Lecture Notes in Applied and Computational Mechanics

Volume 79

About this Series

This series aims to report new developments in applied and computational mechanics—quickly, informally and at a high level. This includes the fields of fluid, solid and structural mechanics, dynamics and control, and related disciplines. The applied methods can be of analytical, numerical and computational nature.

More information about this series at http://www.springer.com/series/4623

Andreas Dillmann · Alexander Orellano
Editors

The Aerodynamics of Heavy Vehicles III

Trucks, Buses and Trains

Springer

Editors
Andreas Dillmann
German Aerospace Center
Göttingen
Germany

Alexander Orellano
Bombardier Transportation
Berlin
Germany

ISSN 1613-7736 ISSN 1860-0816 (electronic)
Lecture Notes in Applied and Computational Mechanics
ISBN 978-3-319-20121-4 ISBN 978-3-319-20122-1 (eBook)
DOI 10.1007/978-3-319-20122-1

Library of Congress Control Number: 2015942617

Springer Cham Heidelberg New York Dordrecht London

Printed on acid-free paper

Springer International Publishing AG Switzerland is part of Springer Science+Business Media
(www.springer.com)

Contents

Part I Train Aerodynamics

Analysis of Flow Structures in the Wake of a High-Speed Train 3
Tomas W. Muld, Gunilla Efraimsson, Dan S. Henningson, Astrid H. Herbst and Alexander Orellano

Experimental Investigation of the Flow Field Underneath a Generic High-Speed Train and the Effects of Ground and Train Roughness 21
Mattias Jönsson and Sigfried Loose

Aerodynamic Drag Reduction of Open-Top Gondola and Hopper Cars in Unit Train Operation and Impact on Train Fuel Consumption and Economics 37
James C. Paul

Sustainable Design of Underground Rail Systems—Aerodynamics at the Interface of Rolling Stock and Civil Construction 61
Andreas Busslinger, Samuel Nyfeler and Peter Reinke

Part II Train Aerodynamics: Cross-wind Effects

Numerical Investigation on the Embankment Configuration for Cross Wind Stability of ICE3 85
Irfan Ali, Arnd Rüter and Carlos-José Rodriguez Ahlert

High-Speed Train Crosswind Analysis: CFD Study and Validation with Wind-Tunnel Tests 99
Carlo Catanzaro, Federico Cheli, Daniele Rocchi, Paolo Schito and Gisella Tomasini

Numerical and Experimental Investigations of the Flow Around a High-Speed Train on an Embankment Under Side Wind Conditions 113
Dan Zhou, Hong-qi Tian, Mark Thompson and John Sheridan

Part III Train Aerodynamics: Experimental Techniques

The Influence of Wind Tunnel Grid Turbulence on Aerodynamic Coefficients of Trains 133
C.N. Nayeri, C. Strangfeld, C. Zellmann, M. Schober, A. Tietze and C.O. Paschereit

Evaluation of the Cross Wind Velocity Through Pressure Measurements on Train Surface 143
F. Cheli, L. Mariano, D. Rocchi, P. Schito and G. Tomasini

Part IV Train Acoustics

A Simplified Model of the Wave Generation Due to Train-Tunnel Entry 161
Sebastian Wagner, Klaus Ehrenfried and Andreas Dillmann

Investigations of Aeroacoustics of High Speed Trains in Wind Tunnels by Means of Phased Microphone Array Technique 175
Andreas Lauterbach, Klaus Ehrenfried and Sigfried Loose

Part V Truck Aerodynamics

Aerodynamics of Commercial Vehicles 195
Thorsten Frank and James Turney

Wheel Housing Aerodynamics on Heavy Trucks 211
David Söderblom, Per Elofsson, Linus Hjelm and Lennart Löfdahl

Part VI Truck Aerodynamics: Drag Reduction

Passive Devices for Reducing Base Pressure Drag in Class 8 Trucks 227
Ramesh Pankajakshan, C. Bruce Hilbert and David L. Whitfield

Numerical and Wind Tunnel Analysis Together with Road Test of Aerodynamic Add-Ons for Trailers 237
G.M.R. van Raemdonck and M.J.L. van Tooren

Computations and Full-Scale Tests of Active Flow Control Applied on a VOLVO Truck-Trailer 253
Mohammad El-Alti, Valery Chernoray, Per Kjellgren, Linus Hjelm and Lars Davidson

Experimental and Full Scale Investigation of Base Cavity Drag Reduction Devices for Use on Ground Transport Vehicles 269
J. Kehs, K. Visser, J. Grossmann, C. Horrell and A. Smith

Part VII Truck Aerodynamics: Experimental Techniques

In Depth Cd/Fuel Economy Study Comparing SAE Type II Results with Scale Model Rolling Road and Non-rolling Road Wind Tunnel Results 287
Adrian Reynard, Mike Camosy, Fritz Marinko, Henri Kowalczyk and Tim Jennings

Towing Tank Experiments for Bluff Body Aerodynamics 303
C.N. Nayeri, J. Glas and C.O. Paschereit

Part VIII Truck Aerodynamics: Computations

What Can LES Do in Vehicle Aerodynamics? 311
Siniša Krajnović

A LES Study of a Simplified Tractor-Trailer Model 327
Jan Östh and Siniša Krajnović

Accurate Drag Prediction Using Transient Aerodynamics Simulations for a Heavy Truck in Yaw Flow 343
Per Elofsson, Guillaume Mercier, Bradley D. Duncan and Samuel Boissinot

Part IX Truck Aerodynamics: Active Flow Control

Analysis of the Active and Passive Drag Reduction Strategies Behind a Square Back Ground Vehicle 363
Charles-Henri Bruneau, Emmanuel Creusé, Delphine Depeyras, Patrick Gilliéron and Iraj Mortazavi

Heavy Trucks Fuel Savings Using the SaOB Actuator 377
A. Seifert, I. Dayan, C. Horrell, J. Grossmann and A. Smith

Part X Bluff Body, Wake and Jet Flows

Comparison of Experimental and Numerically Obtained Flow Properties of a Bluff Body 393
Gandert M.R. van Raemdonck, Peter van Leeuwen and Michel J.L. van Tooren

Quantitative Flow Visualization Applied to a Passive Wake Control Problem 413
A. Landolt, D. Borer, A. Meier and T. Roesgen

Author Index 427

Introduction

This volume contains papers presented at the international conference "The Aerodynamics of Heavy Vehicles III: Trucks, Buses and Trains" held in Potsdam, Germany, during September 12–17, 2010 by Engineering Conferences International (ECI). Leading scientists and engineers from industry, universities and research laboratories, including truck and high-speed train manufacturers and operators were brought together to discuss computer simulation and experimental techniques to be applied for the design of more efficient trucks, buses and high-speed trains in the future.

This conference was the third in the series after Monterey-Pacific Groove in 2002 and Lake Tahoe in 2007. The presentations address different aspects of train aerodynamics (cross wind effects, underbody flow, tunnel aerodynamics and aeroacoustics, experimental techniques), truck aerodynamics (drag reduction, flow control, experimental and computational techniques) as well as computational fluid dynamics and bluff body, wake and jet flows.

We would like to thank ECI for the organization of the conference and the invited speakers, session chairs, organizing committee and authors for the time they spent for a successful meeting.

From the 41 lectures presented at the conference 26 are included in this book.

The editors are grateful to Profs. F. Pfeiffer and P. Wriggers as the Series Editors of "Lecture Notes in Applied and Computational Mechanics" and to Springer-Verlag for the opportunity to publish the results of the conference.

Göttingen — Andreas Dillmann
Berlin — Alexander Orellano
March 2014

Part I
Train Aerodynamics

Analysis of Flow Structures in the Wake of a High-Speed Train

Tomas W. Muld, Gunilla Efraimsson, Dan S. Henningson, Astrid H. Herbst and Alexander Orellano

Abstract Slipstream is the flow that a train pulls along due to the viscosity of the fluid. In real life applications, the effect of the slipstream flow is a safety concern for people on platform, trackside workers and objects on platforms such as baggage carts and pushchairs. The most important region for slipstream of high-speed passanger trains is the near wake, in which the flow is fully turbulent with a broad range of length and time scales. In this work, the flow around the Aerodynamic Train Model (ATM) is simulated using Detached Eddy Simulation (DES) to model the turbulence. Different grids are used in order to prove grid converged results. In order to compare with the results of experimental work performed at DLR on the ATM, where a trip wire was attached to the model, it turned out to be necessary to model this wire to have comparable results. An attempt to model the effect of the trip wire via volume forces improved the results but we were not successful at reproducing the full velocity profiles. The flow is analyzed by computing the POD and Koopman modes. The structures in the flow are found to be associated with two counter rotating vortices. A strong connection between pairs of modes is found, which is related to the propagation of flow structures for the POD modes. Koopman modes and POD modes are similar in the spatial structure and similarities in frequencies of the time evolution of the structures are also found.

T.W. Muld (✉)
Linné FLOW Centre, KTH Aeronautical and Vehicle Engineering, Teknikringen 8,
100 44 Stockholm, Sweden
e-mail: muto02@kth.se

G. Efraimsson
Linné FLOW Centre, KTH Aeronautical and Vehicle Engineering, Stockholm, Sweden

D.S. Henningson
Linné FLOW Centre, KTH Mechanics, Stockholm, Sweden

A.H. Herbst
Aero and Thermodynamics, Specialist Engineering, Bombardier Transportation,
Västerås, Sweden

A. Orellano
Aero and Thermodynamics, Specialist Engineering, Bombardier Transportation,
Hennigsdorf, Germany

A. Dillmann and A. Orellano (eds.), *The Aerodynamics of Heavy Vehicles III*,
Lecture Notes in Applied and Computational Mechanics 79,
DOI 10.1007/978-3-319-20122-1_1

1 Introduction

In today's society there is a trend towards an increased use of faster high-speed trains. As the speed increases, the aerodynamic effects become more prominent, and the safety requirements relating to aerodynamics become more difficult to fulfill. Within Europe, the regulations that specify the safety requirements is the Technical Specification for Interoperability (TSI). The aim is to ensure that trains from different countries within the European Union can operate on the same track and the TSI applies to the Trans-European Rail network.

Passengers standing on platforms waiting for trains experience at the occasion of a passing train wind gusts induced by the train. This is caused by the fact that when the train moves through the air it pulls some of the air with it, due to the viscosity of the fluid. This is known as slipstream and to ensure the safety for passengers on platforms, trackside workers, baggage carts and pushchairs, these gusts need to be weak. Incidents in the United Kingdom between 1972 and 2005 that could have been caused by slipstream is summarized by the Rail Safety and Standards board in the United Kingdom in [6]. In [6], it is reported that there were 4 incidents involving passengers or staff in platforms, 2 incidents involving trackside workers and 13 incidents involving pushchairs. As an example, one child sustained minor head injuries when the pushchair containing the child was affected by the slipstream of a passing passenger train. The TSI states that a train travelling at 200 km/h should not cause wind gusts of above 15.5 m/s at 1.2 m above platforms at a distance 3.0 m from the center of the track. On open tracks the slipstream velocities for trains running at 190–249 km/h should not exceed 20.0 m/s at 0.2 m above top of rail (TOR) and 3.0 m from the center of the track, see [1].

A summary of different experimental work that have performed measurements slipstream is presented in [20]. In addition, 4 regions of the flow around the train were identified, Upstream/nose, Boundary layer, Near wake and Far wake region. In [20] it was observed that the largest slipstream velocities occured in different regions depending on the type of train. For freight trains, the rapid growth of the boundary layer causes high velocities as the train passes, region (2), while for high-speed trains the highest slipstream velocities occur in the near wake, region (3). A schematic view of the different regions and the flow is presented in Fig. 1.

In order to resolve the fluctuations of the slipstream velocities, a time-accurate approach has to be considered for numerical simulation. Such a method is Detached-Eddy simulations (DES), [19], which is the method used throughout this work. DES is a hybrid LES/RANS method that behaves like a Reynolds-Averaged Navier-Stokes (RANS) method close to walls and uses Large Eddy Simulation (LES) far from walls. Examples of numerical work on trains are for instance [5, 7], where DES and LES, respectively, is used to investigate crosswind.

The objective of this paper is to evaluate two different mode decomposition techniques to enable future studies of wake flows behind high-speed trains. First, results are compared for different resolutions of the grid in order to verify the numerical work. The flow in the wake of the ATM are then analyzed using two different tech-

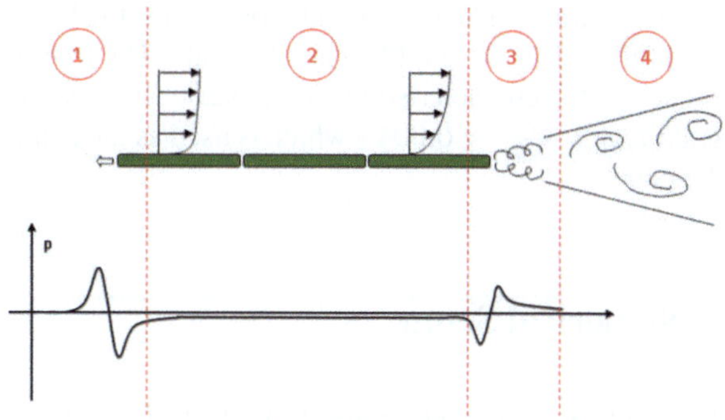

Fig. 1 The different flow regions around a high-speed train

niques of decomposing the flow into modes, Proper Orthogonal Decomposition [8] and Koopman mode decomposition [13], respectively. This is done to enhance the understanding of the flow structures behind the train, which are important for slipstream of high-speed trains. In order to compare the flow results with experimental data, attempt to model a trip wire, that was present in the experimental setup, was presued. This study is presented in Appendix A.

2 Problem Description

The simulated geometry consists of a train, ground and a platform. The train model considered is the ATM with 4 cars. The simulated geometry in this paper also includes simple bogies and 9 car gaps and is shown in Fig. 2. There are more car gaps than cars in order to simulate a long train. For numerical stability the wheels in the bogies is covered with *shoes*, both between body-wheel and ground-wheel. Without the shoes there would be a small gap between the wheels and the car body that would require very small cell sizes as well as very skewed cells between the wheels and the ground.

The results of the numerical study are compared to the results of experimental work performed at German Aerospace Centre (DLR). These experiments were performed in a water towing tank, in which the train model (scale 1:50) was mounted upside down and dragged trough the water. The ground and the platform were stationary, while the train was moving. The velocities around the train were measured with Particle Image Velocimetry (PIV) in a horizontal (xy-) plane at $z = 0.4\ d_h$ above the platform. The geometrical setup in the numerical simulation is made to match this experimental setup.

Fig. 2 Picture of the front half of the train model

The characteristic length scale for ATM is the hydraulic diameter d_h, which is 3 m for a full scale train or 0.06 m in 1:50 scale. The free stream velocity $U_{\text{inf}} = 15$ m/s is chosen as the characteristic velocity scale. It is useful to define a characteristic time scale $T_{ref} = \frac{d_h}{U_{\text{inf}}} \approx 0.004$ s which is used as the characteristic time scale throughout the paper.

3 Numerical Setup

The finite-volume code Star-CD v4 from CD-adapco is used to compute the flow field as described by the incompressible Navier-Stokes equations.

The turbulence model used in all flow simulations presented is DES based on the Spalart-Allmaras (SA) one-equation RANS models, here denoted DES-SA.

In DES-SA the distance to the wall, d, in the SA turbulence model is replaced by the modified wall distance $\tilde{d}$, which is defined as

$$\tilde{d} = \min(C_{DES}\Delta, d), \qquad \Delta = \max(\Delta x, \Delta y, \Delta z). \tag{1}$$

Here C_{DES} is a model constant and Δ corresponds to a filter length in DES, which depends on the grid spacing in each spatial direction.

In [18] a modification to DES-SA, Delayed DES (DDES), was introduced. This modification is done to ensure that the model acts like a RANS model in the boundary layer with any size of the grid cells. The modification is that $\tilde{d}$ is altered as

$$\tilde{d} = d - f_d \max(0, d - C_{DES}\Delta) \tag{2}$$

where f_d is called a shielding function, which has the purpose of shielding the model to act in LES mode inside the boundary layer. The function is designed to be 1 in the LES region and 0 in the rest of the domain.

Three trim-hexa grids were used, denoted Coarse, Medium and Fine Mesh (CM, MM, FM), which were created in Star-CCM+ v3 from CD-adapco. The grids consist of 5 prism layers, with a y^+ of about 1.0, x^+ and z^+ about 600, and 6 different refinements zones. The grids have the same topology but have different reference sizes, where the reference size is the length of the cells in far region. For the trim-hexa grids, this implies that all cells outside the prismlayers scale with the reference length. The ratio between the reference length for CM and FM is $\sqrt{2}$, as suggested by [17] to be a significant difference in size. The wall-normal spacing of the prism layers are not altered between the grids. The total amount of cells are 11, 20 and 28 million cells for CM, MM and FM respectively. A schematic view of the refinement zones and some close up of the grid is shown in Fig. 3.

The numerical scheme used for the convective terms in the Navier-Stokes Equation, is the Monotone advection and reconstruction scheme (MARS), which is a second order accurate scheme available in Star-CD. This is a TVD scheme in which a compression parameter is included, which the user can vary between 0 and 1.

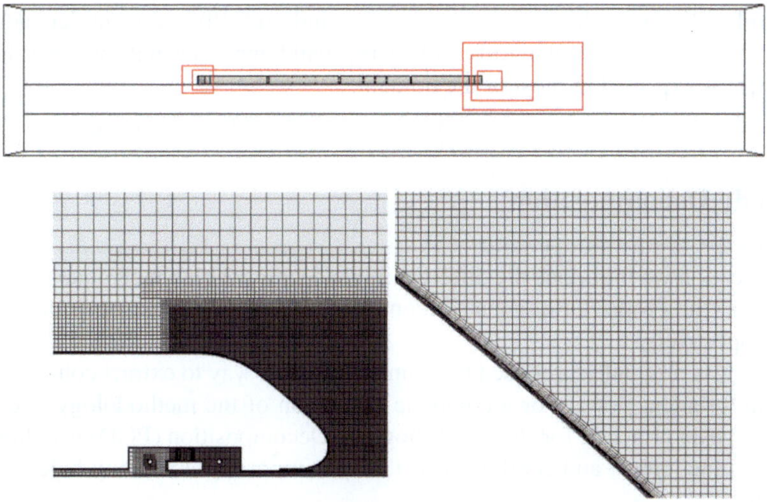

Fig. 3 Grid refinement zones (*upper figure*) and grid close to the rear of the train (*lower figures*)

A higher value increases the sharpness of the solution but the tradeoff is slower convergence. The value of the compression parameter in the computations presented in this paper is set to 0.9, in order to have a high sharpness of the solution but avoide problems with dispersion.

In general, DES simulations are performed using the blended scheme, recommended by [21] in order to reduce the amount of numerical dissipation in the LES-regions. However, when tested on the current test case and grid topology spurious fluctuations appeared close to the interface between refinement zones. It is unclear whether this is due to the solver used, or an inherent feature of the scheme. Oscillations close to interfaces between refinement zones were also found when using the pure central difference scheme. DES-simulations of a train geometry using MARS are presented in [5], where satisfactory results were found.

At the inlet a Dirichlet boundary condition is applied with constant velocity U_{inf} in the streamwise direction. In the experiments the train model is moving into stationary flow, however the fluid is still affected from the last run, which means that some turbulence is still present. The turbulent intensity (TI) is estimated to $TI = 0.3\,\%$ from the experimental results. The turbulent length scale (L_T) at the inlet was estimated from [2] for low turbulence to $L_T = 0.1 d_h$. These are different than the values used in [12]. At the outlet the pressure is prescribed to constant zero gauge pressure. The velocity tangential to the wall is set 0 and the normal component is extrapolated. There are both stationary and moving wall boundary conditions. Stationary wall is applied at the surface of the train, including bogies, wheels and intercar gaps, while moving wall is applied at the ground, platform and water tank walls. The moving walls are moving at a constant speed of U_{inf}. No-slip is applied at the impermeable wall boundaries, which means that all velocity components are zero at the wall, via a hybrid wall approach, which blends a low- and high-Re treatment depending on the y^+ value [15]. Since the grid resolves the wall boundaries with $y^+ \approx 1$, low-Re treatment is expected in most of the domain.

4 Mode Decomposition

In this section we introduce the different mode decomposition techniques applied in this work: the Proper Orthogonal Decomposition and Koopman mode decomposition, respectively.

POD was originally proposed by Lumley, [8], as a way to extract coherent structures in turbulent flows. For a complete derivation of the methodology, see [3] or [9]. The basic principle for Proper Orthogonal Decomposition (POD) is to find a set of basis functions $\boldsymbol{\sigma_i}$ and coefficients a that decomposes the velocity field $u_i(x_j, t)$ such that

$$u_i(x_j, t) \approx \sum_n^{N_M} a^n(t)\sigma_i^n(x_j). \tag{3}$$

Here, N_M is the number of modes. In POD, the minimizing problem,

$$\min_{a,\sigma} I, \quad \text{where} \quad I = \int_T \iiint_\Omega \left(u_i(x_j, t) - \sum_n^{N_M} a^n(t)\sigma_i^n(x_j) \right)^2 d\Omega dt \tag{4}$$

is solved. The characteristics of POD is therefore a maximization of the energy in the lowest modes and that the modes are orthogonal. Thereafter, variational analysis yields an integral eigenvalue problem. This eigenvalue problem is discretized and the integrals are approximated by sums. If the discretized values are arranged into matrices the sums can be written as matrix operation. Rewriting, the resulting eigenvalue problem becomes

$$\Lambda \mathbf{A} = \mathbf{C}\,\mathbf{A}, \qquad \mathbf{C} = \frac{1}{N_T}\mathbf{U}^{\mathrm{T}}\mathbf{W}\mathbf{U}. \tag{5}$$

where $\mathbf{C}$ is the temporal correlation matrix. The matrices $\mathbf{A}$, $\mathbf{U}$ and $\mathbf{W}$ contain the time coefficients, velocity components and volume of each point in the discretized volume, respectively. Solving the eigenvalue problem in Eq. (5) gives the time coefficients, which can then be used to calculate the basis functions.

Koopman mode decomposition [13] decomposes the flow field in a different way than POD. Here the modes are decomposed by the frequency of the motion of each flow structure. Koopman mode decomposition has only recently been introduced in fluid mechanics studies. One first study is presented in [13], where a jet in crossflow is analyzed using Koopman mode decomposition and POD. For a detailed description of Koopman modes, we refer to [13, 16].

To introduce the methodology, consider a discrete dynamical system

$$\mathbf{u}_{k+1} = \mathbf{f}(\mathbf{u}_k), \tag{6}$$

where the function $\mathbf{f}$ shifts the velocity field $\mathbf{u}_k$ from one time step to the next. The Koopman operator (U) is defined as a linear operator such that

$$Ug(\mathbf{u}_k) = g(\mathbf{f}(\mathbf{u}_k)), \tag{7}$$

where g is a scalar valued function. In the following, g, is called an observable, and can be any quantity of interest in the flow. The Koopman operator is hence the operator that shifts any scalar function forward in time. We denote the eigenvalues of U by λ_i and the eigenvectors φ_i, i = 1,2,... Let $\mathbf{g}$ denote any vector observable of the initial flow field $\mathbf{u}_1$. As an example $\mathbf{g}$ can be the force on an object in the flow. In [10] it is shown that $\mathbf{g}$ can be expanded in the eigenvectors of the Koopman operator as

$$\mathbf{g}(\mathbf{u}_1) = \sum_{j=1}^{\infty} \varphi_j(\mathbf{u}_1)\mathbf{v}_j \tag{8}$$

where $\mathbf{v}_j$ is the j-th vector valued expansion coefficients. In a similar manner, the observables at all time instances can be expanded into the same Koopman eigenvectors as for the initial step, since

$$\mathbf{g}(\mathbf{u}_{k+1}) = U^k\mathbf{g}(\mathbf{u}_1) = U^k \sum_{j=1}^{\infty} \varphi_j(\mathbf{u}_1)\mathbf{v}_j = \sum_{j=1}^{\infty} \lambda_j^k \varphi_j(\mathbf{u}_1)\mathbf{v}_j \tag{9}$$

This means that the infinite velocity matrix $\mathbf{U}_\infty$, $\mathbf{U}_\infty = [\mathbf{u}_1\ \mathbf{u}_2\ ...]$, can be represented in terms of the same Koopman modes as in (8), that is

$$\mathbf{U}_\infty = \mathbf{\Phi S}, \qquad \mathbf{\Phi} = \left[\varphi_1(\mathbf{u}_1)\mathbf{v}_1\ \varphi_2(\mathbf{u}_1)\mathbf{v}_2 \ldots\right], \qquad \mathbf{S} = \begin{bmatrix} 1 & \lambda_1 & \lambda_1^2 & \cdots \\ 1 & \lambda_2 & \lambda_2^2 & \cdots \\ \vdots & & \vdots & \ddots \end{bmatrix}, \tag{10}$$

Note that $\mathbf{S}$ is a so called Vandermonde matrix. The eigenvalues of the Koopman operator hence describe the time development of each Koopman mode. Again, the modes can be described as characteristic flow structures. This means that the entire flow can be described by the Koopman modes and the eigenvalues of the Koopman operator.

The algorithm, also referred to as Dynamic Mode Decomposition (DMD), to compute the Koopman modes for the finite dimensional velocity matrix $\mathbf{U}$ is based the Arnoldi method presented in [14], which gives the basis to calculate an approximation of the modes. The algorithm can be found in [13] or [16].

5 Results

5.1 Velocity Profiles

The results of velocity profiles are shown in the horizontal plane $z = 0.4\ d_h$ above the platform, which is the height specified in the TSI, as discussed in Sect. 1. In addition, this height was chosen since the experimental results were only available at this height. For the comparison to experimental data, the reference frame of the numerical results is changed to the reference frame of the experiments. Also, the experimental data only contains two velocity components in the PIV plane. The results are therefore presented in terms of a 2-D velocity magnitude in a ground fixed reference frame. This velocity magnitude is computed from the numerical data as $\| U \| = \sqrt{(U_\infty - U_{CFD})^2 + V_{CFD}^2}$.

First, the results for the three different grids, CM, MM and FM are presented and compared. Most emphasis is put into the comparison of the mean velocity profiles. Comparisons of turbulent quantities between different mesh resolutions for LES type solutions are not straightforward, since finer meshes resolve more of the turbulent quantities. The mean velocity and rms velocity as a function of the streamwise direction along the side of train are shown in Fig. 4, at $0.75d_h$ from the centerline. The results for the two finest grids are very similar, deviating just after the front and close to the rear of the train. The levels of rms velocity seem to be very similar for the grids FM and MM.

The interesting region for slipstream is the wake behind the vehicle. The velocity profiles in this region for the different grids are shown in Fig. 5. Figure 5 show the mean velocity as a function of streamwise position at two different spanwise positions. Close to the train, all three grids yield similar results, but further downstream

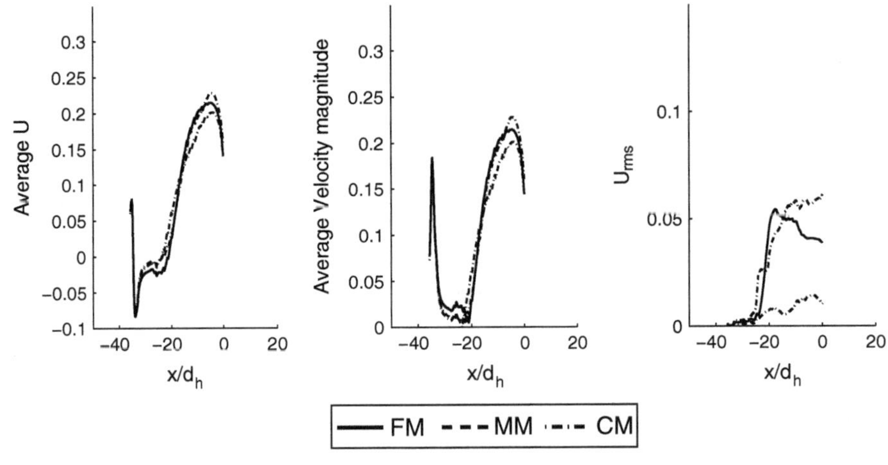

Fig. 4 Mean velocity and u_{rms} on the side of the train with different grids

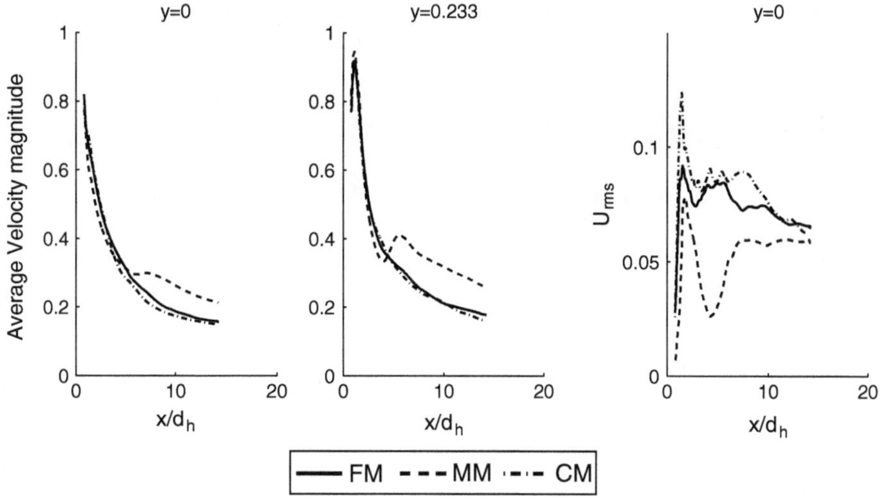

Fig. 5 Mean velocity and u_{rms} in the wake of the train with different grids as a function of streamwise position

the results with the coarse mesh starts to deviate, but the two fine grids show a good agreement along the whole curve, in particular for the mean velocity. The conclusion is therefore that the MM is fine enough and the solution on this grid is decomposed into modes in Sect. 5.2.

5.2 Mode Decomposition

The POD and Koopman modes, respectively are calculated in a subsection of the flow domain. Since the velocity field for all the cells within the decomposition volume has to be stored for the sampled snapshots there is a limitation in amount of cells that can be used. This volume is chosen so that it contain large portions of the near wake region, which is of interest for the slipstream phenomenon. The volume is identified by investigating the instantaneous and rms velocities. The geometric extent of the volume is $8.97d_h \times 4d_h \times 1.83d_h$ and it contains approximately 9.1 million cells, which is almost half of total amount of cells. Part of the volume would contain the last bogie, but since the complicated flow inside the bogie is not directly of interest this part is removed from the domain. Pictures of the instantaneous and rms velocities of the flow field in the region is shown in Fig. 6.

Computing different sets of POD modes using different total sampling times, T_A, and comparing the different sets can be used to investigate convergence as a function of T_A. To compare the different set and to measure the difference the scalar product and L_2-norm of the difference between the sets is computed. The error represents

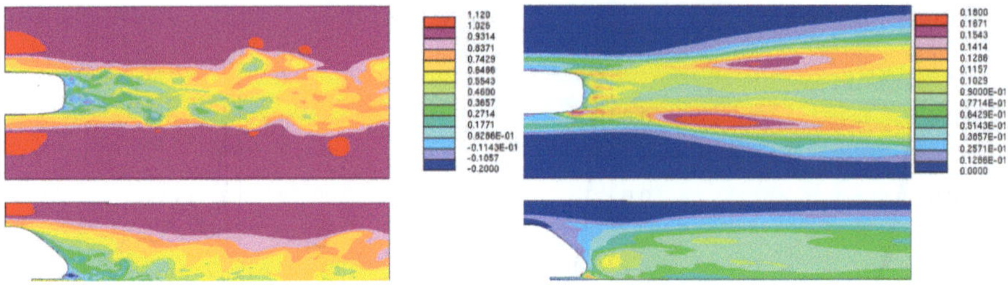

Fig. 6 Instantaneous velocity and u_{rms} behind the train in the volume where mode decomposition is performed

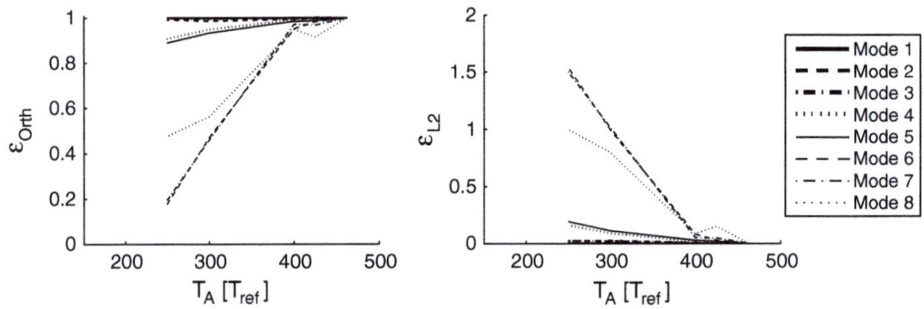

Fig. 7 Convergence of the POD modes as a function of T_A

how much the difference the set of modes are from the set of modes computed with the longest T_A. The errors, ε_{Orth} and ε_{L2}, are defined as

$$\varepsilon_{Orth} = \int\int\int_{\Omega} \sigma_i^{m,1} \sigma_i^{m,2} d\Omega, \qquad \varepsilon_{L2} = \|\sigma_i^{m,1} - \sigma_i^{m,2}\|_2, \tag{11}$$

where 1 and 2 are the different sets of computed modes and m is the number of the mode and i is the coordinate direction. The results for the convergence is then showed in Fig. 7. The first 8 modes have converged after $T_A = 400\ T_{ref}$. From the convergence plot a strong connection between pair of modes can be observed. The convergence of for instance mode 4 follows the convergence of mode 5, which would suggest that these two modes are connected.

The basis functions of the POD modes are shown in Fig. 8, where isosurfaces of positive and negative velocity is shown. The POD mode 1 is the mean flow and the higher modes represent disturbances around this mean. The zero isosurface of streamwise velocity is shown, which shows the intersection between forward and recirculating flow. From the v- and w-component the mean stucture in the wake can be found to be the two counter rotating vortices. Mode 2 and higher are then the flow structures that are related to the perturbation to the counter rotating vortices in the wake. The connection between adjacent modes that was found in the convergence plot is confirmed in the plots for the basis functions. Modes 2–3 and 4–5 show very similar

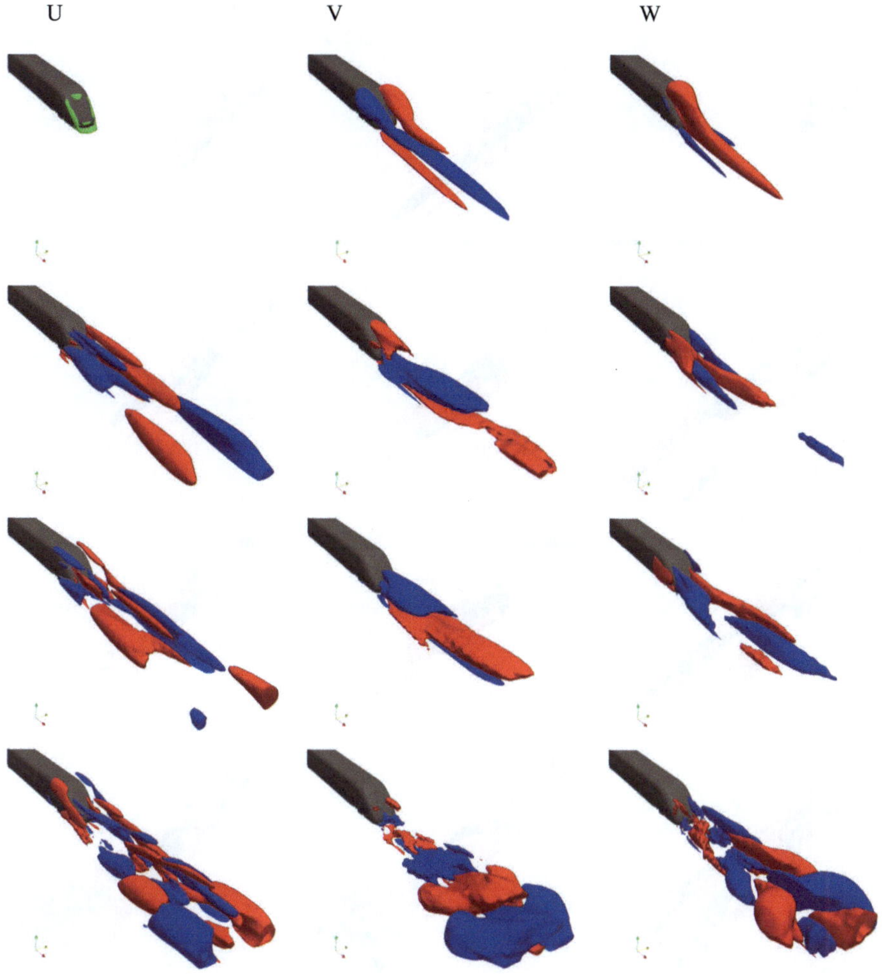

Fig. 8 Isosurfaces of spatial POD modes 1–4

patterns, only with an offset in the streamwise direction. It can be demonstrated that two POD modes, which only differ in phase lag, together represent a propagation of a flow structure.

Similar spatial structures as in the POD modes are found in the Koopman modes, shown in Fig. 9, as they are also dominated by the counter rotating vortices. Mode 1 should per definition be the same as POD mode 1, since the first mode is the mean flow for both methods. For the fluctuating modes the spatial structures in Koopman mode 2, are very similar to those in POD modes 2 and 3. Also the structures in Koopman mode 4 are similar to those in POD mode 4 and 5. However the Koopman modes are not as smooth as the POD modes. If this is due to that Koopman mode

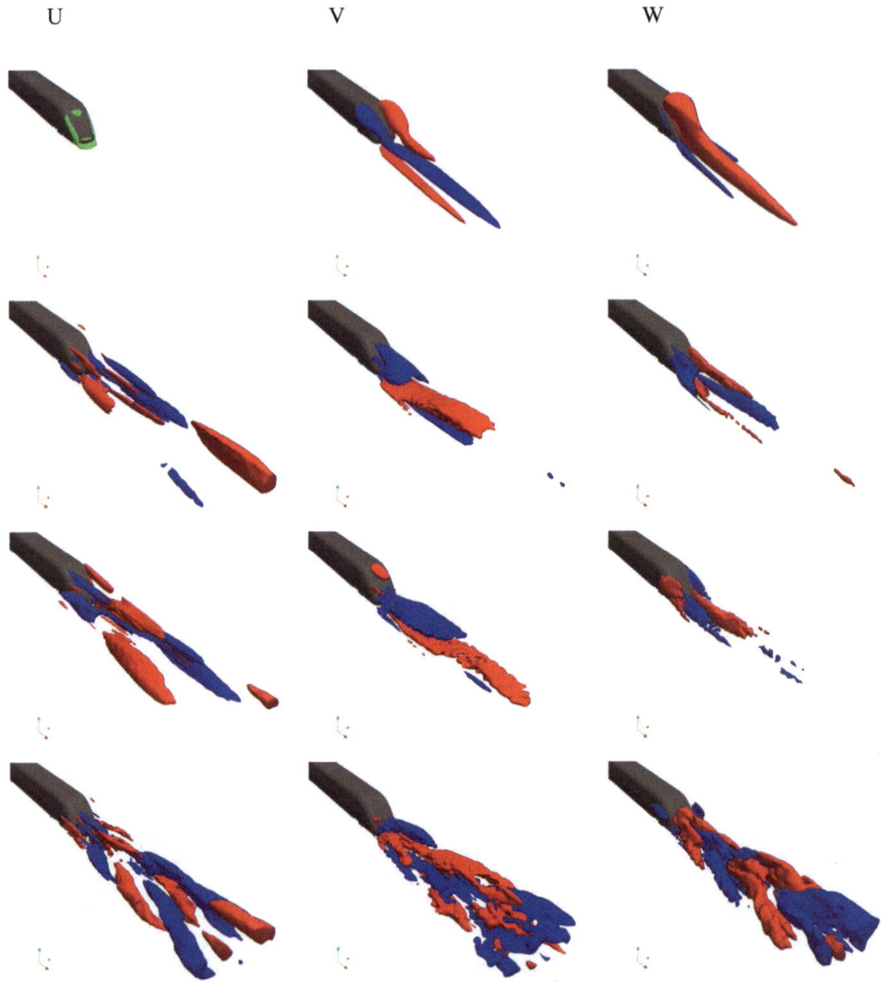

Fig. 9 Isosurfaces of spatial Koopman modes 1–4

decomposition need more snapshots to converge or if it is a inherent property of Koopman modes is at the moment not clear.

To compare the modes extracted by the different decomposition methodologies the frequencies of time coefficients of the POD are compared to the frequencies of the Koopman modes. The frequencies for the POD modes 2-4 and the full spectrum of all Koopman modes are plotted in Fig. 10. For the Koopman mode decomposition the dominant mode has frequency around St = 0.08. POD modes 2 and 3 have very similar frequency spectrum and have a peak close to St = 0.08. This is consistent with that the spatial modes of the two decomposition methods look similar. This is also

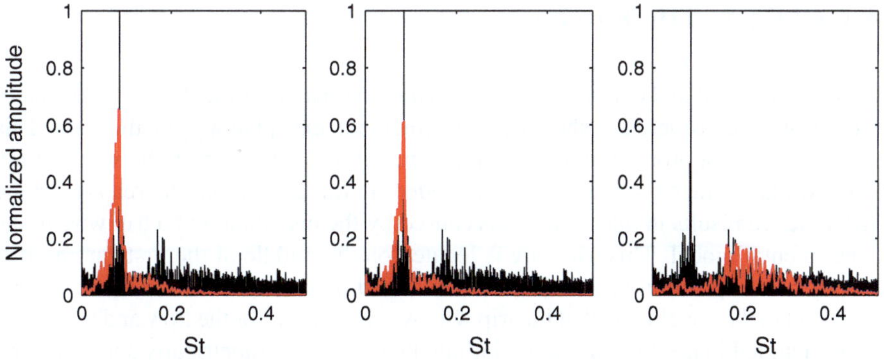

Fig. 10 Spectrum of individual POD modes (*left mode* 2, *middle mode* 3, *right mode* 4) compared to frequencies of all Koopman modes. Each spectrum is normalized by its highest amplitude

true for POD mode 4, which excite higher frequencies in a broad range, including the frequency of Koopman mode 4 that has similar spatial structures.

6 Conclusions

The flow around the Aerodynamic Train Model has been investigated, with a special focus on the wake flow. Grid convergence is proven by comparing velocity profiles for different resolutions of the mesh, where the results for the medium and fine meshes showed little discrepancy, while the results for the coarse mesh did not match the other profiles.

The computed flow is decomposed into POD and Koopman modes. The convergence of the POD modes is investiged by comparing the modes for different numbers of snapshots. From the convergence of the modes and by looking at the basis functions, a strong symmetry in the structures of consecutive modes is found. A further example of the symmetry between the modes is identified from the frequencies of the time coefficients of the POD modes. Modes 2 and 3 are found to have very similar spectral content. These modes are connected to the convection of a flow structure. The first mode demonstrated the presence of two counter rotating vortices and the other modes are perturbations around this mean structure. The Koopman spatial modes show the same type of flow structures as the POD modes. The dominant Koopman mode has frequencies around St = 0.08. The decomposed modes give insight into the flow structures that dominate the wake and further research is needed to identify the behavior and origin of these flow structures.

Appendix A: Trip Forcing

In [12] numerical and experimental results are compared for the ATM. The velocity profiles were considered to show a good agreement, except alongside the train. This was considered a promising first comparison, but the difference on the side of train implied that further investigation was needed. It was found that the reason for the difference in results in this region was caused by the installation of a trip wire in the experimental model. This trip wire is located on the middle of the first car and it is exactly at this position that the experimental and numerical results start to deviate. This lead to the conclusion that the trip wire was important for the flow and that it had to be included in the simulation. Including the trip wire geometrically and resolving the flow around it in the simulation could be very troublesome and computationally demanding. An approach to model the effect of the wire is investigated instead.

Method

The trip forcing model applied in this work is based on the trip forcing presented in [4]. In [4], the flow over a flat wall is considered hence, the y-coordinate is in the wall normal direction. The trip forcing in [4] is modeled via $F_S = (0, F_2, 0)$, where

$$F_2 = exp\left(\{x - t_{x0}/t_{xsc}\}^2 - \{y/t_{ysc}\}^2\right) f(z,t) \tag{12}$$

with

$$f(z,t) = t_{amps}g(z) + t_{ampt}\left((1 - b(t))\, h^i(z) + b(t)h^{i+1}(z)\right) \tag{13}$$

$$i = int(t/t_{dt}), \qquad b(t) = 3p^2 - 2p^3, \qquad p = t/t_{dt} - i. \tag{14}$$

The functions $g(z)$ and $h^i(z)$ are Fourier series with N_{zt} random frequencies, corresponding to steady and transient forcing respectively. Linked together with the Fourier series are the amplitudes t_{amps} and t_{ampt}, s for steady and t for transient. The function $b(t)$ is included in order for the forcing to be continuous in time. The constant t_{x0} is the position of the trip forcing (the wall normal position is $y = 0$) and t_{xsc} and t_{ysc} is the width of the Gaussian function in spanwise and wall normal direction. The variable t_{dt} is the interval between different transient Fourier series for the trip. In this work, we used $N_{zt} = 10$, $t_{amps} = t_{ampt}$, $t_{x0} = 0.1713$, $t_{xsc} = 0.04$, $t_{ysc} = 0.005$ and $t_{dt} = 0.0005$

The original formulation is valid for a flat plate. For simplicity, the curved surface of the train is considered as locally flat, where the y-coordinate is changed to the distance to the wall and z direction is considered to be along the curved surface. For any given point in the fluid, the distance to all the surface points are computed and minimized, in order to find the wall position closest to that cell. The vector from the point to the closest surface point is considered the normal direction and using F_2 from Eq. (12) the forcing in the cartesian coordinate system becomes $F_y = F_2 \times \cos(\theta)$ and $F_z = F_2 \times \sin(\theta)$, where F_y and F_z is the forcing term in the coordinate system of the train, and θ angle between the vector, of the point and the closest surface point, and the y-direction.

Results

The amplitudes of the forcing (t_{amps} and t_{ampt}) are found to be important parameters. Too small amplitudes do not affect the flow significantly while too high amplitudes create a jet from the position of the trip forcing yielding too large impact on the flow on the side of the train. Different amplitudes are tested in order to find an optimum. Amplitudes that are very high or very low can be disregarded only by looking at the instantaneous flow. For these extreme values of the amplitudes, the simulation is stopped before the averaging starts and the results are not presented.

Even tough the results including a trip forcing show a large improvement in the region along side the train, no convergence towards the experimental data were found. There are two regions of the flow where the simulated results does not match experimental results. These two regions are directly after the location of the trip wire ($\approx -30d_h$) and towards the end ($[\approx -20d_h, -5d_h]$) where the velocity seem to be more or less constant with spanwise position. A higher amplitude of the trip forcing match the data just downstream of the trip wire but causes an overprediction of the velocity in the second region. A lower amplitude would give a good estimate towards the rear of the train, but underpredict the velocity close the trip wire. In addition, trying to match the u_{rms} values is even more challenging. Results for different amplitudes are shown in Fig. 11. The black vertical line represents the position of the trip wire. In

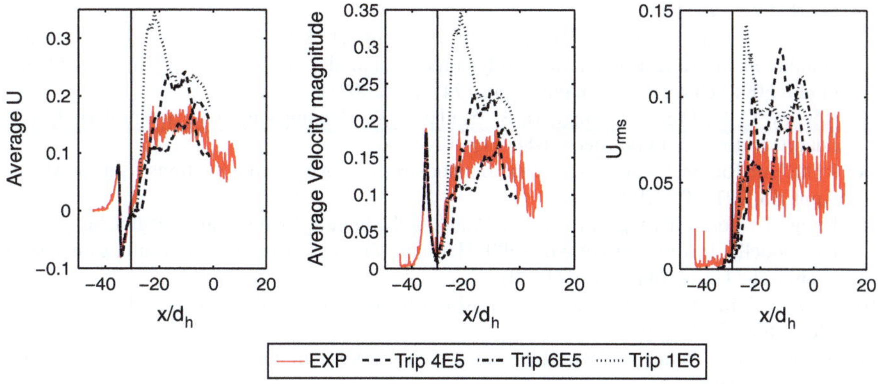

Fig. 11 Velocities and u_{rms} alongside the train with three different amplitudes of trip forcing (4E5,6E5,1E6) compared to experimental results. *Vertical line* is the position of the trip wire

this study only the amplitudes are varied. Changing more parameters of the volume forcing would increase the computational cost extensivly. From the results in this section, we conclude that the the trip wire is needed in the numerical simulation in order to compare to experimental results and that the volume forcing improves the results but was unsuccessful to reproduce the experimental results.

References

1. Barrot, J.: Commission decision of 21 February 2008 concerning a technical specification for interoperability relating to the rolling stock sub-system of the trans-European high-speed rail system. Official J. Eur. Union **L84**(51) (2008)
2. Casey, M., Wintergerste, T. (eds.): ERCOFTAC special interest group on quality and trust in industrial cfd best practice guidelines. ERCOTFTAC (2000)
3. Cazemier, W., Verstappen, R.W.C.P., Veldman, A.E.P.: Proper orthogonal decomposition and low-dimensional models for driven cavity flows. Phys. Fluids **10**(7) (1998)
4. Chevalier, M., Schlatter, P., Lundbladh, A., Henningson, D.: Simson- a pseudo-spectral solver for incompressible boundary layer flows. Technical Report KTH Mechanics (2007)
5. Diedrichs, B.: Unsteady aerodynamic crosswind stability of a high-speed train subjected to gusts of various rates. In: Proceedings Euromech Colloquium 509: Vehicle Aerodynamics, pp. 39–50 (2009)
6. Figura-Hardy, G.I.: RSSB Slipstream Safety—Analysis of existing experimental data on train slipstreams including the effects on pushchairs. Rail Safety and Standards Board (2007)
7. Hemida, H., Krajnović, S., Davidson, L.: Large-eddy simulation of the flow around a simplified high speed train under the influence of a cross-wind. In: Proceedings of the 17th AIAA Computational Fluid Dynamics Conference (2005)
8. Lumley, J.L.: The structure of inhomogeneous turbulent flows. In: Yaglom, A.M., Tatarsky, V.I., (eds.) Atmospheric Turbulence and Radio Wave Propagation, pp. 166–178 (1967)
9. Manhart, M., Wengle, H.: A spatiotemporal decomposition of a fully inhomogeneous turbulent flow field. Theor. Comput. Fluid Dyn. **5**, 223–242 (1993)
10. Mezić, I.: Spectral properties of dynamical systems, model reduction and decompositions. Nonlinear Dyn. **41**, 309–325 (2005)
11. Morel, T.: Effect of base slant on flow in the near wake of an axisymmetric cylinder. Aeronaut. Q. **31**, 132–147 (1980)
12. Muld, T.W., Efraimsson, G., Henningson, D.S., Herbst, A.H., Orellano, A.: Detached eddy simulation and validation on the aerodynamic train model. In: Proceedings Euromech Colloquium 509: Vehicle Aerodynamics, pp. 174–187 (2009)
13. Rowley, C.W., Mezić, I., Bagheri, S., Schlatter, P., Henningson, D.S.: Spectral analysis of nonlinear flows. J. Fluid Mech. **641**, 115–127 (2009)
14. Ruhe, A.: Rational Krylov sequence methods for eigenvalue computations. Linear Algebra Appl. **58**, 391–405 (1984)
15. Rung, T.: Formulierung universeller Wandrandbedingungen für Transportgleichungsturbulenzmodelle. Institutsbericht Nr. 02/99, Hermann-Föttinger-Institut für Strömungsmechanik, Technische Universität Berlin (1999)
16. Schmid, P.J.: Dynamic mode decomposition of numerical and experimental data. J. Fluid Mech. **656**, 5–28 (2010)
17. Spalart, P.R.: Young-person's guide to detached-eddy simulation grids. NASA CR 2001-211032 (2001)
18. Spalart, P.R., Deck, S., Shur, M.L., Squires, K.D., Strelets, M., Travin, A.: A new version of detached-eddy simulation, resistant to ambiguous grid densities. Theor. Comput. Fluid Dyn. **20**, 181–195 (2006)

19. Spalart, P.R., Jou, W-H., Strelets, M., Allmaras, S.R.: Comments on the feasibility of LES for wings, and on a hybrid RANS/LES approach. In: First AFOSR International Conference on DNS/LES (1997)
20. Sterling, M., Baker, C.J., Jordan, S.C., Johnson, T.: A study of the slipstreams of high-speed passenger trains and freight trains. Proc. Inst. Mech. Eng. Part F: J. Rail Rapid Trans. **222**, 177–193 (2008)
21. Travin, A., Shur, M., Strelets, M., Spalart, P.R.: Detached-eddy simulations past a circular cylinder. Flow Turbul. Combust. **63**, 293–313 (1999)

Experimental Investigation of the Flow Field Underneath a Generic High-Speed Train and the Effects of Ground and Train Roughness

Mattias Jönsson and Sigfried Loose

Abstract Results obtained from two component Particle Image Velocimetry (PIV) measurements on three different 1:50 generic high-speed train configurations hauled through a water towing tank over a smooth (Plexiglas) and a rough (grinding belt) ground at a speed of 4 m/s are presented. Principally, the three different generic high-speed train configurations are based on the same model (front car, two cars and tail car). The smooth generic high-speed train configuration (smooth GHSTC) reflects no bogies and covers the bogie cut outs and inter car gaps, the rough generic high-speed train configuration (rough GHSTC) is obtained by removing the bogies and leave the bogie cut outs and inter car gaps open and for the generic high-speed train configuration (GHSTC) the bogies are not removed but the inter car gaps are left open. A PIV set-up was chosen that the light sheet defines a vertical plane (XZ) between the ground and the train in the symmetry line of the train. Comparing the PIV results obtained for the GHSTC with full scale measurements, it was found that the same flow structures develop in the vicinity of the head and the tail of the train. But, the measured underfloor U-velocity of the downscaled model measurements did not reach the same value as those measured for the full scale high-speed train. The reason which is ascribed to the better aerodynamic underframe of the train model in the downscaled model measurements. The flow field underneath the GHSTC was fully developed at the beginning of the second car, in agreement to the full scale measurements. For the three train configurations three different flow fields underneath the train were obtained. The lowest velocities were found for the smooth GHSTC and the highest for the rough GHSTC. Further, the ground roughness changed the flow fields underneath the different train configurations.

M. Jönsson (✉) · S. Loose
German Aerospace Center, Bunsenstraße 10, 37073 Göttingen, Germany
e-mail: Mattias.Joensson@dlr.de

S. Loose
e-mail: Sigfried.Loose@dlr.de

A. Dillmann and A. Orellano (eds.), *The Aerodynamics of Heavy Vehicles III*,
Lecture Notes in Applied and Computational Mechanics 79,
DOI 10.1007/978-3-319-20122-1_2

1 Introduction

Ballast projection which is also known as ballast flight takes place whenever one or more stones start to move or fly underneath the train. The flying stones can damage the underframe equipment of the train. They can also damage trackside objects when the ballasts are reflected to the side. Ballast projection has been considered as a problem in train aerodynamics since the 1980s. Back then it was correlated with snow and ice accumulation on the train underframe and it's dropping during travel. This problem could be minimized e.g. by lowering the ballast [9]. Since the beginning of the 21st century ballast projection incidents were reported also in the absence of winter conditions [1]. This indicated that the aerodynamic loads induced by high-speed trains are high enough to move the ballasts on the track bed [7]. To gain more knowledge regarding this observation many interesting studies were made within the DeuFraKo project [5], which included a full scale measurement [2]. The full scale measurement techniques used have been installed in the track bed. They have to be below Top Of Rail (TOR) for security reasons. Therefore it was only possible to measure aerodynamic loads and the flow field between the ground and TOR. To investigate the entire gap between the ground and the train, two component Particle Image Velocimetry (PIV) [4] was conducted in the water towing tank at DLR Göttingen. The obtained results are discussed below.

2 Experimental Setup

2.1 Water Towing Tank Göttingen

The PIV measurements which are presented in this paper were all conducted in the water towing tank at DLR Göttingen. The water towing tank is 18 m long open steel tank with a cross-section of 1.1×1.1 m equipped with thick glass windows allowing the use of optical measurement techniques and flow visualization, see Fig. 1. The moving model rig is hauled with the help of a steel cable and an electrical motor (installed outside the tank) and travels on two rails installed on the upper edges of the tank. For the presented results a rig speed of 4 m/s was used, which corresponds to a Reynolds number of approximately 0.24 Mio based on the reference length of 0.06 m of the model and the kinematic viscosity of water [6]. The 1:50 train model configuration (front car, two cars and tail car) is mounted on the moving model rig with the help of eight NACA profiles, i.e. two NACA profiles per car. The NACA profiles press the train model into the water. The NACA profiles were chosen in a way that they minimize the influence of the mounting on the flow field. They were mounted on the roof to avoid disturbing the flow field underneath the train model. The effects of the NACA profiles on the flow field underneath the train model can not completely be neglected but they are considered as very small. For the correct train ground simulation a ground plate was installed into the tank. The ground plate

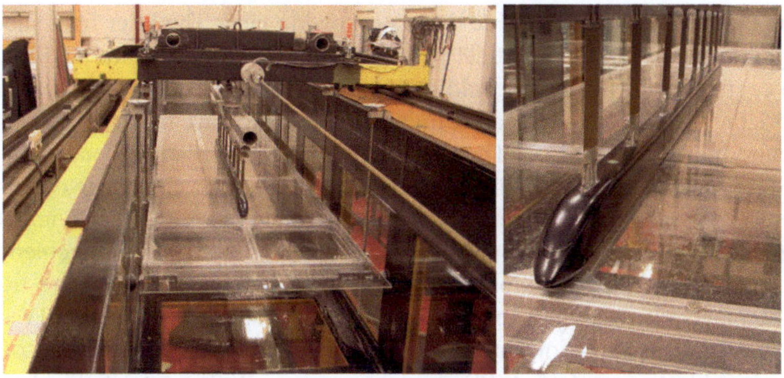

Fig. 1 A train model mounted on the moving model rig with the installed ground plate in the water towing tank at DLR Göttingen

was held by eight threaded rods allowing to adjust the ground clearance to any desired value. For this measurements a ground clearance between the wheel and the ground of 0.047 mm (235 mm in full scale) was chosen according to the standard ground clearance used in cross-wind measurements [6]. The ground plate consists of Plexiglas plates mounted on an aluminum frame. The plates had to be transparent for the laser light sheet needed for the PIV.

2.2 PIV Setup and Measurement for the Water Towing Tank

The sketch in Fig. 2 shows the two component PIV set-up used for the measurements in the water towing tank. The laser, laser light sheet optics, light sheet mirror and

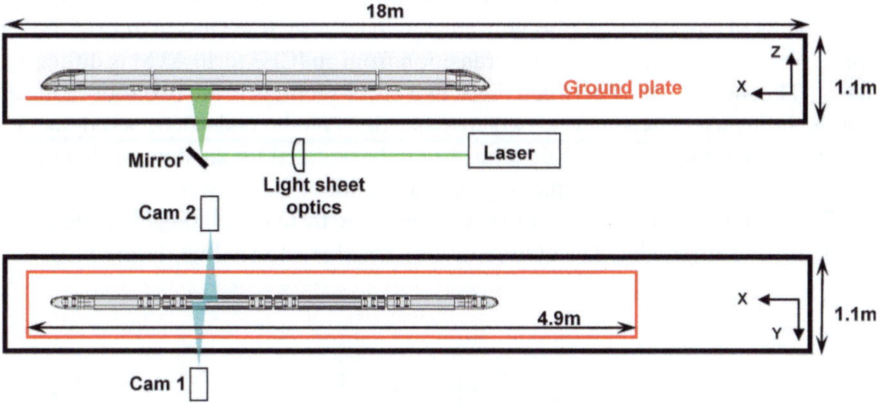

Fig. 2 Sketch of the two component PIV set-up for the water towing tank

cameras were all installed outside the water towing tank. The laser light sheet was arranged in a way that the vertical plane (XZ) between the ground and the train, at the centre line of the train, could be measured. This plane was chosen because the horizontal velocity underneath a train is expected to have their highest values in the centre line of the train [7]. The two cameras were installed perpendicular to the laser light sheet at both sides of the water towing tank. Two cameras were installed to increase the measurement field (also known as field of view). The complete field of view covered 0.065×0.025 m (W $\times$ H). Two PCO 1600 digital cooled 14 bit CCD cameras with a resolution of 1600×1200 pixels (W $\times$ H) were used, so that 450 pixels resolved the gap between the ground and the train. The PIV system had a maximum acquisition rate of 10 Hz. For the considered rig speed of 4 m/s it was possible to sample 5–6 PIV images per run. With a total field of view of 0.065×0.025 m (W $\times$ H) and a distance of 0.4 m between the images, the position had to be shifted 8 times to cover the entire train length ($\sim$2.1 m). At every position 10 runs were recorded and ensemble averaged to determine the mean flow field underneath the train. Additionally for a specific position 80 runs were recorded to be able to analyze the statistical convergence.

2.3 Train Model and Train Model Configurations

The used train model for the PIV measurements was a 4 unit (front car, 2 cars and tail car) 1:50 generic high-speed train model configuration. It is not possible to use more than 4 units owing to the dimensions of the water towing tank. The generic high-speed train model is a combination of the German high-speed train Inter City Express 3 (ICE3) and the Aerodynamic Train Model (ATM) which is similar to the ICE2. In Fig. 3 a close up view of the front is shown. The front of the train model has the shape of an ICE3 and shortly after the first bogie the cross-section smoothly changes to the cross-section of an ATM. A similar transition existed for the tail car since the ATM cross-section smoothly changes back to an ICE3 just in front of the last bogie. For the underframe itself the transition from an ICE3 to an ATM is difficult to detect. The most obvious between these two cross-sections is the two roof radiuses. But the influence between these two cross-sections on the underfloor aerodynamics is insignificant. Further, simplifications had to be realized for the downscaled model geometry. They are visible at the bogies where a lot of details were neglected.

Three different train configurations were realized by manipulating the underframe on the same train model. The configurations are characterized by different degrees

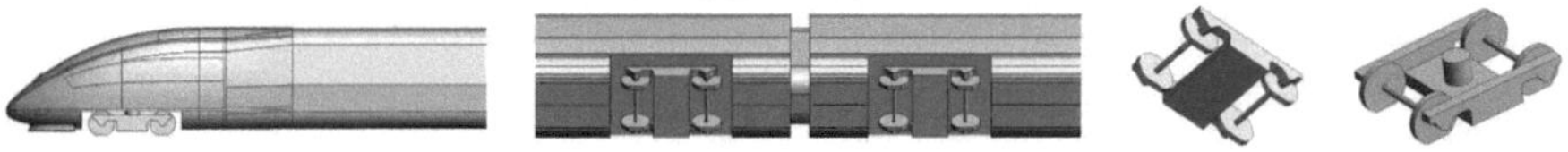

Fig. 3 The 1:50 generic high-speed train model: Front, underbelly and simplified bogie

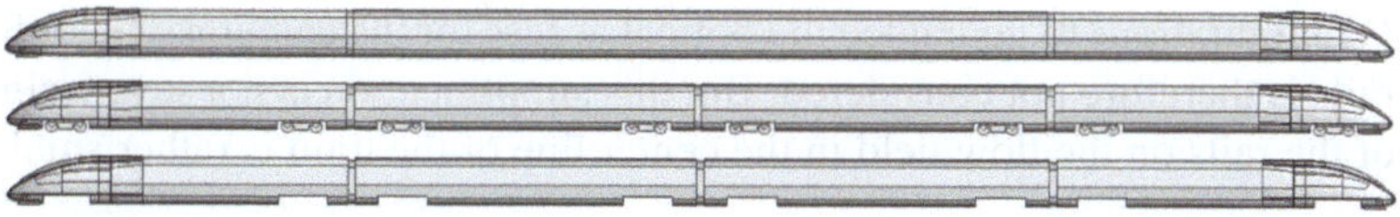

Fig. 4 The three train configurations: The smooth GHSTC (*upper*), the GHSTC (*middle*) and the rough GHSTC (*lower*)

of roughness. The generic high-speed train configuration (GHSTC) is a conventional high-speed train with inter car gaps and two bogies per car. For the smooth generic high-speed train configuration (smooth GHSTC) the bogies were removed and the bogie cut outs and the inter car gaps were covered. Hence a completely smooth underframe without protruding objects or gaps was realized. For the third configuration i.e. the rough generic high-speed train configuration (rough GHSTC), the bogies were removed and the bogie cut outs and inter car gaps were left open. CAD models of the three train configurations are presented in Fig. 4.

2.4 Ground Configurations

Each of the three different train configurations were all measured with the smooth and the rough ground. Thus in total six test cases were investigated. The smooth ground was realized with a smooth Plexiglas plate. For the rough ground additional Plexiglas plates were mounted on the ground plate and a grinding belt was glued on these additional plates. For the PIV measurements a small part of the grinding belt was removed, to guarantee optical accessibility for the laser light sheet, see Fig. 5. The roughness of the grinding belt was chosen such that the grain size corresponds to a 1:50 ballast stone. According to the European standard [3] the ballast stones selected for track beds are limited by holes of sieves with a width of 22.4–63 mm. This allows stones of sizes between 31.5 and 50 mm or 31.6 and 63 mm. The applied grinding belt had a grain size of less than 0.764 mm. No rails were mounted on both

Fig. 5 The two ground configurations, rough ground (*left*) and smooth ground (*right*)

ground configurations to guarantee the optical access for the cameras. The influence of the rails is therefore not considered. But this simplification is not severe since the effect of the rails on the flow field in the center line of the train is rather small.

3 Data Evaluation

The major steps in the evaluation of the PIV data are listed and described below.

(1) Calculate the flow field with the help of the cross-correlation technique [8] between the PIV images.
(2) Find the shift between the runs to allocate the exact position of the train.
(3) Shift the data into its right position and average the overlap between the cameras.
(4) Interpolate the data on a global grid for the calculation of the ensemble average and the standard deviation, as in Eqs. 1 and 2.
(5) Use the ensemble average flow field at every position to construct the entire flow field underneath the train.

$$C(x_j, z_k) = \frac{1}{n} \sum_{i=1}^{n} C_i(x_j, z_k) \tag{1}$$

$$C_{STD}(x_j, z_k) = \sqrt{\frac{1}{n-1} \sum_{i=1}^{n} (C_i(x_j, z_k) - \overline{C}(x_j, z_k))^2} \tag{2}$$

For the cross-correlation of the PIV images a multigrid interrogation (grid refinement) was used with a final interrogation size of 24 × 24 pixels (0.522 × 0.522 mm) with an overlap of 75 %. With this configuration the gap between the ground and the train was resolved with 75 data points.

The shift between the runs was calculated to enhance the accuracy of the ensemble average. Reference lines were painted on the side of the train to identify the shift in the middle of a car or for the smooth GHSTC. For the later no characteristics such as the bogies of the train existed.

The velocity field from both cameras was then shifted into its right position and the overlap between the cameras averaged. With the data at its right position the data could be interpolated on a new global equidistant grid ($\Delta x = 0.13$ mm, $\Delta z = 0.13$ mm) for statistical averaging. The outer edges of the mean flow field were obtained with a lower sample rate due to the shift between the runs. Therefore data points generated in less than 8 runs (out of total of 10 runs) were excluded. For the measurements for which the position was fixed and than around 80 runs were realized the data fields of all of the realized runs were averaged. The last step was to shift the ensemble averages into the correct positions underneath the train and to average the overlaps between the different positions. With this principle the flow field underneath the entire train was reconstructed.

4 Results

4.1 Velocities Along the Train Length

The mean flow field underneath the train can be divided into the three regions, the flow around the head, along the train body and around the tail. This can be seen in Fig. 6, where the normalized horizontal component U of the mean velocity field is plotted along the length of the train for the three considered trains and both ground types. The flow field in the region close to the head and before the first train axle, reflects a positive U-velocity maximum which is followed by a negative U-velocity minimum which is the head effect. In front of the head the flow is pushed in the traveling direction of the train. Close to the head displacement causes the flow to draw aside in perpendicular to the traveling direction. The result is a reversed flow region directly underneath the head (the negative U-velocity minimum). Further downstream the train drags fluid in traveling direction. Thus the U-velocity is increasing with the train length until the 2nd car is reached. There the U-velocity settles around a constant value. The flow field changes then at the tail of the train (at the last bogie) where the third and last region is observed. Part of the third region is the near wake, reflected

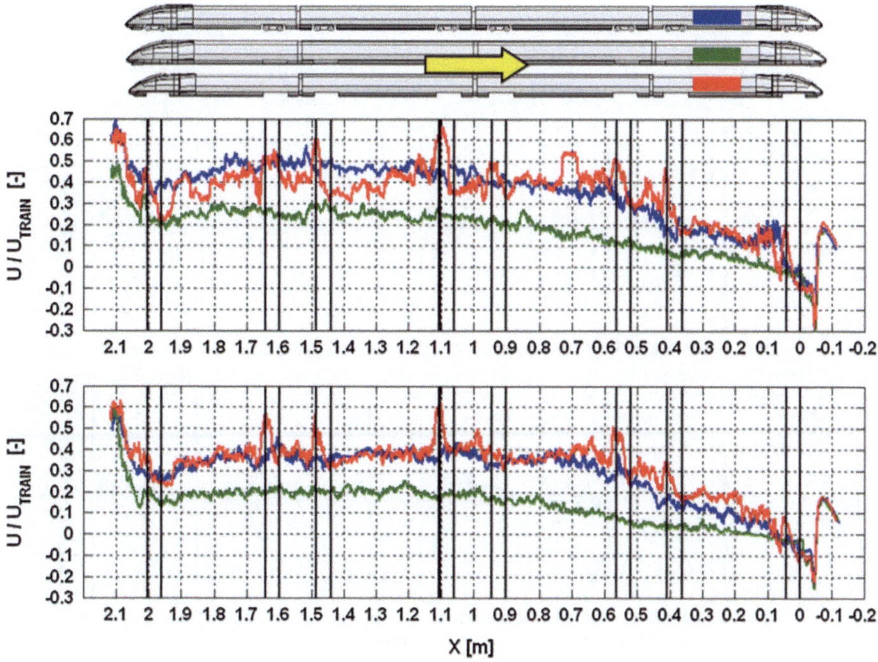

Fig. 6 Normalized U-velocity at TOR ($Z/H_{GAP} = 0.4753$) along the train length for the GHSTC (*blue*), the smooth GHSTC (*green*) and the rough GHSTC (*red*) for the smooth (*upper plot*) and the rough (*lower plot*) ground. The *black vertical lines* are the positions of the train axles

by a tiny decrease of the U-velocity just before the tail and directly after a strong acceleration of the flow. The smooth GHSTC shows the lowest U-velocity for both ground types. This is not surprising since this configuration has a smooth underframe with no protruding objects. The behavior for the GHSTC and the rough GHSTC is different due to ground roughness. For the rough ground the rough GHSTC produces a higher U-velocity at the front car than for the GHSTC. Thereafter the distribution of the U-velocity of both train configurations is almost similar. Only in the vicinity of the bogies the rough GHSTC produces higher U-velocity than the GHSTC. However for the smooth ground the U-velocities obtained for the GHSTC and the rough GHSTC are the same at the front car and the 1st car. For the second half of the train the U-velocity obtained for the rough GHSTC are lower than those measured for the GHSTC. Also for the smooth ground higher U-velocity values are observed around the bogies for the rough GHSTC. The effect of the ground roughness on the velocity distribution over the gap between the ground and the train is discussed more in detail in Sect. 4.2.

In Fig. 7 the normalized vertical component W of the mean velocity field is plotted along the train length for the three train configurations and both ground types. As expected the vertical velocity values for the smooth GHSTC are small except around the head and the tail due to the absence of protruding objects. The small

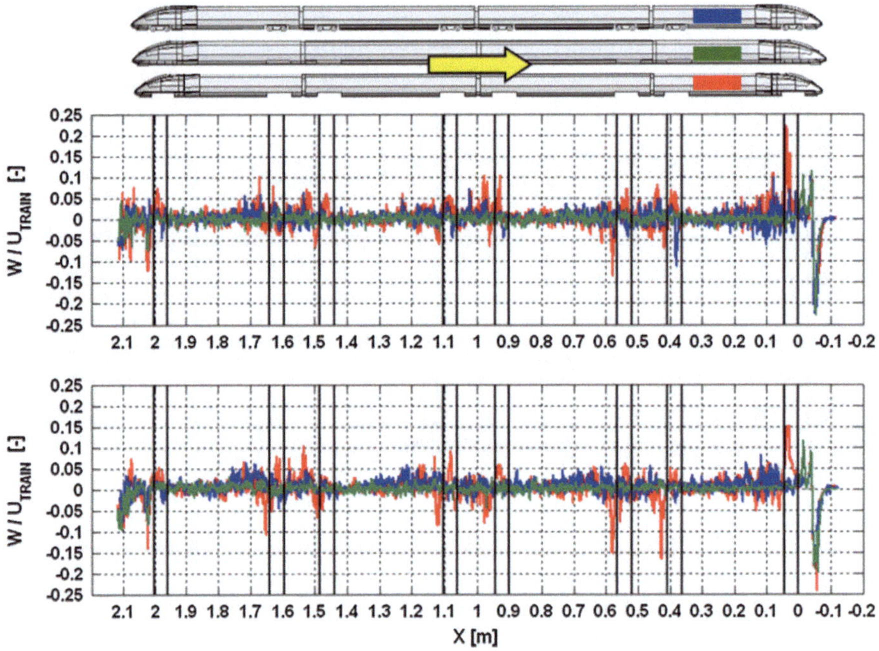

Fig. 7 Normalized W-velocity at TOR ($Z/H_{GAP} = 0.4753$) along the train length for the GHSTC (*blue*), the smooth GHSTC (*green*) and the rough GHSTC (*red*) for the smooth (*upper plot*) and the rough (*lower plot*) ground. The *black vertical lines* are the positions of the train axles

W-velocity fluctuations around zero are considered to be noise. For the two other train configurations the effects of the bogies and the open bogie cut outs are reflected in the W-velocity. The highest W-velocities in both directions are found for the rough GHSTC. The effects from the bogie areas are stronger at the beginning of the train than at the end. This is due to the boundary layer growth underneath the train. A maximum W-velocity value with a direction towards the ground are observed at every bogie followed by a W-velocity variations around zero which settles itself around the middle of every car. For the open bogie cut outs a maximum in the W-velocity towards the train appears first as the flow approaches the bogie cut out. Then a strong maximum of the W-velocity with a direction towards the ground appears where the flow leaves the bogie cut out. This is also followed by W-velocity variations with amplitudes which decrease in the middle of the car. The GHSTC and the rough GHSTC produce the same W-velocity value as the smooth GHSTC at the middle of the car, for every car, and the W-velocity stay constant until the next downstream bogie. This is interpreted as a W-velocity for the GHSTC and the rough GHSTC which is zero at this position just as that obtained for the smooth GHSTC.

4.2 Velocity Profiles

To obtain a better overview of the development of the flow field behavior, additionally velocity profiles over the gap between the ground and the train were extracted. The velocity profiles shown are from the measurements for a selected position and statistical averaging from 80 runs, instead of the 10 runs used for the results along the train length. In Fig. 8 the three train configurations are compared with each other for five different positions underneath the train. The positions underneath the train are indicated by the red vertical lines on the train configurations above the velocity profile plots. The front car velocity profile is the plot to the right and the tail car velocity profile is the plot to the left. The U-velocity is normalized with the train speed and plotted against the height normalized with the gap distance between the ground and the train underbelly. The ground is at $Z/H_{GAP} = 0$ where $U/U_{TRAIN} = 0$ and the train underbelly is at $Z/H_{GAP} = 1$ where $U/U_{TRAIN} = 1$.

Studying the velocity profiles from the right to the left plot the flow field development underneath the train is analyzed. The flow develops along the entire gap, especially at the ground, until it reaches the 2nd car where the flow is fully developed. Comparing the last two velocity profiles (two velocity profiles to the left) no difference can be observed, hence the flow is fully developed. The smooth GHSTC produces the lowest U-velocity over the entire gap for both ground types. Comparing the results for GHSTC and the rough GHSTC in Fig. 8 it can be seen that the flow of the two train configurations develops differently for the two ground types. The first three velocity profiles (front car, downstream of the 3rd and 4th bogie) of the GHSTC and the rough GHSTC are nearly the same for both ground types. Differences are observed at the end of the 2nd car and later at the tail car. For the rough ground the

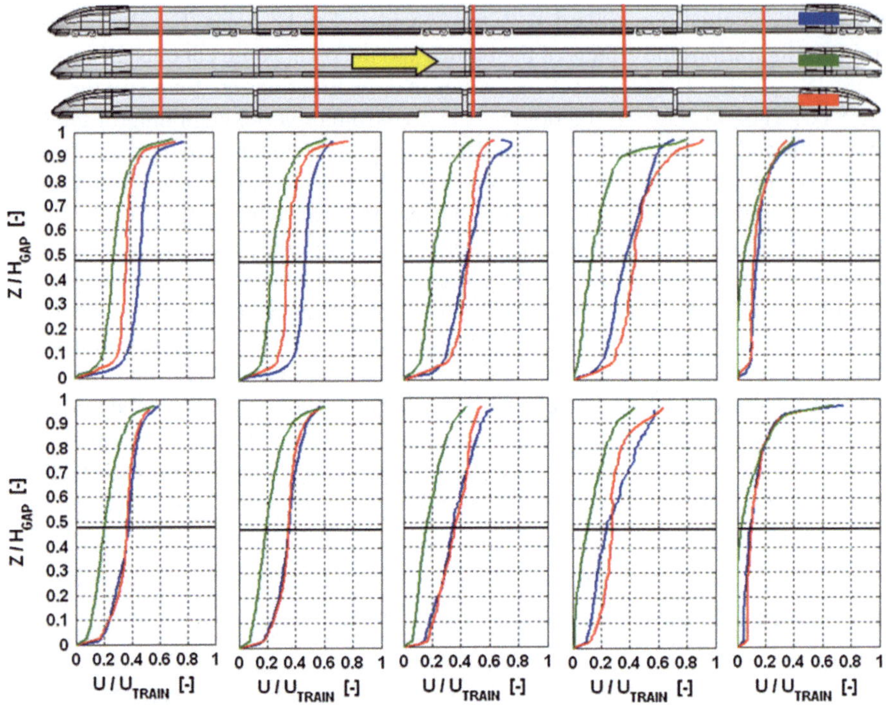

Fig. 8 Velocity profiles over the gap between the ground and the train underframe for the GHSTC (*blue*), the smooth GHSTC (*green*) and the rough GHSTC (*red*) for the smooth (*upper row*) and the rough (*lower row*) ground. The *red vertical lines* on the train configurations indicate the velocity profile positions and the *black horizontal line* is the TOR

velocity profiles are similar. But for the smooth ground the velocities of GHSTC overtake those of the rough GHSTC finally leading to higher U-velocity values.

Regarding ballast projection the velocity below TOR and the velocity gradient at the ground are two important factors among others. The velocity gradient at the ground is proportional to the shear force on the ground i.e. the ballasts. If ballast starts to move or elevates the velocity close to the ground has a big impact on the forces on the moving or flying ballast. Ranking the train configurations regarding the risk for ballast projection, by looking at the mean U-velocity component, the smooth GHSTC would be less sensible than the two other train configurations. The reason is that it has the lowest U-velocity and also the lowest velocity gradient close to the ground for both ground types. The ranking of the most susceptible train configuration is more difficult to develop due to the ground roughness effect between the GHSTC and the rough GHSTC. For the rough ground the rough GHSTC is the most susceptible due to the higher U-velocities around the bogies. However for the smooth ground it is believed that ballast projection is more an issue for the GHSTC. The rough GHSTC generates higher U-velocity values in the vicinity of the bogies but the U-velocity values between the bogies for the 2nd car and the tail car are lower. The difference between the velocity profiles measured for the GHSTC and the rough GHSTC for the 2nd car and the tail car is larger than the difference of the velocity profiles behind the fourth bogie. Therefore the GHSTC on the smooth ground is considered to be more susceptible for ballast projection.

All velocity profiles obtained for the 1st car and the tail car are plotted in Fig. 9. The effect of the ground roughness can be extracted by comparing the velocity profiles of the smooth ground (full lines) with the velocity profiles of the rough ground (dashed lines). For the smooth GHSTC the difference between the smooth and the rough ground types increases along the train length as the boundary layers develops. The ground roughness slows down the flow in the traveling direction underneath the train and with increasing U-velocities the difference is also increasing between the results of the ground types. The increasing difference between the ground configurations along the train length could not be identified for the GHSTC, where a clear difference was found from the beginning until the end of the train. For the GHSTC and the smooth GHSTC a difference of the flows for the two ground types was found over the entire gap. This was only observed for the front car and around the bogies for the rough GHSTC, see the plot to the right in Fig. 9. Further downstream the train the difference could only be found below TOR. The train roughness of the rough GHSTC is large enough that the ground roughness only influences the flow field close to the ground and not the flow field close to the train. The GHSTC is more sensitive to the ground roughness than the rough GHSTC, or the GHSTC is not as rough as the rough GHSTC.

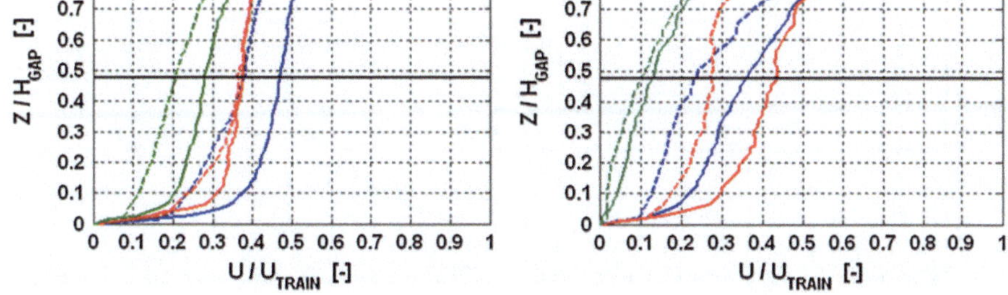

Fig. 9 Velocity profiles after the 3rd bogie and in the *middle* of the tail car for the GHSTC (*blue*), the smooth GHSTC (*green*) and the rough GHSTC (*red*) for the smooth (*full lines*) and the rough (*dashed lines*) ground. The *red vertical lines* on the train configurations indicate the velocity profile positions and the *black horizontal line* is the TOR

4.3 Standard Deviation Profiles

Profiles of the standard deviation of the U-velocity (U_{STD}) over the gap between the ground and the train underframe are plotted in Fig. 10 for the smooth (upper row) and for the rough (lower row) ground type. Comparing the three train configurations for both ground types it is easy to observe that the rough GHSTC leads to the highest U_{STD} values for all positions. This train configuration is also the roughest one and therefore it should lead to higher U_{STD} values. The lowest U_{STD} values was reached by the smooth GHSTC, this was also expected due to the completely smooth underframe. The value of the U_{STD} for the smooth GHSTC lies around 0.05–0.06 for the region above TOR at the beginning of the train and with the train length the U_{STD} value below TOR develops to a constant value over the gap of 0.05–0.06 for both ground types. The relation for the GHSTC to the other two train configurations changes with the train length. At the beginning of the train the behavior of the GHSTC is very close to that of the rough GHSTC and far away from the smooth GHSTC. For the next two downstream positions, i.e. downstream region of the 3rd and 4th bogie, the GHSTC profiles lies between those of the two other train configurations. This indicates that the open bogie cut outs are able to produce higher additional

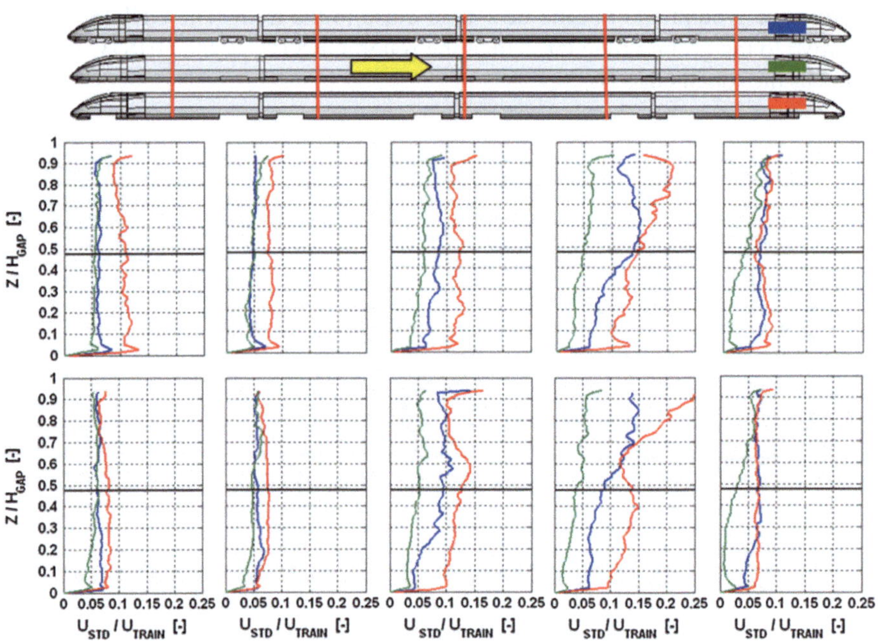

Fig. 10 U_{STD} profiles over the gap between the ground and the train underframe for the GHSTC (*blue*), the smooth GHSTC (*green*) and the rough GHSTC (*red*) for the smooth (*upper row*) and the rough (*lower row*) ground. The *red vertical lines* indicate the velocity profile positions and the *black horizontal line* is the TOR

ΔU-velocity to the mean than for the bogies and of course the smooth wall. For the two last positions the U_{STD} values of GHSTC are similar to the smooth GHSTC, except for the rough ground type where a difference was found in the region below TOR. The reason might be, that when the flow is fully developed the disturbances from the upstream bogie has a minor influence on the flow from the middle of the car until the next bogie. This is not the case for the rough GHSTC which produces a higher U_{STD} value, the disturbances from the open bogie cut out have not settled as for the GHSTC.

Summarizing the velocity profiles results in Sect. 4.2 and the results from the U_{STD} profiles the less sensible for ballast projection is the smooth GHSTC for both ground types. The smooth GHSTC are characterized by the lowest value of both quantities. For the ballast projection ranking of the train configurations the Gaussian distribution (or normal distribution) was assumed for the velocity distribution underneath the train. Applying the 2σ rule, that 95 % of all values extracted from the normal distribution are within 2σ of the average (in this case 2 U_{STD} of the mean U-velocity). Ranking the train configurations after the upper limit of the U-velocity distribution the rough GHSTC is the most susceptible for ballast projection on both ground types. The rough GHSTC reflects the highest U-velocity for all five positions for the rough ground. For the smooth ground the rough GHSTC leads to the highest U-velocity values for the front car and after the 3rd and 4th bogie. However at the end of the 2nd car the GHSTC produces a larger U-velocity values and for the tail car the GHSTC and the rough GHSTC had the same U-velocity. The U-velocity of the GHSTC was only higher at one position underneath the train and therefore the rough GHSTC was considered to be more susceptible for ballast projection.

The ground roughness also had an effect on the U_{STD} values. In Fig. 11 the U_{STD} profiles are plotted for both ground types for three different positions underneath the train. The ground roughness effect differs for the three train configurations. For the smooth GHSTC a small difference between the ground types was found for the

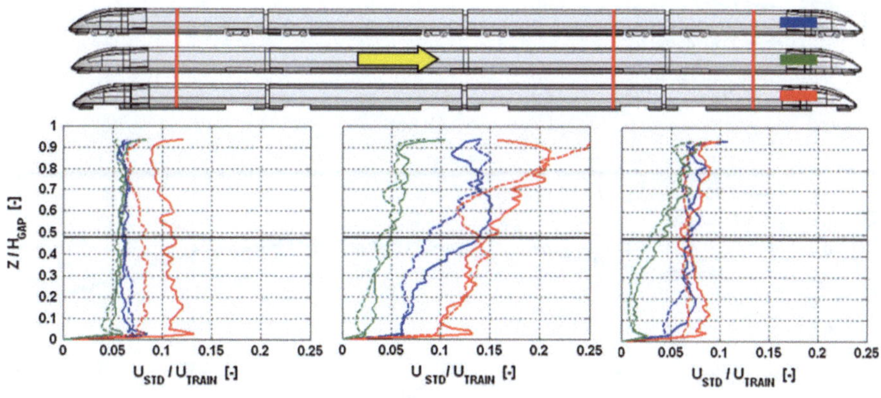

Fig. 11 U_{STD} profiles for the GHSTC (*blue*), the smooth GHSTC (*green*) and the rough GHSTC (*red*) for the smooth (*full lines*) and the rough ground (*dashed lines*). The *red vertical lines* indicate the velocity profile positions and the *black horizontal line* is the TOR

first half of the train where the smooth ground leads to a higher U_{STD} value. For the second half of the train the difference decreases and it's hard to differentiate the results obtained for the two ground types. The GHSTC showed the same ground roughness effect like the smooth GHSTC for the first half of the train. For the second half of the train the relation between the ground types changed and the rough ground generates a higher U_{STD} value than the smooth ground. The tendency for the ground roughness effect is different for the rough GHSTC than for the two other train configurations. The difference between the ground types increases with the train length. The U_{STD} value measured for the smooth ground was a bit higher than those for the rough ground measurements of the front car. Around the bogies the ground types showed little influence and a clear difference at the end of the train where the smooth ground lead to higher U_{STD} values than for the rough ground.

4.4 Comparison with Full Scale Measurements

The PIV results from the water towing tank were also compared to full scale measurements [2] to identify differences and similarities between full scale and downscaled model measurements. For the comparison the length of the train model was multiplied with the model scale factor. Two comparisons were conducted due to the different lengths of the train sets (14 units for the full scale and 4 units for the downscaled model measurements), one where the first axle and one where the last axle of the trains were aligned, see Fig. 12. The two curves do not fit perfectly, due to the

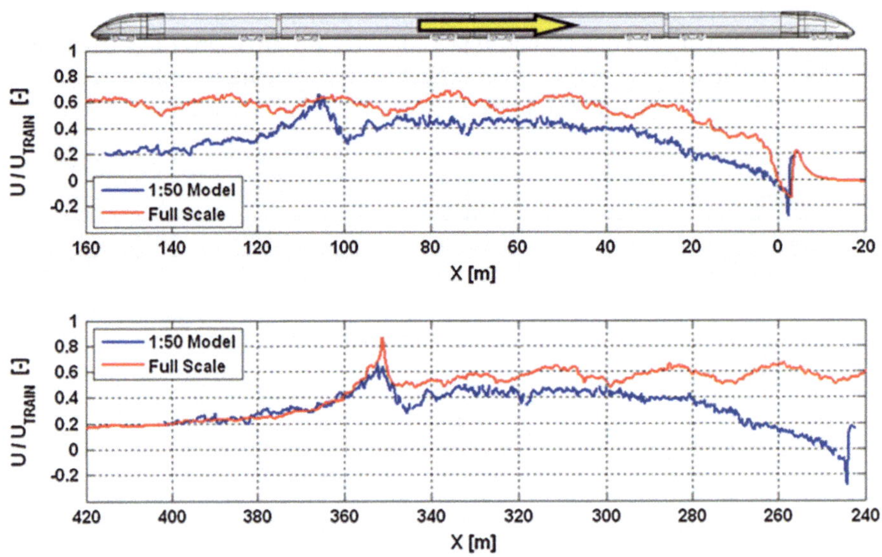

Fig. 12 Comparison to full scale measurements. In the *upper plot* the first axles of the trains are aligned and in the *lower plot* the last axle of the trains are aligned

different geometrical shapes of the trains. For the downscaled model measurements a generic high-speed train configuration was used and for the full scale measurements the Italian high-speed train ETR 500 was measured. However the same flow field characteristics were found around the head and the tail of the train. The flow field was also fully developed at the beginning of the 2nd car, which was also observed for the full scale measurements. The U-velocity distribution obtained for the downscaled model measurements is always lower than that of the full scale measurements. The reason is that the train model in the downscaled model measurements is aerodynamically better than the ETR 500. The simplified bogies, in Fig. 3, used for the water towing tank are completely smooth on the surface facing the ground. This is not the case for a full scale bogie on a train where a lot of protruding objects exists such as gearboxes, electrical motors, cables etc. There are also a lot of cavities where the air can flow into and be accelerated by the train, the simplified bogies shields the bogie cut out and makes the train underframe smoother than for the ETR 500. The bogie differences is also one of the reasons that the downscaled model measurements does not have a self-similar or wave pattern which was found for the full scale measurements. Another reason is that for a full scale train there are also air outlets (cooling air) on the underframe of the train which gives a momentum into the flow field. The comparison showed that the normalized U-velocity distribution underneath the train for the downscaled model measurements is comparable to that of a full scale high-speed train cruising at a speed of 250 km/h.

5 Conclusions

- It was possible to create three different flow fields underneath the train by manipulating the underframe on the same train model.
- The flow fields were fully developed at the beginning 2nd car. This was also observed for the full scale measurements.
- Considering the different train geometries and the simplifications that were done for the downscaled model measurements a comparable U-velocity distribution to the full scale measurements was achieved.
- The ground roughness changed the flow fields underneath all three train configurations.
- The smooth GHSTC is considered to be the less sensitive and the rough GHSTC the most susceptible for ballast projection.
- Important information can be extracted from downscaled model measurements, however to extrapolate the results into full scale should be done with caution.

References

1. Claus, P.: WP 1.2 ballast projection: prestudy of ballast projection incidents. In: DeuFraKo Project Aerodynamics in Open Air (AOA) (2007)
2. Deeg, P., Jönsson, M., Kaltenbach, H.J., Schober, M., Weise, M.: Cross-comparison of measurement techniques for the determination of train induced aerodynamic loads on the trackbed. In: Proceedings of the BBAA VI, Milano, Italy, 20–24 July 2008. http://bbaa6.mecc.polimi.it/uploads/treni/BPR04.pdf
3. Din, E.N.: 13450:2002: Aggregates for Railway Ballast. Gesteinskörnung für Gleisschotter, Deutsche Fassung (2002)
4. Eckelmann, H.: Einführung in die Strömungsmeßtechnik. B.G Teubner, Stuttgart (1997)
5. Kaltenbach, H.J.: WP1 underfloor aerodynamics summary report. In: DeuFraKo Project Aerodynamics in Open Air (AOA) (2008)
6. prEN 14067–6:2006: Railway Applications—Aerodynamics—Part 6: Requirements and Test Procedures for Cross Wind Assessment (2006)
7. Quinn, A.D., Hayward, M., Baker, C.J., Schmid, F., Priest, J.A., Powrie, W.: A full-scale experimental and modelling study of ballast flight under high-speed trains. In: Rai, F-J. (ed.) Proceedings of Institute of Mechanical Engineering (2009). doi: 10.1243/09544097JRRT294
8. Raffel, M., Willert, C.E., Wereley, S.T., Kompenhans, J.: Particle Image Velocimetry, 2nd edn. Springer, Berlin (2007)
9. Shinojima, K.: Study on the phenomena of snow adhering to and dropping from shinkansen train, and the countermeasures. Q. Rep. **25**(2), 41–44 (1984)

Aerodynamic Drag Reduction of Open-Top Gondola and Hopper Cars in Unit Train Operation and Impact on Train Fuel Consumption and Economics

James C. Paul

Abstract A three-part study was conducted to determine the effectiveness of retrofit aerodynamic drag reducing devices on the fuel consumption and economics of open-top, bulk commodity, gondola and hopper rail cars in unit train service. Specific applications included trains transporting coal from mines to power plants. During the first part of the study wind tunnel testing and computational fluid dynamics were combined with an extensive literature search to rank the drag reducing effectiveness of a variety of devices including covers, internal baffles, end treatments, gap fillers, car side geometry, and underbody modifications. During the second portion of the study, three approaches were utilized to determine fuel savings associated with each aerodynamic retrofit device. These included two classical methods and a newly-developed train energy model. Results were validated using fuel consumption data provided by a U.S. Class I railroad. During the third portion of the study, an economic analysis of the candidate devices was completed which included the following parameters: weights of the retrofit devices, manufacturing and installation costs, drag reduction effectiveness, and projected return on investment. The study predicted round trip fuel savings, due to the addition of aerodynamic modifications to open-top rail cars in unit-train service, ranging from 2.7 to 19.9 % and return on investment durations as short as 2 years, depending upon the type of device, route, and car utilization. It was shown that economic viability of car modifications depends only partly on aerodynamic performance. Some of the modifications exhibiting high levels of drag reduction were eliminated from additional consideration due to high associated costs and negative impact on payload capacity.

1 Introduction

The current study was conducted to determine the effectiveness of retrofit aerodynamic drag reducing devices on the fuel consumption and economics of open-top gondola and hopper rail cars in unit train operation, including cars in dedicated

J.C. Paul (✉)
Airflow Sciences Corporation, 12190 Hubbard Street, Livonia 48150, USA
e-mail: jpaul@airflowsciences.com

A. Dillmann and A. Orellano (eds.), *The Aerodynamics of Heavy Vehicles III*,
Lecture Notes in Applied and Computational Mechanics 79,
DOI 10.1007/978-3-319-20122-1_3

coal-hauling service. The study was divided into three parts: (1) wind tunnel testing and computational fluid dynamics were combined with an extensive literature review to identify and rank candidate retrofit drag reducing modifications; (2) three methods were employed to determine fuel savings associated with each aerodynamic retrofit device; and (3) an economic analysis of the candidate retrofit devices was performed based on capital and operational cost data.

An initial task involved definition of the North American coal car fleet. The purpose of the task was to identify the potential market for aerodynamic drag reduction retrofit devices. The research involved a number of information sources including the Railinc UMLER database [72, 73], the U.S. DOE Energy Information Agency [13, 14], the U.S. Surface Transportation Board [71], the Association of American Railroads [88, 93], industry publications [7, 9, 50, 59, 70, 80], and interviews with utility company representatives [85–87, 89, 90]. In addition, calculations were performed based on utility industry statistics (number of coal cars per megawatt multiplied by the total generating capacity of U.S. coal-fired power plants—megawatts) and coal car utilization statistics (coal car loads per year divided by average coal car utilization—trips per car per year). The data sources and calculations provided a consistent result for the composition of the fleet and the results are summarized in Fig. 1.

Another initial task involved selection of evaluation methods for defining the aerodynamic behavior of the candidate retrofit aerodynamic devices. Methods were sought that provided acceptable levels of accuracy while meeting project time and budget constraints. Both wind tunnel testing and computational fluid dynamics (CFD)

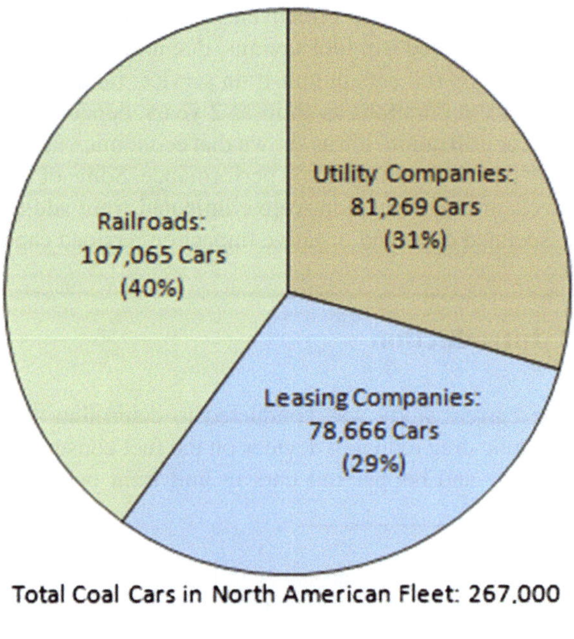

Fig. 1 Composition and ownership of North American coal car fleet

were employed [68]. To match full-scale relationships between momentum and viscous aerodynamic forces, as well as complex bluff body wake structures, the scale model railroad cars were evaluated at Reynolds numbers corresponding to those of their full-scale counterparts. Several researchers have documented the relationship between Reynolds number and drag coefficient, reinforcing the requirement for choosing wind tunnel model scales and air speeds that replicate the full-scale Reynolds number [15, 20, 30, 43, 44, 77, 103].

Significant research and development is underway in the area of CFD codes and their application to bluff body flows [1, 2, 5, 29]. These include Unsteady Reynolds-Averaged Navier-Stokes (URANS), Large Eddy Simulation (LES) [49], Detached Eddy Simulation (DES), Direct Numerical Simulation (DNS), spectral methods, vortex methods, and Lattice-Boltzmann methods [6, 101]. All of these have been applied to the evaluation of heavy vehicle aerodynamics [1–64].

For the current study, appropriate tools were sought to provide practical engineering solutions within commercial timeframes and budgets. Because of the many configurations to be modeled, computationally-intensive, and/or Reynolds-number-limited methods such as LES, DES, and DNS could not be accommodated. Two 3-D Reynolds-Averaged Navier-Stokes (RANS), finite-volume simulation codes were selected: (1) a proprietary code, VISCOUS [58, 60, 61], and (2) a commercial code, FLUENT [19]. The Reynolds stress tensor was addressed using the K-ε equations [52, 53]. Although this method does not model detailed turbulent structures, it does predict average surface pressures and force differences with sufficient accuracy for rail car design and cross-wind-induced tipping moment determination [4]. The CFD simulations and wind tunnel tests were supplemented and validated with field tests.

2 Section I: Identify and Rank Candidate Retrofit Coal Car Modifications

In addition to wind tunnel testing and CFD simulations, an extensive literature review was completed to assist in evaluating candidate coal car retrofit drag reduction modifications. Categories of drag reduction methods include: covers, internal baffles, car end treatments, inter-car gap controls, side wall treatments, and underbody modifications.

2.1 Review of Past Work

Results of research on train resistance components, including aerodynamic drag, are included in papers by Davis [12], Hay [36], Hammitt [32–35], Hickey [37–41], Johansen [45], Johnson [46], Joshi [47], Klemin [48], MacFadyen [55], Mailoux [56], Miller [57], Schito [78], and Paul [62–64, 66]. Railroad freight car aerodynamics research has been supported by the U.S. railroads, equipment manufacturers, and the

Association of American Railroads ([16, 21, 24, 26, 27, 102]). Recent increases in fuel prices have once again generated interest in reducing train resistance.

2.2 Results of Current Research

Covers have been shown to be very effective for reducing coal car aerodynamic drag ([8, 11, 25, 54, 65]). Designs range from simple fabric covers to elaborate automatic-opening fiberglass structures. All covers act to reduce the entry of high velocity air to the car interior (see Fig. 4). This high velocity air has been shown to impact the downstream end wall of the car where it increases the static pressure and thus acts to resist forward motion. Flat covers reduce aerodynamic drag by over 60 % for empty gondola cars and 40 % for empty hopper cars. In addition the covers act to reduce the drag of loaded cars between 6 and 10 % (Fig. 2).

Internal Baffles provide significant drag reduction without imposing the operational restrictions of full and partial covers. The baffles produce a "trapped vortex" that acts to direct the free-stream air along the top of the car and reduce the impact pressure on the downstream vertical end-wall of the open-top cars as shown in Figs. 3 and 4. Many baffle designs have been investigated, including solid and perforated partitions, partial-height designs, and triangular-shaped inserts that fit within the existing internal bracing [18, 21, 65, 75]. It is interesting to note that the internal

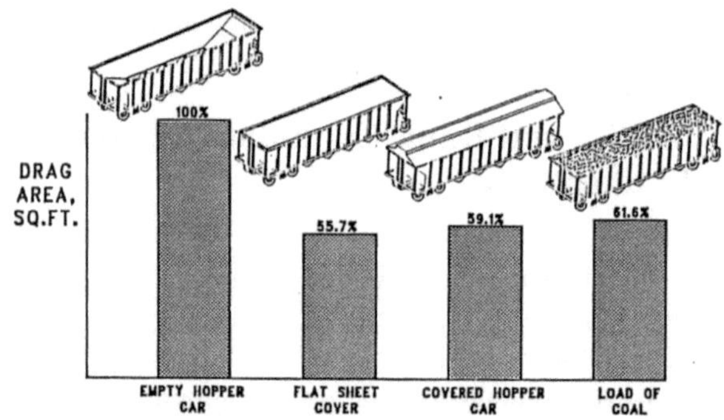

Fig. 2 Effect of coal load and covers on gondola car aerodynamic drag (for example, the drag of a gondola car with a flat cover is 55.7 % of the drag of the same car with no cover or load)

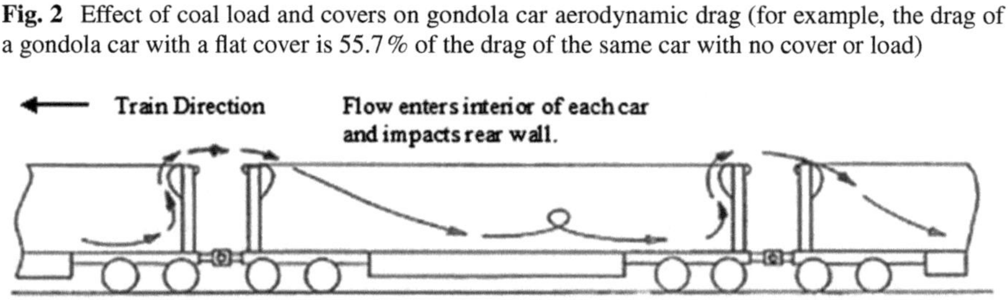

Fig. 3 Observed flow behavior in vicinity of open-top rail cars

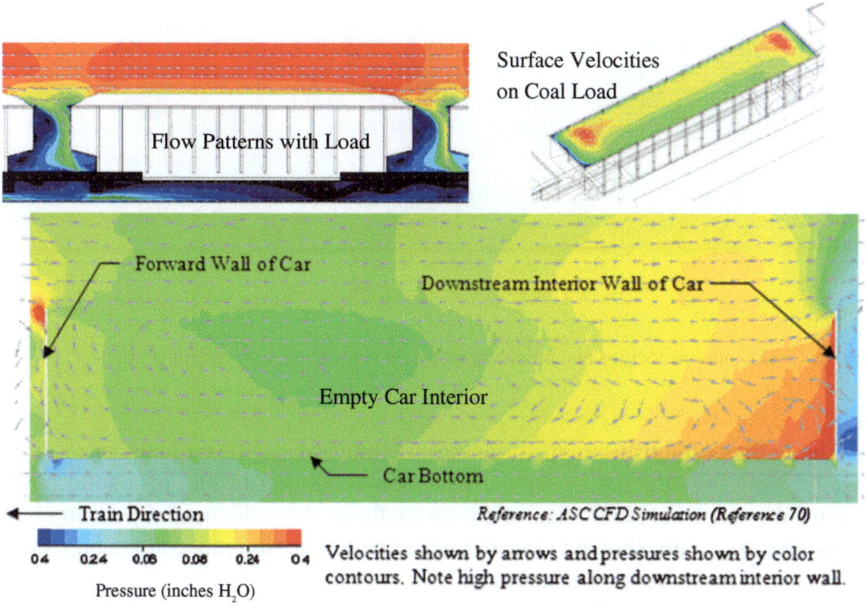

Fig. 4 CFD results: velocities and pressures along car centerline (no baffles)

Fig. 5 Example coal car with internal baffles

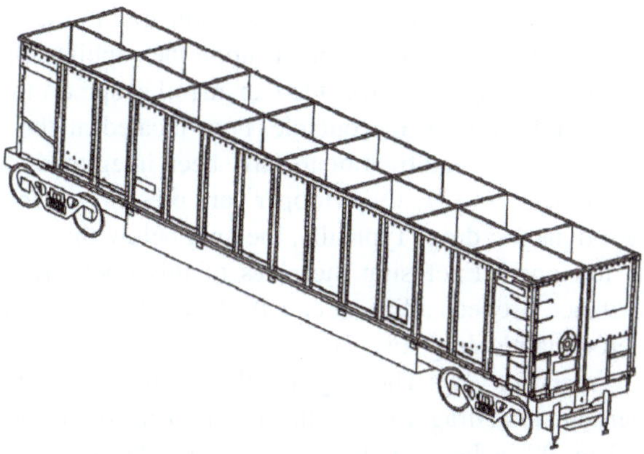

bracing alone acts to reduce aerodynamic drag 5–10 % by reducing the free-stream air from entering the car cavity. CFD modeling and wind tunnel testing indicate that internal baffles reduce aerodynamic drag between 16 and 60 % depending upon the car design and cross-wind conditions. One of the most effective designs identified during the study includes a full-height, longitudinal baffle along the car centerline and three to seven transverse full-height, solid baffles installed perpendicular to the car centerline, as shown in Fig. 5. This agrees with the results obtained by other researchers, which indicate the optimum cavity opening width and length are on the same order as the cavity vertical dimension [75, 82–84, 96].

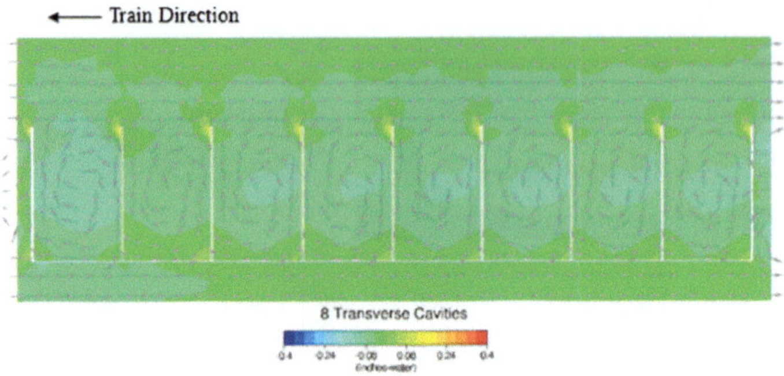

Fig. 6 CFD results: velocities and pressures along car centerline (with internal baffles)

The baffles establish “trapped vortices” within the cavities which act to direct exterior flow over the top of the car in a manner similar to that of a solid cover. The results of a CFD simulation of the cavity flows along the central portion of the open-top coal car are shown in Fig. 6. Interior baffles act to reduce drag under cross wind conditions as well ([28] through [76]).

Car End Treatments: The end geometry of both gondola and hopper cars offer another area where modifications can be made to reduce aerodynamic drag. Adding airfoils or fairings to the tops of the vertical end walls has been shown to reduce aerodynamic drag from 5 to 31 % compared to the uncovered, empty car [65]. The level of drag reduction depends upon the sophistication of the add-on designs. The most effective design included an airfoil shape on the inside of the car and a semi-circular fairing on the outside, both located at the top of the vertical end walls. Some of these end treatments have been integrated into coal cars offered by several manufacturers [31, 69]. Hopper cars present yet another opportunity for reducing aerodynamic drag. Typically, the area below the slope sheet at each end of the car is left open. Enclosing the sides of this open area effectively decreases the gap distance between adjacent cars [98], which, as noted below, produces a reduction in aerodynamic drag (Fig. 7).

Inter-Car Gap: The length of the gap between adjacent cars has an effect on train aerodynamic drag. In fact, the drag associated with wide inter-car gaps (greater than 2.3 m, 90 inches) has been measured to be over 55 % greater than that associated with narrow inter-car gaps (less than 0.8 m, 30 inches) (Ref. [23], p. 10; Ref. [99], pp. 157–158, and Ref. [25], pp. 165–170). Closely-spaced cars appear to the air stream as one long body with a relatively uninterrupted boundary layer whereas cars separated by large gaps appear as distinct objects [42, 74]. Reducing the inter-car gap is effective at reducing aerodynamic drag for hopper and gondola cars with and without covers, loaded and empty, and under cross wind conditions [100]. A plot of drag effects due to changes in inter-car gap distances is shown in Fig. 8 for the case of gondola cars equipped with flat covers. It is noted that reducing the inter-car gap to 0.64 m (25 inches) from the standard gap distance of 1.10 m (43.5 inches), reduces the drag area (drag coefficient multiplied by reference area, see Eq. 3) by 0.65 m^2

for hopper cars and 0.56 m^2 for gondola cars [65]. Employing gap fillers (flexible devices that block the flow of free stream air into the gap) reduces the aerodynamic drag of hopper and gondola cars by 7–9 % for empty, uncovered cars and 19 % for covered cars, based on standard car spacings.

Side Wall Treatments: Exposed, external ribs associated with the sides of standard-design hopper and gondola cars have been shown to disrupt the air flow along the vehicle and increase the aerodynamic drag in a manner similar to that of the open tops [17, 23, 65]. For ribs at standard spacings between 0.7 m (27 inches) 1.02 m (40 inches), free stream flow has been observed entering the region between the ribs and striking the upstream rib faces. Placing smooth side sheets over the exposed ribs reduces the aerodynamic drag on the empty, uncovered cars by 9–13 % [17]. For the loaded and covered cars, drag reductions of 24–30 % were measured after addition of the smooth sides [17, 65]. The application of aerodynamic fairings to the individual ribs, or providing aerodynamic shaping to the rib cross sectional geometry, reduced the drag by 6 % for the empty, uncovered cars and by 15 % for the loaded or covered cars. Rounding the square-edged vertical corners at the car ends exhibited minimal drag reduction, likely due to the boundary layer effect along the sides of the train.

Underbody Modifications: During the wind tunnel test program, car models were equipped with removable underbody components, including wheel trucks, drop bottoms, and hoppers, to allow measurement of the drag contribution for the components individually and collectively, Wheel trucks were found to have a higher drag contribution at non-zero yaw angles. Based on the empty, uncovered car drag measurements, the wheel trucks represent from 2.5 to 3.3 % of the car drag at zero degrees yaw and from 5.5 to 12 % of the car drag at 10° yaw [17, 21, 65]. Drop bottoms and hoppers, on the other hand, act to decrease the drag of the cars under cross-wind conditions. At zero degrees yaw, the drop bottoms and hoppers have little effect on aerodynamic drag, but at 10° yaw, the drag for the drop bottom and hopper cars is 11 % lower than the flat bottom cars. Side skirts are also effective under yaw conditions, reducing drag between 5 and 8 %, depending upon baseline car geometry and yaw angle [68].

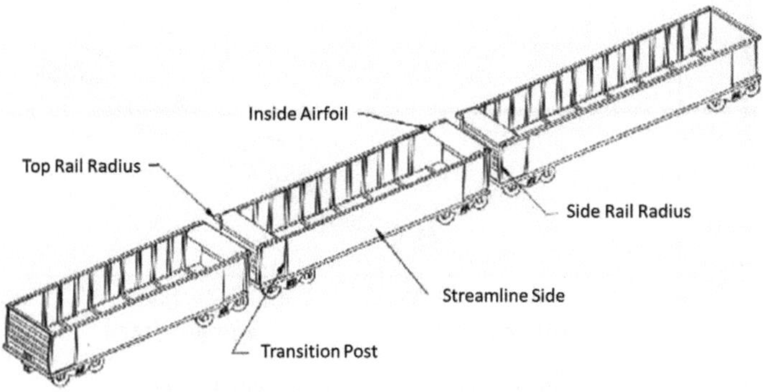

Fig. 7 Aerodynamic end treatments and side wall modifications for gondola car

Ranking of Candidate Coal Car Aerodynamic Devices: To facilitate comparisons of the various devices, wind averaged drag areas will be utilized. The definitions of these terms follow:

The formula for aerodynamic drag is:

$$D = C_d \, A \, q \tag{1}$$

where:

$$\begin{aligned} D &= \text{Drag Force (N)} \\ C_d &= \text{Drag Coefficient (non-dimensional)} \\ A &= \text{Cross Sectional Area of Coal Car } (m^2) \\ q &= \text{Dynamic Pressure of Air} \end{aligned}$$

The dynamic pressure is a function of air density and speed as shown in the following formula:

$$q = 1/2 \, \rho V^2 \tag{2}$$

where:

$$\begin{aligned} \rho &= \text{Density of Air } (kg/m^3) \\ V &= \text{Air Velocity } (m/sec) \end{aligned}$$

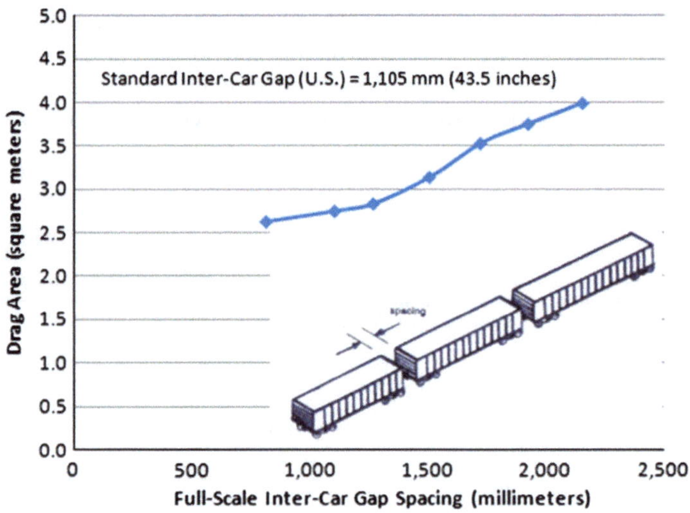

Fig. 8 Effect of inter-car gap on drag area of gondola cars with flat roofs

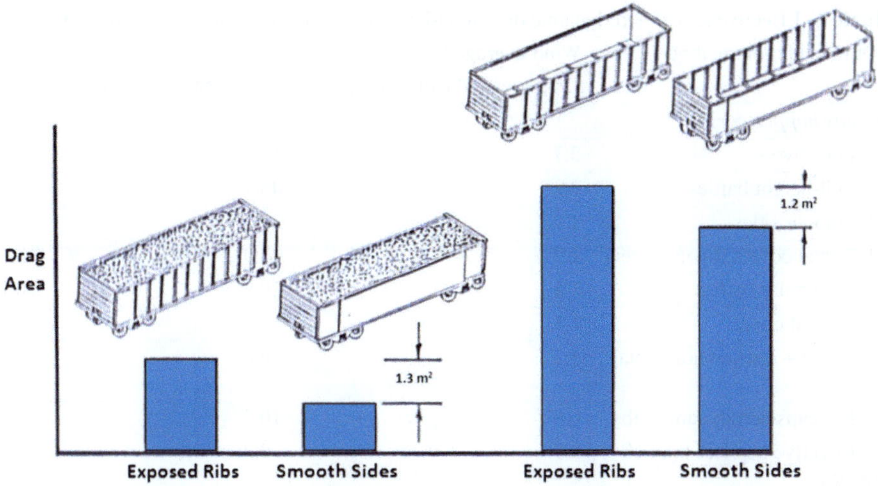

Fig. 9 Effect of adding smooth sides to loaded and empty open top gondola cars

It is convenient to group the drag coefficient and cross sectional area together and refer to them collectively as the drag area. This eliminates any confusion regarding the selected value for the cross sectional area (Fig. 9).

$$\text{Drag Area} = C_d\, A \tag{3}$$

Aerodynamic drag varies with yaw angle (cross wind effects). To rank the effectiveness of the drag reducing devices, the concept of Wind-Averaged Drag (WAD) Coefficient is introduced. The wind-averaged drag coefficient is associated with a yaw angle based on average wind speeds and train routes in North America. It is defined in SAE Publication J1252 [79] and in the report prepared by Frank Buckley of the University of Maryland [10]. Statistical methods are utilized to define the wind average yaw angle. The American Wind Energy Association (www.awea.org) and the National Climatic Data Center (NCDC) of the U.S. Department of Commerce (www.ncdc.noaa.gov) publish meteorological data for the United States. The representative average wind speed is 11 km/h. Based on an average train speed of 70 km/h, the wind averaged yaw angle is 5.5°, from which the wind-averaged drag value can be determined.

Selected retrofit aerodynamic modifications have been ranked by effectiveness and are summarized in Table 1. These data are referenced to the baseline wind average drag areas for empty and loaded gondola and hopper cars summarized in Table 2 [see, for example, Ref. [17], pp. VII–28].

Table 1 Effectiveness of aerodynamic drag modifications on coal car wind-averaged drag

Drag reduction method	Wind-averaged	
	Drag area reduction (m^2)	Drag area reduction (%)
Empty hopper cars		
1. Flat covers	3.7	51.7
2. Full-height baffles	3.2	45.1
3. Smooth sides	1.9	27.4
4. Partial covers (open center)	1.7	23.9
5. Triangular baffles	1.5	20.9
6. Domed covers	1.5	2.0
7. Inside + outside horizontal airfoils	1.2	16.8
8. Rib caps/aerodynamic ribs	0.8	10.7
9. Effective gap less than 10 inches	0.7	9.2
10. Cross bracing	0.6	8.8
11. End enclosure	0.5	7.0
12. Side skirts	0.4	5.4
Empty gondola cars		
1. Flat covers	3.5	44.1
2. Full-height baffles	3.3	40.5
3. Triangular baffles	2.0	25.1
4. Domed covers	1.8	22.3
5. Partial covers (open center)	1.6	20.4
6. Inside + outside horizontal airfoils	1.6	19.9
7. Smooth sides	1.2	15.5
8. Cross bracing	0.8	8.7
9. Rib caps/aerodynamic ribs	0.7	8.7
10. Effective gap less than 10 inches	0.6	6.9
11. Side skirts	0.4	4.7

2.3 *Results and Conclusions: Section I: Aerodynamic Drag Reduction*

Wind tunnel testing and CFD simulations were utilized to evaluate retrofit aerodynamic modifications to open-top gondola and hopper cars. The devices exhibiting the largest reductions in drag were covers (20–50 %), baffles (20–45 %), end treatments (7–19 %), and side modifications (16–27 %).

Table 2 Baseline wind averaged drag areas for coal cars (drag at 5.5° Yaw angle)

Car	Configuration	Drawing	Wind-averaged drag area (m^2)
Hopper	Loaded	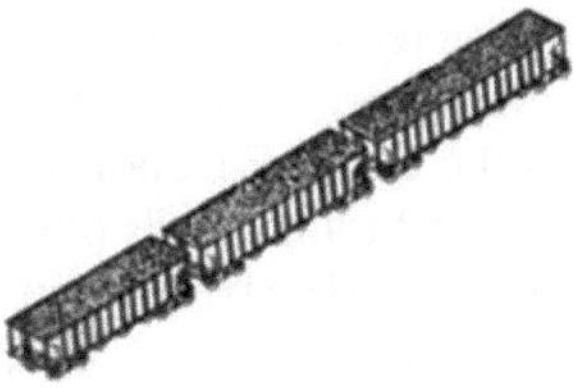	4.6
Hopper	Empty	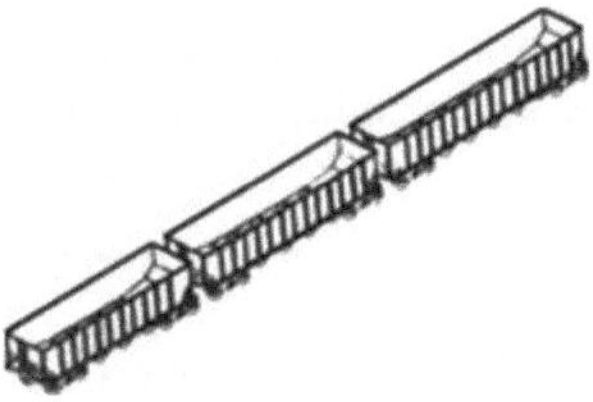	7.1
Hopper	Loaded	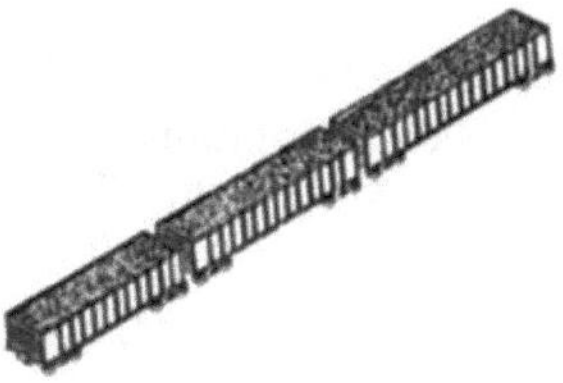	5.1
Hopper	Empty	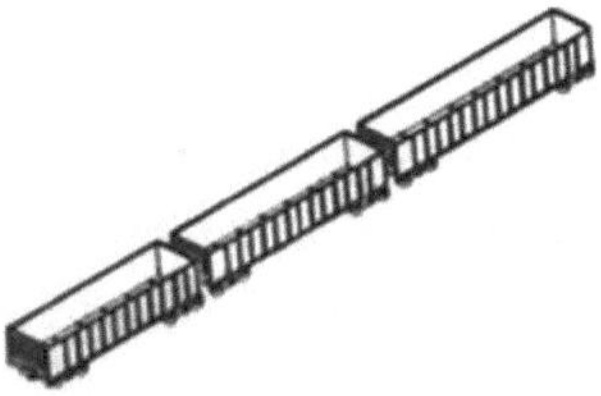	8.0

3 Section II: Fuel Savings Calculations

An analysis was performed to determine the impact of the various aerodynamic modifications on train energy requirements. This study also included calculations of fuel consumption for representative unit trains and the economic impacts associated with installing the retrofit aerodynamic devices. Economic parameters calculated during the analysis included fuel savings, revenue penalties of reduced payload (due to added weight of aerodynamic devices), fabrication costs for the devices, and return on investment.

3.1 Review of Past Work

To determine the impact on locomotive fuel usage resulting from implementation of the identified aerodynamic drag reducing devices, representative train consists were simulated using three calculation procedures: (A) trains operating on straight level track: calculation based on the Davis Equation for train power requirements, the energy content of diesel fuel, and train speeds obtained from a major U.S. railroad ([12, 36, 62], and [94] through [97]), (B) trains operating on both level and hilly routes using the AAR AERO Subroutine and associated fuel usage equation [22], and (C) a spreadsheet-based fuel calculation procedure developed specifically for ATP, as part of the current study, for unit coal trains traveling specific routes between a Powder River Basin (Wyoming) mine and power plants located in the US Midwest and US Southeast [68].

3.2 Train Energy and Fuel Economy Calculations

Fuel consumption for the straight, level track condition was defined using the following equation [62] and train speed histograms (distance traveled within each speed category) supplied by a Class I U.S. railroad.

$$\Delta FC = [E_{df}/\eta][\Delta D/q][\Sigma V_i^2 L_i] \tag{4}$$

where

$$\begin{aligned} \Delta FC &= \text{fuel consumption (liters)} \\ E_{df} &= \text{energy content of diesel fuel (joules/liter)} \\ \eta &= \text{locomotive propulsion system efficiency} \\ \Delta D/q &= \text{change in wind averaged drag area } (m^2) \\ \Sigma V_i^2 L_i &= \text{velocity (V) and distance (L) from histogram} \end{aligned}$$

This method was shown to over-predict the portion of fuel savings attributed to the coal car aerodynamic modifications, likely because it does not properly represent actual trip conditions, such as elevation changes, flange friction (curving), and train acceleration events.

The second method used to calculate train fuel usage was based on the train aerodynamics subroutine (AERO: available from the Association of American Railroads) in conjunction with an empirically-derived (English units) fuel calculation equation (with options for specifying hilly or level routes) [22].

$$FC = K(0.0015\,W + 0.00256\,S_d\,V^2 + C\,W) \tag{5}$$

where:

$$
\begin{aligned}
\mathrm{FC} &= \text{fuel consumed (gallons per 100,000 miles)} \\
\mathrm{K} &= \text{fuel consumption factor (fuel consumed per distance} \\
&\quad\ \text{traveled per unit of tractive resistance)} \\
&= 20.38\ \text{gallons/100,000 miles/lbf} \\
\mathrm{S_d} &= \text{train consist drag area (ft}^2) \\
\mathrm{V} &= \text{train speed (miles/hour)} \\
\mathrm{W} &= \text{consist weight (lbf)} \\
\mathrm{C} &= \text{Hill Climbing Factor} \\
&= 0.0 \text{ for level routes} \\
&= 0.0007 \text{ for hilly routes}
\end{aligned}
$$

This method predicts lower savings than those calculated by the straight, level track method described above. Universal application of the method has been questioned because of the average empirical values employed.

A third approach for calculating train fuel usage was developed during the current study and incorporates a more detailed evaluation of the train resistance forces, including: (1) gravity (elevation changes) [36], (2) rolling resistance (including journal bearing friction) ([36, 104] through [91]), (3) force required to accelerate the train [92], (4) flange resistance (curving) [91, 92], and (5) aerodynamic drag. If F(x) is the total resistance force acting on the train (axial direction), the energy required to overcome this force is given by the following formula:

$$
Energy\ Required = E = \int_0^L F(x)dx \tag{6}
$$

To determine the energy required for a given train journey, the track survey data is divided into segments and the forces acting along each segment are calculated using equation (7):

$$
E \approx \sum_{i=0}^{i=N} F_i X_i \tag{7}
$$

where

$$
\begin{aligned}
i &= \text{trip segment} \\
F_i &= \text{forces acting over segment } X_i
\end{aligned}
$$

The results for a typical unit coal train round trip (mine to power plant and return) is shown in Table 3 for the case of a relatively high-speed operation. The results of the train energy model spread-sheet calculations were compared to fuel usage data and transportation efficiency (gross-tonne-kilometers per liter, gross-ton-miles per gallon) data supplied by Class I U.S. Railroads for three routes (mine-to-power

Table 3 Baseline calculations for unit coal train, mine-to-power plant round trip, without aerodynamic modifications, high-speed route, distance of each segment = 1,500 km

Resistive force component	Energy expended			
	Outbound (loaded)		Return (empty)	
	(MJ)	% of Total	(MJ)	% of Total
Elevation changes	–	0.0	20,385	5.2
Rolling resistance	272,437	53.2	74,429	18.9
Acceleration: starting resistance	4,029	0.8	550	0.1
Speed changes	44,552	8.7	12,171	3.1
Flange resistance (Curves)	6,880	1.3	1,880	0.5
Aerodynamic drag	184,564	36.0	285,118	72.2

plant and return trip). The results are shown in Fig. 10 and verify the accuracy of the calculation procedure.

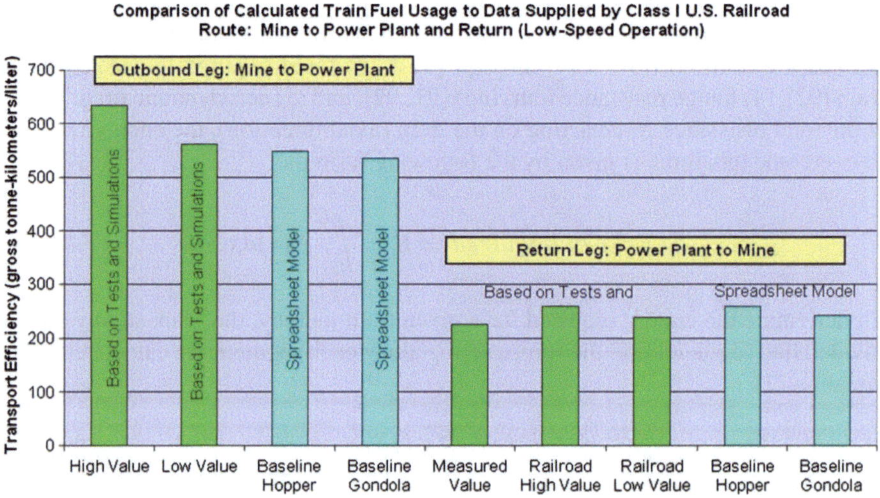

Fig. 10 Comparison of measured and predicted train fuel consumption

A comparison of the three energy calculation procedures is shown in Fig. 11. As noted above, methods 1 (straight, level track), and 2 (AAR Aero subroutine) tend to provide overly-optimistic predictions of fuel savings, while method 3 (spreadsheet-based train energy model) provides good correlation with actual test data.

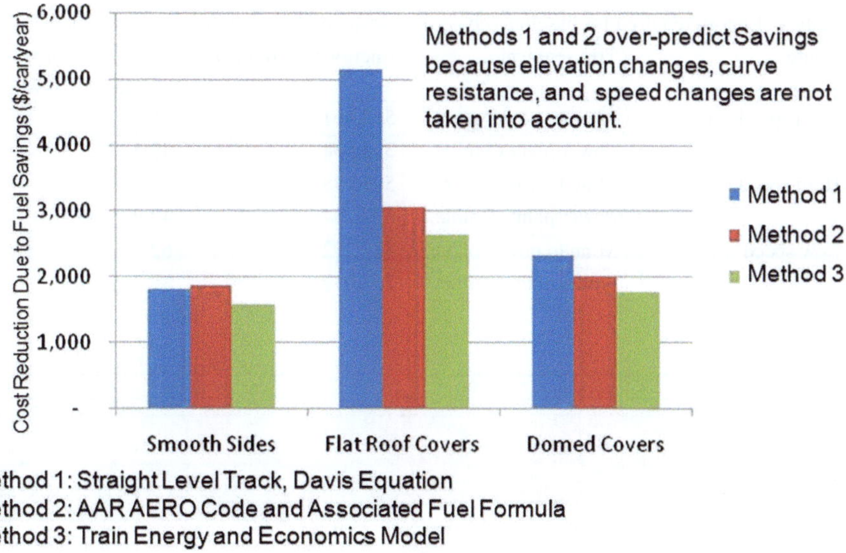

Fig. 11 Comparison of train energy and fuel consumption calculation methods, high speed unit train, three aerodynamic modifications

3.3 Results and Conclusions, Section II: Fuel Savings Calculations

The three histograms evaluated during the study highlight the effect of train speed on energy requirements. Total energy expended during the low-speed round trip (average speed 42 km/h), normalized by distance, is 25 % less than that expended for the high speed unit train (average speed 64 km/h) as illustrated in the Table 4 and the graph in Fig. 12. For these examples, it is interesting to note that the energy required to transport empty coal cars back to the mine requires between 60 and 80 % of the energy required to transport loaded cars from the mine to the power plant. It is also worth noting that aerodynamic drag is one of the largest contributors to train resistance, accounting for between 57 and 72 % of the total energy required for transporting the empty cars and between 22 and 36 % of the total energy required for the loaded cars.

4 Section III: Economic Analysis

The verified spread-sheet-based train energy and economics calculation method was employed to evaluate the cost effectiveness of several retrofit coal car aerodynamic drag reducing devices. Criteria employed for selecting the candidate devices

Table 4 Energy required for baseline (no aerodynamic modifications) unit coal trains

Route	Trip segment	Energy required (MJ)	Portion of round trip energy required (%)
High speed route 1	Mine to power plant	512,461	56.5
	Power plant to mine	394,534	43.5
High speed route 2	Mine to power plant	501,198	57.1
	Power plant to mine	377,294	42.9
Low speed route	Mine to power plant	417,392	62.3
	Power plant to mine	253,057	37.7

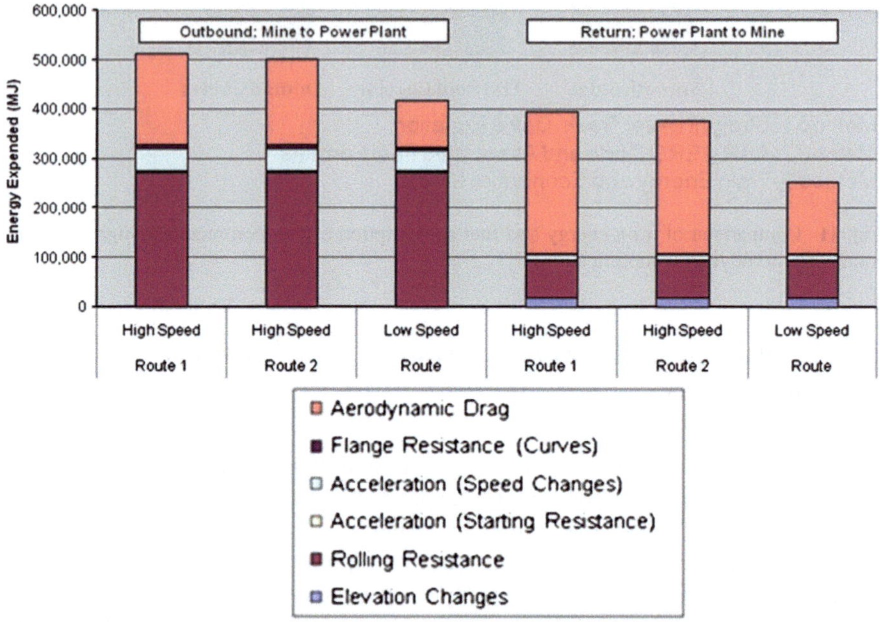

Fig. 12 Energy analysis: baseline gondola cars (100 cars, 4 locomotives, trip distance = 1,500 km each way)

included: (1) drag reduction effectiveness, (2) durability, (3) ease of manufacture and installation, (4) weight, (5) impact on loading and unloading operations, and (6) expected acceptance by the railroads. The candidate devices include (a) smooth sides, (b) side skirts, (c) end closures, (d) combinations of side and end treatments, and (e) roof covers and baffles.

Economic Calculations: Fuel usage can be determined based on the calculated energy required and the efficiency of the locomotive propulsion system. Propulsion system efficiency was evaluated based on information from three sources: (A) locomotive manufacturer, (B) calculations involving efficiencies of the diesel engine, generator, and traction motors, based on engineering handbooks (Ref. [3], p. 257),

High Speed Route 1									
Car Type	Aerodynamic Modification	Weight of Add-On Aerodynamic Devices (lbs)	(tons)	Fuel Used Per Car Per Year (gallons/car/year)	Fuel Savings (%)	Fuel Cost Savings ($/car/year)	Lost Revenue ($/car/year)	Net Cost Reduction $/car/year	ROI (years)
Gondola	None (Baseline)			7,781					
	Smooth Sides	2,000	1.00	7,052	9.4	$ 1,589	$ 490	$ 1,099	5.0
	Flat Roof Covers	2,500	1.25	6,566	15.6	$ 2,649	$ 613	$ 2,036	3.9
	Domed Covers	3,000	1.50	6,861	11.8	$ 2,006	$ 735	$ 1,271	9.4
Hopper	None (Baseline)			7,333					
	Smooth Sides	2,000	1.00	6,356	13.3	$ 2,130	$ 490	$ 1,640	4.9
	Hopper End Enclosure	1,000	0.50	7,069	3.6	$ 576	$ 245	$ 331	9.1
	Flat Roof Covers	2,500	1.25	6,403	12.7	$ 2,027	$ 613	$ 1,414	7.1
	Domed Covers	3,000	1.50	6,555	10.6	$ 1,696	$ 735	$ 961	12.5

Fig. 13 Example output from train energy and economics model for several representative retrofit aerodynamic modifications: hopper cars, high-speed route

and (C) locomotive propulsion system information contained in the text by Kutz [51]. Efficiency is given in units of kilowatt-hours/gallon (kW-h/l). These locomotives produce, on average, 3.0 kW-h of useful work for each liter of diesel fuel consumed (11.3 kW-h/gal). Thus, projected fuel associated with a particular trip can be determined by:

$$\text{Fuel required} = \frac{\text{(Total energy required, kilowatt-hours)}}{\text{(Locomotive efficiency, kilowatt-hours per liter)}} \tag{8}$$

The spread sheet program also includes fuel costs, freight rates ($/net delivered ton), train speed history, aerodynamic characteristics of the locomotives and cars, car weights, load weights, position-in-train effects, and fuel price. The penalty associated with weights of the aerodynamic devices is included in the economic calculations. The lost revenue due to carrying the added weight (hence less coal) was subtracted from the projected fuel savings to obtain the net cost savings per car per year. Fuel savings are highest during the empty part of the trip (power plant to mine) where aerodynamic drag accounts for the largest portion of tractive resistance. For both legs of the trip, rolling resistance and aerodynamic drag together represent approximately 90 % of the train resistance.

An example output of the economic section of the spreadsheet-based train energy model is shown in Fig. 13. For this example, the diesel fuel price was assumed to be US$2.18/gallon, the freight through rate was entered as US$14.00/net delivered ton mile, and the car utilization was assumed to be the industry average, 35 trips per car per year.

To calculate return on investment (ROI), the predicted fuel savings for each retrofit aerodynamic device were multiplied by the fuel price. The lost revenue associated with the added weight of the aerodynamic device was subtracted from the fuel savings to obtain the net cost reduction. The cost of the installed aerodynamic device was then divided by the net cost reduction to obtain the ROI. It is noted that some devices exhibiting significant aerodynamic drag reductions, notably domed roof covers, have unacceptably long ROIs due to their high weights (reduced freight revenue) and high fabrication costs.

4.1 *Results and Conclusions, Secttion III: Economic Analysis*

The current study indicates that round trip fuel savings due to the addition of aerodynamic modifications to open-top coal cars (which includes several designs not shown in Fig. 13) range from 2.7 to 19.9 %. Based on industry publications, transportation of coal within the United States requires on the order of 1.5 billion gallons of diesel fuel each year. Thus, a 5 % fleet-wide fuel savings would be 75 million gallons, or 2 % of all Class I railroad fuel consumption.

4.1.1 Potential for Product Development

The current study indicates there is significant potential for reducing aerodynamic resistance of open-top railroad cars. The economic benefit to the railroads associated with installation of these devices is also favorable. Although many retrofit modifications were evaluated, the research indicates additional development will produce more efficient and cost-effective designs. Recommended tasks include application of design-for-manufacturing methods, materials selection (light weight, durable), improved installation procedures, design for extended product life, cost reduction, consideration of maintenance requirements, reduce impacts of weather conditions (e.g. frozen coal), and elimination of interference with operation of railcar mechanical equipment. Once the retrofit designs have been finalized, prototype devices will be fabricated and tested on a selected unit train. The goal of the pending test program will be to define fuel savings associated with selected retrofit coal car aerodynamic devices and to verify their applicability.

It is noted that several of the aerodynamic drag reduction devices can also serve as passive coal dust suppression systems [67, 68]. Development of these systems can be integrated into the aerodynamic drag reduction program.

Acknowledgments This work was funded by Aero Transportation Products, Inc., which also approved release of the results. Preparation of this paper was funded by Airflow Sciences Corporation.

References

1. McCallen, R., Browand, F., Ross, J. (eds.): The Aerodynamics of Heavy Vehicles: Trucks, Buses, and Trains. Springer, New York (2004)
2. Browand, F., McCallen, R., Ross, J. (eds.): The Aerodynamics of Heavy Vehicles II: Trucks, Buses, and Trains. Springer, Berlin (2009)
3. Adler, U., Bazlen, W. (eds.) Automotive Handbook, p. 216. Stuttgart, Germany, Robert Bosch, GmbH (1976)
4. Ahmed, S.R.: Some applications of RANS methods. In: Hucho, W.H. (ed.) Sect. 15.6.3.2, Computational Fluid Dynamics, Chap. 15, Aerodynamics of Road Vehicles, 4th edn. Society of Automotive Engineer, Warrendale, Pennsylvania (1998)

5. Ali, I.: Numerical investigation of the embankment configuration for cross wind stability of ICE3. In: Aerodynamics of Heavy Vehicles III: Trucks, Buses, and Trains, Potsdam, Germany, 12–17 Sept 2010
6. Anderson, D.A., Tannehill, J.C., Pletcher, R.H.: Computational Fluid Dynamics and Heat Transfer. Hemisphere Publishing Corporation, Washington (1984)
7. Annual Report, FreightCar America, p. 2
8. Baranowski, K.: Rail car covers for bottom discharge cars. In: National Coal Transportation Association. Western Logistics and Planning Committee Meeting, Omaha, Nebraska, 23 July 2008
9. Bossert, A.: FreightCar America: Coal Orders Will Pick Up Soon. Alpha Investment News, 12 Aug 2008. http://seekingalpha.com/article/90507-freightcar-america-coal-car-orders-will-pick-up-soon
10. Buckley, F.T., Marks, C.H., Walston, W.H.: A study of methods for improving truck fuel economy. Prepared for the National Science Foundation under Grant SIA-74-14843, University of Maryland (1978)
11. Dahlan, W.G., Pladson, W.S.: Cover for open top rail car. U.S. Patent 5,762,002, 9 June 1998
12. Davis Jr, W.J.: The Traction resistance of electric locomotives and cars. Gen. Electr. Rev. **29**, 685–707 (1926)
13. U.S. Department of Energy, Energy Information Administration, EIA-860 Data. Inventory of Electric Utility Power Plants in the United States. http://www.eia.doe.gov/cneaf/electricity/ipp/ipp_sum.html and Existing Capacity by Energy Source. http://www.eia.doe.gov/cneaf/electricity/epa/epat2p2.html
14. U.S. Department of Energy, Energy Information Administration. Weekly U.S. Coal Production Overview. http://www.eia.doe.gov/cneaf/coal/weekly/weekly_html/wcpweek.htm
15. Elofsson, P.: Effect of transient aerodynamics on uncertainty in drag predictions using aerodynamic simulations for a heavy truck in yaw flow. In: Aerodynamics of Heavy Vehicles III: Trucks, Buses, and Trains, Potsdam, Germany, 12–17 Sept 2010
16. Engdahl, R., Gielow, R.L., and Paul, J.C.: Train Resistance—Aerodynamics, Volume I of II, Intermodal Car Aerodynamics. Association of American Railroads (1987)
17. Engdahl, R., Gielow, R.L., Paul, J.C.: Train Resistance—Aerodynamics, Volume II of II, Open Top Car Applications. Association of American Railroads (1987)
18. Engdahl, R.: Opening cover for railroad cars. U.S. Patent 4,524,700, 25 June 1985
19. FLUENT 6.3 User's Manual. Lebanon, New Hampshire, ANSYS Inc. (2007)
20. Frank, T.: Aerodynamics of commercial vehicles.In: Aerodynamics of Heavy Vehicles III: Trucks, Buses, and Trains, Potsdam, Germany, 12–17 Sept 2010
21. Furlong, C.F., Gielow, M.A., Gielow, R.L., Paul, J.C., Nelson, R.L.: Results of Wind Tunnel and Full-Scale Tests Conducted from 1983 to 1987 in Support of the Association of American Railroad's Train Energy Program. Livonia, Michigan, Airflow Sciences Corporation Report Number R-88-010, AAR Report R-685, 30 Dec 1988
22. Furlong, C.F., Nelson, R.K., Gielow, R.L., Paul, J.C.: Aerodynamic Subroutine Users Guide, Train Aerodynamics Simulator, Livonia, Michigan, Airflow Sciences Corporation Report R-88-003, July 1988 (also published as AAR Report R-683)
23. Gielow, M.A., Furlong, C.F.: Results of Wind Tunnel and Full-Scale Tests Conducted from 1983–1987 in Support of the Association of American Railroads' Train Energy Program. AAR Report R-685, 30 Dec 1988
24. Gielow, R.L., Holmes, A.E., Paul, J.C.: Aerodynamic Drag Analysis of the Ortner 100-Ton Rapid Discharge Car, Livonia, Michigan, Airflow Sciences Corporation Report R-82-014, 15 Dec 1982
25. Gielow, R.L., Keeler, D.G., Polihonki, D.R., deJager, D.E., Paul, J.C.: 16 % Scale Gondola, Locomotive, and 89 Foot Car Wind Tunnel Test. Livonia, Michigan, Airflow Sciences Corporation Report R-85-011, 31 Jan 1986
26. Gielow, R.L., Paul, J.C., Nelson, R.K.: A summary of the 1980 ASC effort for ford motor company. Livonia, Michigan, Airflow Sciences Corporation Report R-80-009 (1980)

27. Gielow, R.L., Paul, J.C., Nelson, R.K.: F-150 Light Duty Truck Pre-Test Report, University of Maryland Wind Tunnel. Livonia, Michigan, Airflow Sciences Corporation Report R-82-012, Nov 1982
28. Gielow, R.L., Paul, J.C.: Aerodynamically structured railway car with corner, air flow guides. U.S. Patent 4,738,203, 19 April 1988
29. Gosman, D.: Aspects of CFD applications to heavy vehicle aerodynamic design. In: Aerodynamics of Heavy Vehicles III: Trucks, Buses, and Trains, Potsdam, Germany, 12–17 Sept 2010
30. Götz, H., Mayr, G.: Optimization in the wind tunnel—problems with reduced scale models. In: Hucho, W.-H. (ed.) Sect. 9.5, Aerodynamics of Road Vehicles, Warrendale, Pennsylvania, USA, Society of Automotive Engineers, pp. 425–427 (1998)
31. Haentjens, S., Chappet, P., Tessier, J.L., Ardissone, J.P., Guibergia, J.P.: Device for reducing the aerodynamic drag of a cavity in a flow, and a vehicle, in particular a rail vehicle, provided with such a device. U.S. Patent 5,546,865, 20 Aug 1996
32. Hammitt, A.G.: Aerodynamic Forces on Freight Trains, Volume 1—Wind Tunnel Tests of Containers and Trailers on Flatcars. U.S. Department of Commerce, NTIS PB-264304, Dec 1976
33. Hammitt, A.G.: Aerodynamic Forces on Freight Trains, Volume III, Correlation Report, Full Scale Tests of Trailers on Flat Cars and Comparison with Wind Tunnel Results. U.S. Department of Commerce, NTIS PB-288137, Sept 1978
34. Hammitt, A.G.: Aerodynamic Forces on Freight Trains, Volume IV, Wind Tunnel Tests of Freight Cars and New Trailer and Container Car Designs. U.S. Department of Commerce, NTIS PB80-174899, June 1979
35. Hammitt, A.G.: Aerodynamic Forces on Various Configurations of Railroad Cars for Carrying Trailers and Containers. Wind Tunnel Tests of Six Scale Model Configurations. U.S. Department of Commerce, NTIS PB80-174881, Jan 1979
36. Hay, W.W.: Railroad Engineering, pp. 69–89. John Wiley and Sons, New York (1982)
37. Hickey, P.K., Gielow, R.L., Holmes, A.E., Paul, J.C.: Aerodynamic Drag Analysis of the Lo-Pac 2000 Intermodal Car. Livonia, Michigan, Airflow Sciences Corporation Report R-82-011, 15 Sept 1982
38. Hickey, P.K., Gielow, R.L., Holmes, A.E., Paul, J.C.: Aerodynamic Modifications to the Budd Lo-Pac 2000 Intermodal Car. Livonia, Michigan, Airflow Sciences Corporation Report R-83-001, 31 Jan 1983
39. Hickey, P.K., Gielow, R.L., Paul, J.C.: An Aerodynamic Treatment Package for the Ortner Four Pocket Rapid Discharge Coal Car. Livonia, Michigan, Airflow Sciences Corporation Report R-83-013, 5 Aug 1983
40. Hickey, P.K., Holmes, A.E., Paul, J.C., Gielow, R.L.: Aerodynamic Design Study of an Aluminum Gondola Car for Pullman Standard. Livonia, Michigan, Airflow Sciences Corporation Report R-83-005, 1 April 1983
41. Hickey, P.K., MacFadyen, A., Holmes, A.E., Paul, J.C., Gielow, R.L.: An Aerodynamic Analysis of the Trailer Train Company 4-Runner Intermodal Car Using Numerical Flow Simulation Methods. Livonia, Michigan, Airflow Sciences Corporation Report R-82-005, 7 May 1982
42. Hoerner, S.F.: Drag of Railroad Vehicles. Fluid Dynamic Drag, Sect. XII, Land-Borne Vehicles, pp. 12–10 to 12–15. Brick Town, New Jersey, Published by the Author (1965)
43. Hucho, W.-H., Jansse, L.J., Emmelmann, H.-J.: The Optimization of Body Details—A Method for Reducing the Aerodynamic Drag of Road Vehicles. SAE Paper No. 760185, Society of Automotive Engineers, Warrendale, Pennsylvania, USA (1976)
44. Hucho, W.-H.: Tests with reduced-scale models. In: Hucho, W.-H. (ed.) Sect 13.4, Aerodynamics of Road Vehicles, Warrendale, Pennsylvania, USA, Society of Automotive Engineers, pp. 680–694 (1998)
45. Johansen, F.J.: Air resistance of passenger trains. Inst. Mech. Eng. Proc. **134**, 91–208 (1936)
46. Johnson, R.W., Nelson, R.K.: Wind tunnel test results of 16 % scale partitioned gondola car. Livonia, Michigan, Airflow Sciences Corporation [Report and Funding Organization Information Proprietary], 15 Dec 2004

47. Joshi, P.B.: Aerodynamic Forces on Freight Trains, Volume III, Full-Scale Aerodynamic Validation Tests of Trailer-On-A-Flat Car (Series II). U.S. Department of Commerce, NTIS PB-281823, March 1978
48. Klemin, A.: Aerodynamics of Railway Trains. Railway Mechanical Engineer, Part II, Sept 1934, Part III, Oct 1934
49. Krajnovic, S.: What Can LES do for aerodynamics of heavy vehicles? In: Aerodynamics of Heavy Vehicles III: Trucks, Buses, and Trains, Potsdam, Germany, 12–17 Sept 2010
50. Kruglinski, A.D.: The Coal Car Fleet: Sizing for the "90s—Equipment Financial Trends", Railway Age, Oct 1989
51. Kutz, M.: Handbook of Transportation Engineering, Chap. 23. The McGraw-Hill Companies Inc., New York (2004)
52. Launder, B.E., Spalding, D.B.: Mathematical Models of Turbulence. Academic Press, New York (1972)
53. Launder, B.E., Spalding, D.B.: The numerical computation of turbulent flow. Comp. Methods Appl. Mech. Eng. **3**, 269 (1974)
54. Luckring, A.K.: Slidable cover assembly for gondola railroad car. U.S. Patent 6,250,233, 26 June 2001
55. MacFadyen, A., Widjaja, D., Holmes, A.E., Paul, J.C., Gielow, R.L.: An Aerodynamic Drag Comparison of Streamlined and Conventional Locomotives Utilizing Numerical Airflow Simulation Methods. Livonia, Michigan, Airflow Sciences Corporation Report R-81-011, June 1981
56. Mailoux, C.O.: Problems of heavy electric traction. Trans. A.I.E.E. **23**, 723 (1905)
57. Miller, R.W.: Cover for rail car. U.S. Patent 5,181,474, 26 June 1993
58. Nelson, R.K., Banka, A.L.: VISCOUS Users Guide. Michigan, Airflow Sciences Corporation, Livonia (1993)
59. Norfolk Southern Railway Coal Car Fleet. http://www.nscorp.com. 2002 NS Coal Fleet Summary
60. Pankajakshan, R.: Passive devices for reducing base pressure drag in class 8 trucks. In: Aerodynamics of Heavy Vehicles III: Trucks, Buses, and Trains, Potsdam, Germany, 12–17 Sept 2010
61. Patankar, S.V.: Numerical Heat Transfer and Fluid Flow. Hemisphere Publishing Corporation, New York (1980)
62. Paul, J.C., Gielow, R.L., Holmes, A.E., Hickey, P.K.: Reduction of Intermodal Car Aerodynamic Drag through Computerized Flow Simulation. American Society of Mechanical Engineers, Paper 83-RT-4 (1983)
63. Paul, J.C., Gielow, R.L.: Aerodynamic Enhancements of the Trailer Train Front Runner Single Platform Rail Car. Livonia, Michigan, Airflow Sciences Corporation Report R-84-009, 1 June 1984
64. Paul, J.C., Johnson, R.W., Yates, R.G.: Application of CFD to rail car and locomotive aerodynamics. In: Browand, F., McCallen, R., and Ross, J. (eds.) The Aerodynamics of Heavy Vehicles II: Trucks, Buses, and Trains, pp. 259–297. Springer, Berlin (2009)
65. Paul, J.C., Nelson, R.L., Groesbeck, P.D., Polihonki, D.R., Banka, A., Keeler, D., Gielow, R.L.: 03/10 Scale Hopper and Gondola Rail Car Wind Tunnel Test. Livonia, Michigan, Airflow Sciences Corporation Report R-84-019, 28 Dec 1984
66. Paul, J.C., Smith, T.W., Gielow, M.A., Gielow, R.L.: 3/10 Scale Wind Tunnel Test of Skeletonized and Well-Type Intermodal Rail Cars. Livonia, Michigan, Airflow Sciences Corporation Report R-83-019, 1 Oct 1983
67. Paul, J.C.: Computer Modeling and Wind Tunnel Testing to Determine Coal Loss from Open Top Railroad Cars. Airflow Sciences Corporation, Livonia, Michigan, Report R-09-L3C-08, 17 July 2009
68. Paul, J.C.: Final Report: Coal Car Aerodynamics and Coal Dust Suppression, Phases I, 2, and 3. Livonia, Michigan, Airflow Sciences Corporation Report R-08-ATP-03. Rev. 3, 15 Jan 2010

69. Przbylinski, P.G., Bodinger, D.C., Wille, H.S.: Fuel Efficient Coal Gondola. American Society of Mechanical Engineers, Paper 84-WA/RT-18, Dec 1984
70. Rahm, R.: PRB Coal Users' Group Enjoys Growing Interest In Its Concerns, pp. 68–73. Power, July 2008
71. Rail Rates Analysis. The Surface Transportation Board, Office of Economics, Environmental Analysis, and Administration (OEEAA), pp. 14–19, Dec 2000
72. RailincUMLER, E.M.I.S.: Equipment, Registration and Information System, Database of Information on Essentially all Rail Cars and Locomotives Moving on the General Railway System in North America, UMLER Data Specification Manual, January 1. Railinc. Cary, North Carolina (2005)
73. Railroad Facts 2007. Association of American Railroads, Policy and Economics Department. Freight Cars in Service, pp. 51–52
74. Roshko, A. Koenig, K.: Interaction effects on the drag of bluff bodies in tandem. In: Sovran, G., Morel, T., Mason, W.T. (eds.) Aerodynamic Drag Mechanisms of Bluff Bodies and Road Vehicles, pp. 253–273. Plenum Press (1978)
75. Ross, J.C., Mehta, R.: Heavy vehicle drag reduction: experimental evaluation and design. Presentation to U.S. Department of Energy, NASA Ames Research Center, Experimental Aero-Physics Branch, 18 April 2006
76. Saunders, J.W., Watkins, S., Cassar, R.J.: Vortex optimisation of slotted tops and cavities of two different open rail wagons. J. Wind Eng. Ind. Aerodyn. **49**, 421–430 (1993)
77. Schewe, G.: Reynolds number effects in flow around bluff bodies. In: Aerodynamics of Heavy Vehicles III: Trucks, Buses, and Trains, Potsdam, Germany, 12–17 Sept 2010
78. Schito, P.: Hi-speed train cross wind analysis: CFD study and validation with wind-tunnel tests. In: Aerodynamics of Heavy Vehicles III: Trucks, Buses, and Trains, Potsdam, Germany, 12–17 Sept 2010
79. Society of Automotive Engineers: Wind Tunnel Test Procedure for Trucks and Buses. SAE J1252 (1979)
80. SourceWatch Encyclopedia. Existing U.S. Coal Plants, includes ownership and generating capacity information for the 1,522 coal-fired generating units located within the U.S. http://www.sourcewatch.org/index.php?title=Existing_U.S._Coal_Plants
81. Stachowiak, G.W., Batchelor, A.W.: Engineering Tribology, Elsevier Publisher, Butterworth (2000)
82. Stevens, S.G.: Car tarpaulin. U.S. Patent 894,759, 28 July 1908
83. Storms, B.L., Ross, J.C., Dzoan, D.: Reducing the aerodynamic drag of coal cars and class 8 trucks. Presentation to the U.S. Department of Energy, Office of Heavy Vehicle Technology, NASA Ames Research Center, Experimental Aero-Physics Branch, 22 Feb 2006
84. Storms, B.L., Ross, J.C., Dzoan, D.: reducing the aerodynamic drag of empty coal cars. In: Proceedings of JRC2005 Joint Rail Conference, pp. 139–145. Pueblo Colorado, ASME RTD2005-70056, 16–18 Mar 2005
85. Teleconference with Harry Mullins: Manager. Railcar Maintenance, Southern Company, Birmingham, Alabama, 9 Dec 2008
86. Teleconference with Janet Gellici, CEO, American Coal Council, Washington, DC, 9 Dec 2008
87. Teleconference with Randy Rahm: President. CoalTech Consultants Inc., Topeka, Kansas, 9 Dec 2008
88. Teleconference with Ron Lang: Engineer. Association of American Railroads, Transportation Technology Center Inc, Pueblo, Colorado, 3 Sept 2008
89. Teleconference with Thomas C. Canter. Executive Director, National Coal Transportation Association, Littleton, Colorado, 9 Dec 2008
90. Teleconferences and e-mail correspondence with Larry Rose: Manager. Operations and Logistics, The Detroit Edison Company, Detroit, Michigan, 9–12 Dec 2008
91. The American Railway Engineering and Maintenance-of-Way Association (AREMA), Manual for Railway Engineering (2007)

92. The American Railway Engineering and Maintenance-of-Way Association (AREMA). Practical Guide to Railway Engineering, Chap. 2, p. 57 (2003)
93. The Rail Transportation of Coal. Volume No. 3, Association of American Railroads, Policy and Economics Department, Rail Carloads, Length of Haul, pp. 22–26 and 47, Jan 2007
94. Tietjens, O.G., Ripley, K.C.: Air resistance of high-speed trains and interurban cars. Trans. ASME **54**(APM-54-23), 235–251 (1932)
95. Tuthill, J.K.: High Speed Freight Train Resistance: Its Relation to Average Car Weight. University of Illinois Engineering Bulletin. No. 376 (1948)
96. van Dyke, M.: An Album of Fluid Motion, p.15. The Parabolic Press, Stanford, California (1982)
97. Vincent, H.S.: Air resistance of railroad equipment. ASME Trans. **59**(RR-59-4), 617–640 (1937)
98. Walker, D.E., Early, S.R.: Aerodynamic end closures for railway hopper cars. U.S. Patent 4,909,154, 20 March 1990
99. Watkins, S., Saunders, J.W., Kumar, H.: Aerodynamic drag reduction of goods trains. J. Wind Eng. Ind. Aerodyn. **40**, 147–178 (1992)
100. Watkins, S., Vino, G.: The effect of vehicle spacing on the aerodynamics of a representative car shape. J. Wind Eng. **96**, 1232–1239 (1998)
101. Wesseling, P.: Principles of Computational Fluid Dynamics. Springer, Heidelberg (2000)
102. Widjaja, D., Scofield, J.W., Holmes, A.E., Gielow, R.L.: An Application of Airflow Modeling Methods to the Santa Fe 10-Pack Fuel Foiler. Livonia, Michigan, Airflow Sciences Corporation Report R-80-004, 1 Oct 1980
103. Wiedemann, J., Ewald, B.: Turbulence manipulation to increase effective reynolds numbers in vehicles aerodynamics. AIAA J. **27**, 763–769 (1989)
104. Williams, J.A.: Engineering Tribology. Cambridge University Press, New York (2005)

Sustainable Design of Underground Rail Systems—Aerodynamics at the Interface of Rolling Stock and Civil Construction

Andreas Busslinger, Samuel Nyfeler and Peter Reinke

Abstract In high-speed rail tunnels and heavy duty underground systems, aerodynamic issues have a substantial impact on the consumption of energy and resources. Together with the interrelated thermal conditions and the tunnel ventilation, various parameters of tunnel and vehicle design affect the resulting life-cycle costs and consequently the sustainability of these systems. For example, the power demand of metro systems is determined by the rolling stock features, the civil layout of tunnels and stations as well as by the way of operating the mechanical and electrical systems, including tunnel ventilation. Costs for traction power are influenced by vehicle design but equally by the choice of cross-sections and arrangement of shafts in the tunnel. Ventilation and cooling costs are caused by on-board systems of trains as well as by the equipment in tunnels and in stations. Thus, aero-thermal features of both, rolling stock and civil construction should be optimized together. During the design process, the above topics are commonly addressed separately. An overall system optimization covering for example rolling stock, civil design, track layout, tunnel ventilation and station ventilation is often missing. Awareness of rolling stock and infrastructure designers of the individual impact of the various factors affecting the energy demand of underground systems could be improved by more data for decision making. This paper aims at triggering more profound research work for a better understanding of the impact of the various design parameters on tunnel aerodynamics and the closely linked ventilation and cooling. Life-cycle costs would be reduced and a sustainable design shall be promoted.

1 Introduction

Transportation systems need to become faster, more frequent, more comfortable, safer and more economical. At the same time, our and future generations are facing constraints such as limited natural resources, required environmental protection,

A. Busslinger · S. Nyfeler · P. Reinke (✉)
HBI Haerter Ltd., Thunstrasse 32, 3005 Bern, Switzerland
e-mail: peter.reinke@hbi.ch

A. Dillmann and A. Orellano (eds.), *The Aerodynamics of Heavy Vehicles III*,
Lecture Notes in Applied and Computational Mechanics 79,
DOI 10.1007/978-3-319-20122-1_4

global warming and financial limitations. Conflicting demands are not new in engineering. However, the current diverging demands are rather fundamental. Every opportunity for improvement should be considered. Design and operation of transportation systems need to become more sustainable.

Sustainability has become a wide-ranging term that is applied to almost every aspect of life. In the context of engineering and design of transportation systems, the following design principles are in focus:

- Energy efficiency: develop processes/products which require less energy
- Long-term consideration: achieve quality and durability by longer-lasting and better-functioning systems considering the whole life-cycle of system
- Holistic view: improve in parts only if leading to an overall improvement
- Low-impact materials: use of small quantities of materials which require little energy to process and which have a low impact on the environment

These principles of sustainable design are a challenge for all technical systems. In the context of underground rail systems, the improvement of aerodynamic conditions is one key element to achieve an overall progress here.

In underground rail systems, aerodynamic phenomena affect both, the design of vehicles and of civil infrastructure. Particularly as the velocity and/or frequency of trains increase, the effects of tunnel aerodynamics on the design of tunnels and vehicles are becoming more important. In urban, heavy-duty underground systems or in tunnels of high-speed rail systems, several of the following aerodynamic phenomena need to be considered:

- Traction power requirements of trains
- Pressure loads on the vehicle, tunnel structure and equipment
- Pressure comfort and health limits related to pressure changes
- Micro-pressure waves and resulting noise (sonic boom, vibration)
- Loads due to air velocity acting on vehicle, tunnel structure and equipment
- Comfort and safety related to elevated flow velocities of air
- Air-exchange or climate in vehicle and tunnels/stations and related power demand of equipment in infrastructure

Aerodynamic issues (pressure, air velocity) are closely linked to the tunnel climate (air-exchange rate, temperature, humidity, air quality). This is important because many underground projects for urban transportation are located in warm and tropical regions of the world and tunnel aerodynamics allows improving the tunnel climate. In addition, the climate in very long tunnels becomes an important factor to assure system functionality. Therefore, aerodynamic issues have to be considered increasingly together with tunnel climate as well as ventilation and cooling of vehicles and tunnels including possible underground stations. The phenomena mentioned above have in common that all of them can be influenced by both, the design of the rolling stock and by the layout of the tunnels including possible underground stations. The interface of civil construction and of rolling stock offers the potential for optimization of the aero-thermal conditions.

2 Objectives

This paper shall present some challenges of sustainable design of underground rail systems at the interface of vehicle and tunnel aerodynamics. Examples are provided illustrating that the aero-thermal conditions of underground rail systems can be influenced simultaneously by both, the rolling stock and the civil design. Shortcomings of current projects shall be illustrated to motivate further research.

3 Aerodynamics Phenomena in Underground Rail and Metro Systems

The order of magnitude of pressure deviations from normal pressure in a high-speed rail tunnel may reach, for example, more than $+/-5$ kPa. Other pressure changes and air velocities might be created for the following reasons:

- *wind at portals*: Wind or storm may lead to substantial pressures acting on the air in the tunnel.
- *meteorological pressure differences*: Across tunnels connecting different valleys or mountain ridges substantial pressure difference might occur.
- *meteorological pressure changes*: The meteorological pressure in a region changes with time. During a year and even during a day the change of the meteorological pressure might be quite substantial, i.e. more than train-induced pressure fluctuations. However, the time scales of the changes are much larger (hours, days) than those of trains (seconds).
- *thermostatic pressure*: Temperature differences between the tunnel air and the outside may lead to pressure differences between portals.
- *tunnel ventilation*: The pressures created by fans may lead to considerable forces particularly acting on the walls of air ducts and shafts.

The typical magnitude of pressure changes of the above phenomena is smaller than those created by trains. Thus, the train-induced pressure fluctuations and resulting aerodynamic conditions are dominating in heavy-duty underground systems or in tunnels of high-speed rail systems.

The train-induced pressure fluctuations cause or affect the phenomena as listed in Table 1.

4 Measures to Control the Aero-Thermal Phenomena in Underground Systems

The aero-thermal phenomena mentioned above can be influenced and handled by a variety of measures. Several examples are given in Table 2.

Table 1 Major questions regarding aero-thermal phenomena in high-speed rail tunnels and/or heavy duty underground systems

Phenomena and examples of related questions	Location of impact	Stations (platform, mezzanine level	Other rooms (technical, social, vending offices)	Rail tunnel (incl. cross-passages, stairwells, shafts)	Vehicles
Traction power	How does a tunnel increase or affect the power demand of trains moving in a tunnel?				X
Pressure loads	What are the mechanical forces acting on the tunnel structure, the equipment in the tunnel, the vehicle, the load on freight trains, the passengers and staff?	X	X	X	X
Pressure comfort	How to avoid the unpleasant perception of pressure changes by hearing organs, i.e. aural discomfort?	X		X	X
Health limits	How to avoid harming pressure on hearing organs?	X		X	X
Micro-pressure waves	How to avoid the detonation-like, loud bang at portals upon train passage at opposite portal (sonic boom)?	X		X	
Loads due to air flow	What are the forces acting on signs/signals in a tunnel due to high air velocities?	X		X	X
Comfort and safety related to air flow	How to provide an acceptable, low velocity of air on the platform/access ways of an underground station?	X		X	X

(continued)

Table 1 (continued)

Phenomena and examples of related questions	Location of impact	Stations (platform, mezzanine level	Other rooms (technical, social, vending offices)	Rail tunnel (incl. cross-passages, stairwells, shafts)	Vehicles
Climate	How does heat release from vehicles and air-exchange influence the temperature, the humidity and the air-quality in a tunnel?	X	X	X	X
Ventilation of tunnel and related power consumption	How can tunnel ventilation (passive and mechanical) provide adequate temperature, humidity and air-quality? How does the vehicle and tunnel design increase or affect the power demand of the infrastructure?	X	X	X	X
Ventilation/air-conditioning of vehicle	How can train-bound HVAC-systems (passive and mechanical) provide adequate temperature, humidity and air-quality? (HVAC = heating, ventilation, air-conditioning)	X		X	X

Table 2 Examples of technical measures to cope with aerodynamic phenomena in tunnels

Phenomena	Examples of measures at vehicle	Examples of measures at tunnel
Traction power	• Apply aerodynamic streamlining	• Increase cross-sections (e.g. by single-tube, double-track tunnel)
	• Use advanced drive systems and components for control of traction power	• Introduce shafts
	• Implement light-weight construction	• Allow re-circulation of air within twin-tube systems (pressure relief ducts, no full-height platform screen doors)
		• Choose better vertical alignment
Pressure loads	• Apply aerodynamic streamlining	• Increase cross-sections of tunnels and particularly near portals
	• Implement robust design	• Introduce shafts/robust design
Pressure comfort/Health limits due to pressure	• Apply aerodynamic streamlining	• Increase cross-sections of tunnels, particularly near portals
	• Seal vehicles	• Introduce shafts
	• Implement passive or active pressure control at heating, ventilation and air-conditioning units (HVAC)	
Micro-pressure waves (sonic boom)	• Apply aerodynamic streamlining at train head and tail	• Design portals in favorable manner
		• Use active/passive measures to mitigate pressure waves inside of tunnel (portal design, damping of tunnel)
Loads due to air flow	• Apply aerodynamic streamlining	• Implement robust design
Comfort and safety related to air flow	• Apply aerodynamic streamlining	• Introduce pressure relief or draught relief shafts or shield-off platform region by baffle blades, locks, guide vanes, etc.
		• Create aerodynamic decoupling between platforms and trackway
		• Reduce open gap between vehicle and platform screen doors

(continued)

Table 2 (continued)

Phenomena	Examples of measures at vehicle	Examples of measures at tunnel
Climate and power for HVAC of tunnel/stations	• Apply aerodynamic shape	• Support natural ventilation
	• Release less heat, e.g. use regenerative breaking	• Support train-induced ventilation (e.g. by double-tube, single-track tunnels; platform screen doors)
	• Release heat such that efficient heat removal is possible (high enthalpy air at appropriate location)	• Remove heat in efficient manner by extracting air of high enthalpy
	• Use fire resistant materials or on-board fire extinguishing systems (i.e. relief requirements for tunnel ventilation)	• Use energy-efficient control schemes to provide ventilation according demands only
Climate and HVAC power of vehicle	• Reduce losses of cold-air during stops at stations	• Keep tunnel cool by preventing air re-circulation at portal/shaft
		• Allow for good air-quality by frequent air-exchange

Table 2 indicates several ways to influence the aerodynamic conditions in underground systems and to control the impact of tunnel aerodynamics on vehicles. Some of the examples exhibit inherent design conflicts. On the one hand, for example, choosing a double-tube, single-track tunnel is of advantage with respect to controlled air-exchange, heat removal and climate control in heavy-duty underground systems or in very long rail tunnels. Due to the piston effect of trains, warm air is being carried out of the tunnel system in an efficient manner due to mono-directional traffic. On the other hand, single-tube, double-track tunnels are preferable regarding other aerodynamic phenomena. In comparison to double-tube tunnels, these offer a larger free cross-sectional area. Larger cross-sections lead to smaller pressure forces, better pressure comfort (ignoring oncoming traffic), reduced traction power demand and better suppression of non-acceptable micro-pressure waves.

In summary, a design might be good for the tunnel aerodynamics but might be bad for the climate or other system features. Therefore, all aero-thermal phenomena have to be taken into account and further, possibly conflicting demands need to be considered:

- safety (for example related to fire/smoke control)
- operational and maintenance flexibility
- reliability and availability of system
- systems harmonization with other parts of network

- environmental protection
- cost efficiency

Obviously, there is a wide range of possible perspectives and scopes to influence and/or to cope with aerodynamic phenomena in tunnels. The focus here is on the particular interface of civil construction and rolling stock. Some examples of possible aerodynamics optimization at the interface of rolling stock and civil construction shall be addressed.

5 Life-Cycle Costs of System Balancing Between Vehicle and Tunnel and Between Construction and Operation

Sustainable investment goes along with reaching the required system performance while keeping lowest life-cycle costs. Reduced life cycle costs are a prerequisite for sustainable design.

Figure 1 shows examples of improving the aerodynamic conditions in a tunnel with respect to pressure comfort, i.e. reduced pressure fluctuations inside trains:

- pressure relief shafts at tunnels
- streamlined trains with active pressure control inside coaches.

The relief shafts create partial reflections of the pressure waves. This weakens the pressure waves without the need to increase the free cross-sectional area of the tunnel (see [1, 2]). For certain project settings it might be a reasonable opportunity to improve the pressure comfort by using shafts and keeping the free cross-sectional tunnel area comparatively small (e.g. at tunnels of short and medium length; trains with no or limited sealing; small overburden and/or cut-and-cover tunnels; no residential area in immediate vicinity of shaft for noise reasons). Aerodynamic streamlining of trains is beneficial for various other aerodynamic reasons (less traction power, reducing pressure fluctuations, reduced micro-pressure waves, improved cross-wind stability).

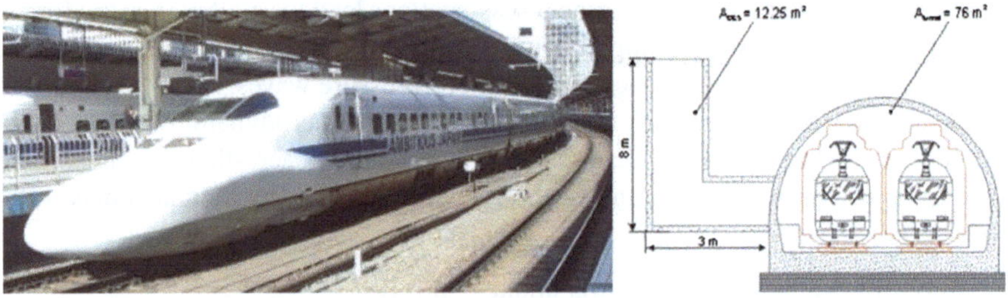

Fig. 1 Aerodynamically streamlined Shinkansen 700 with active pressurization system of cabin (*left*) and shaft to improve pressure comfort in a tunnel (*right*)

Table 3 Balancing expenditures between construction and operation and trains and tunnels

Expenditures	Rolling stock	Tunnel
	Measure:	Measure:
	• Apply aerodynamic streamlining	• Increase cross-sections of tunnels, e.g. near portals
	• Seal vehicles	• Introduce shafts or open cross-connections between adjacent tunnels
	• Implement passive or active pressure control of A/C units	
Investment/construction costs	• Costs for raw materials and construction, e.g., for a more robust vehicle structure, measures for sealing, space requirements for streamlining and pressure control	• Costs for raw materials and construction
		• Costs for land and space for shafts
Operation and replacement costs	• Maintenance costs of additional equipment and sealing	• Costs for operation of additional infrastructure (cleaning, regular refurbishment, handling of vandalism, safety, security, etc.)
	• Traction power costs due to additional weight of vehicle	
	• Costs due to reduced space for commercial utilization due to streamlining of train ends, larger HVAC units, etc.	

Table 3 provides an example of the task to properly balance costs. The example deals with achieving pressure comfort inside of high-speed trains. Expenditures are possible:

- for measures at rolling stock and/or for measure at tunnels
- during construction and/or during operation

In order to minimize life-cycle costs in the context of pressure comfort, several aspects have to be considered to balance measures at vehicles and tunnels. Spending on tunnels such as larger cross-section and/or shafts becomes preferable for the following boundary conditions:

- small share of high-speed tunnels within whole rail network
- high frequency of trains
- expected strong increase in future energy costs

- emphasis on simple, reliable systems with high availability
- measure beneficial for other aerodynamic purposes as well, such as limiting traction power demand and suppression of micro-pressure waves
- in case of shafts: combination of measure with other functions possible, e.g., emergency exit, access to technical rooms, smoke control during fire emergency, etc.
- in case of shafts: small overburden of tunnel and sufficient distance to residential areas

Spending on vehicles becomes preferable if the above boundary conditions are less dominating. In addition, measures to improve aerodynamic streamlining of trains will lead in most cases to smaller pressure fluctuations in tunnels. Therefore, reducing aerodynamic drag and friction to reduce energy consumption, to improve cross-wind stability, to reduce creation of critical micro-pressure waves will improve pressure comfort as well.

Expenses for construction costs should be preferred in contrast to operation costs in case of expected strong increase in future energy costs and expected strong increase in future maintenance costs. The above example on pressure comfort shows that various boundary conditions need to be considered. Optimization from a technical or physical point of view might not be feasible for operational or legal reasons. Responsibilities regarding balancing of costs for underground systems and optimization of life-cycle costs and sustainability are illustrated simplified in Fig. 2.

Figure 2 shows that optimizing life-cycle costs requires co-ordinated activities of the involved parties. On the one hand, the potential of possible optimization is limited for the following unchangeable reasons:

- boundary conditions for design of tunnels and stations (space limitations especially in urban areas, geo-technical constraints, access and operational demands, etc.)
- high-level restrains (norms, laws, safety and availability requirements, network-wide or manufacturer-wide standards, etc.)
- uncertainties with respect to future developments (customer demands, costs for electricity, future vehicle/drives and further technology, etc.)

On the other hand, the potential for improvements is not fully exploited yet. The awareness of designers and decision-makers could be improved. Knowledge about

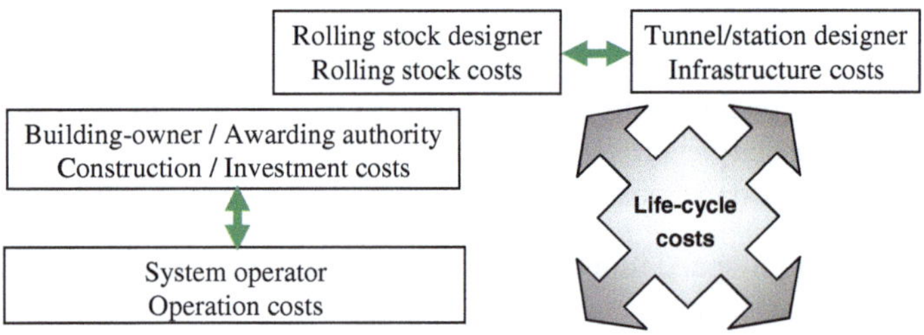

Fig. 2 Responsibilities regarding allocation and balancing of costs of underground systems

mutual dependencies of design parameters of vehicles and infrastructure and their consequences for system emergency consumption and life-cycle costs are limited. In typical projects, time and resources are missing elaborate and optimize a more sustainable design. Easy accessible, best-practice recommendations and more data as basis of decision-making would be beneficial. Selected case studies would allow to present opportunities and lead to more sound design decisions.

6 Choosing Appropriate Measures for Improvements

In Table 4, selected examples are provided in order to highlight the challenge of balancing measures at vehicle with measures at tunnels/stations and the task to balance costs between construction and operation. Table 4 is of qualitative nature only. For application in projects, data and a quantitative analysis on the sensitivity of the different measures are needed.

7 Selected Examples of Measures

Three further examples regarding the challenge of choosing appropriate measures for improving the aerodynamic conditions in underground rail systems shall be presented:

- Choice of free-cross-sectional area: Consequences for life cycle costs
- Provision of pressure relief between tubes: Balancing costs for traction power and for tunnel cooling
- Definition air exchange in tunnel: Balancing measures for control of temperature in vehicles and tunnels

7.1 Adapting the Free-cross-sectional Area of Tunnel

Figure 3 shows the power requirements for a passenger train passing through the Swiss Gotthard Base Tunnel at 250 km/h (GBT; operational in 2017). The figure illustrates:

- Aerodynamic resistance is a major contributor to the power consumption of the trains, both, on open track and in the tunnel.
- In the tunnel the aerodynamic resistance leads to significantly higher power requirements than on the open track particularly at high velocity (in example of Fig. 3 by a factor of 3 more than on open track).

Table 4 Measures at tunnel/station and rolling stock to influence aero-thermal conditions

Major impact on (XX major impact, X slight impact, – no impact) Measures	Traction power	Pressure loads	Pressure comfort/ Health	Micro-pressure waves	Loads due to air flow	Comfort and safety of air flow	Tunnel climate and HVAC	Vehicle climate and HVAC
Measures at vehicle (selected examples only)								
Aerodynamic shaping	XX	XX	XX	XX	XX	X	–	–
Regenerative braking	XX	–	–	–	–	–	XX	X
Energy efficient design of drives and HVAC system with reduction of heat release	XX	–	–	–	–	–	XX	XX
Light-weight structures	XX	–	–	–	–	–	X	X
Passive/active pressure control of passenger compartments	–	–	X	–	–	–	–	X
Fire resistant material	–	–	–	–	–	–	XX	–
On-board fire extinguishing systems	X	–	–	–	–	–	XX	–
Measures at tunnel/station								
Choice of tunnel free cross-section al area	XX	XX	XX	XX	XX	XX	XX	X
Shape of portal, i.e. gradual change of cross-section	–	X	X	XX	–	–	–	–
Alignment of portals of double-tube systems	–	–	–	–	–	–	XX	–
Number, location, features of pressure relief shafts	X	XX	XX	X	X	–	–	–
Number, location, features of draught relief shafts	X	X	X	X	X	X	XX	

(continued)

Table 4 (continued)

Major impact on (XX major impact, X slight impact, – no impact) Measures	Traction power	Pressure loads	Pressure comfort/ Health	Micro-pressure waves	Loads due to air flow	Comfort and safety of air flow	Tunnel climate and HVAC	Vehicle climate and HVAC
Adjustment friction and damping characteristic of tunnel	X	X	X	XX	X	X	–	–
Open doors of cross-passages/crossovers	XX	X	X	X	X	X	X	X
Optimized overtrack exhaust (OTE) and underplatform exhaust (UPE)—more efficient removal of heat from vehicle	–	–	–	–	–	–	XX	XX
Vertical alignment to reduce traction power for acceleration and losses due to breaking	X	–	–	–	–	–	X	X
Full-height platform screen doors	X	–	–	–	–	XX	XX	X
Design of gaps at stations with platform screen-doors	–	–	–	–	–	X	X	X
Release of heat from technical equipment in tunnel	–	–	–	–	–	–	XX	XX
Avoidance of unnecessary heat ingress at portals/stations	–	–	–	–	–	–	XX	XX
Support natural or train-induced ventilation (shafts, openings)	X	X	X	–	–	–	XX	XX

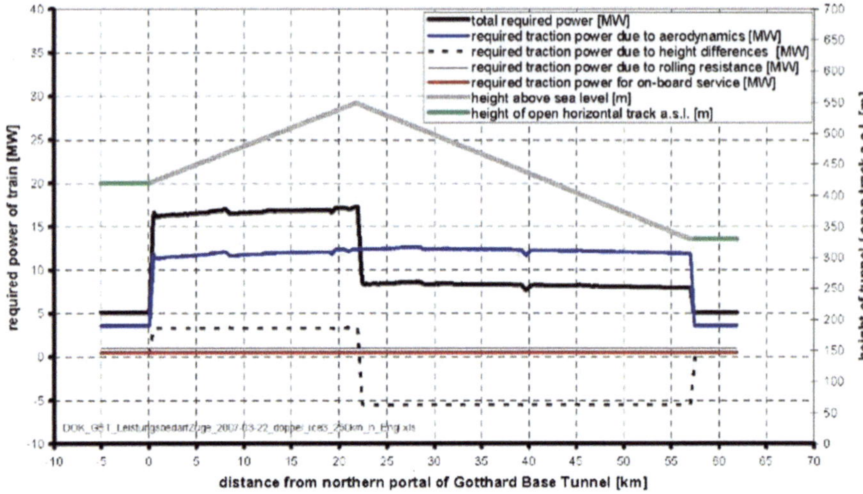

Fig. 3 Required power of high-speed passenger train for run from north to south through Gotthard Base Tunnel at 250 km/h with train length of approx. 400 m

Despite the inherent higher aerodynamic resistance in tunnel, the GBT is an example of saving energy by tunnel infrastructure. Simulations have been made comparing the existing shorter Gotthard tunnel and its access ramps with the new longer and lower Gotthard Base Tunnel (GBT). It was found that the expected 300 trains a day with a share of about 15 % of passenger trains will consume 20–30 % less power using the new and wider GBT (including energy for operation of the tunnel). At the same time, travel times are reduced by 40 % for passenger trains (81 min rather than 138 min for reference journey) and by 25 % for freight trains (145 min rather than 190 min for reference journey) [7].

For a generic tunnel, the influence of the free cross-sectional area on investment costs and the costs for traction power are shown in Fig. 4.

Increasing the free cross-sectional area of the tunnel reduces traction power and increases the costs for construction. Figure 4 indicates that for given boundary conditions an optimal tunnel size exists. Figure 5 highlights the impact of the costs for energy and the train velocity and for a certain train frequency on the optimal cross-sectional area of the tunnel. Increased energy costs lead to larger optimal tunnel cross-section.

First generation high-speed tunnels were mostly built as single-tube, double-track systems. The current trend of building more twin-tube, single-track rather than single-tube, double-track tunnels leads to reducing the free cross-sectional area from about 80–50 m^2 for typical European high-speed rail tunnels (see Fig. 6). According to Figs. 4 and 5, this change may lead to sub-optimal cross-sections in terms of tunnel aerodynamics and life-cycle costs. The aerodynamics, rather than the kinetic envelope of trains or the tunneling method might become the determining factor for the size of tunnels of high-speed rail connections. Considering other beneficial aero-

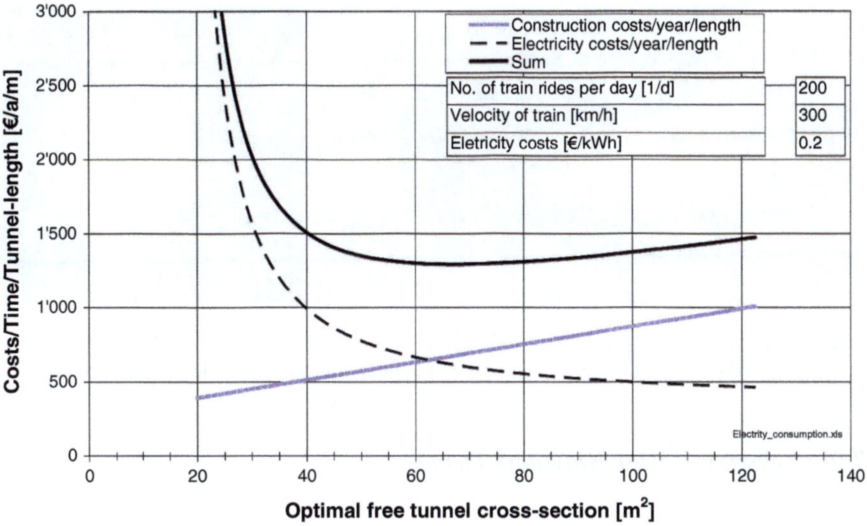

Fig. 4 Specific investment and operation costs for a rail tunnel for different sizes of cross-sectional areas; illustration of principal relationship only

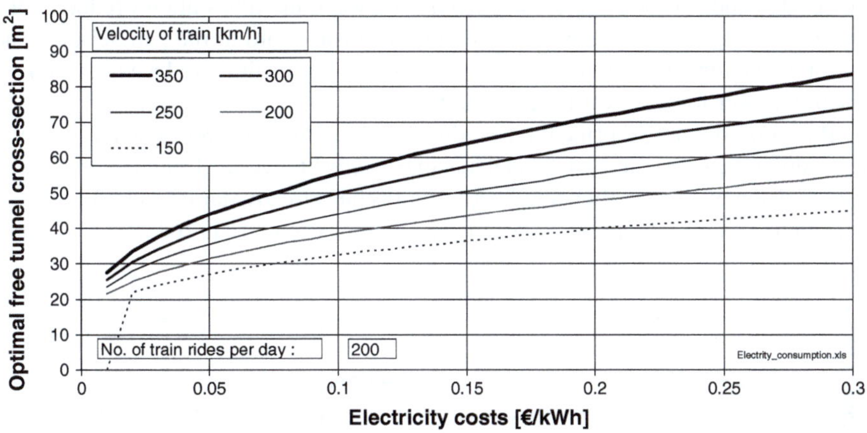

Fig. 5 Optimal size of cross-sectional area of tunnel depending on costs for electrical power; illustration of principal relationship only

dynamic aspects of large cross-sections (pressure comfort, micro-pressure waves, traction power, smoke control), the choice of new high-speed tunnel systems should be evaluated with care.

Apart from reducing energy consumption, improvements may allow to reduce the required maximum traction power to be installed (for example with respect to number of locomotives) power supply of rail network. Alternatively, maximum allowable train length or train weight can be adapted.

Fig. 6 Portal of tunnel Richthof of German high-speed rail line Hanover-Wuerzburg and of Loetschberg Base Tunnel; cross-sectional area of tunnels approx. 82 and 45 m^2, respectively

7.2 Pressure Relief Between Tubes

In very long tunnels, the piston effect of trains is required for air-exchange and temperature control of the tunnel. Twin-tube, single-track tunnels lead to an efficient air-exchange. Simultaneously, these tunnel systems lead to an increased traction power demand. Introducing cross-openings between the tubes allows reducing the tractive effort, since air in front of the train is pushed into to the parallel tube and re-enters the tube behind the train. Figure 7 shows the resulting reduced demand for traction power for a 20-km-long rail tunnel.

Based on the above principle to reduce the maximum power demand, pressure relief ducts were installed in Channel tunnel between France and England (see Fig. 8). The decision was made after construction works had already started and rolling stock specified. The traction power consumption, the maximum required power of trains

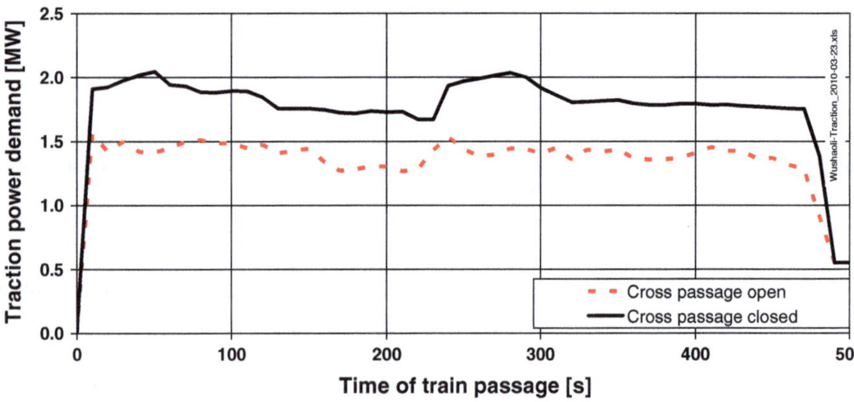

Fig. 7 Traction power demand in a 20 Km long twin-tube, single-track tunnel with open and closed cross-passages at same train speed

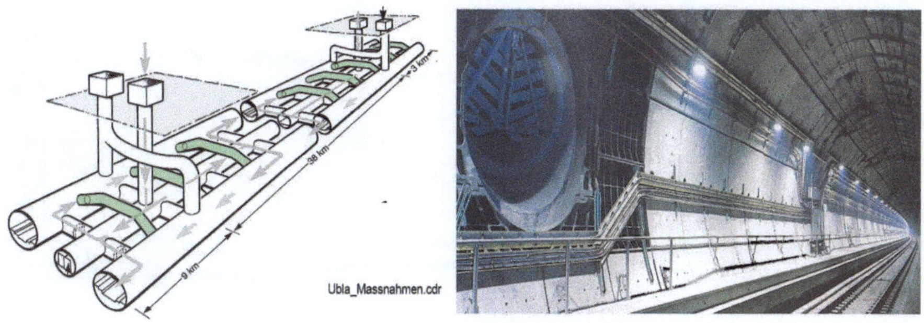

Fig. 8 Pressure relief ducts in Channel Tunnel to reduce traction power demand of trains (spacing of 500 m)

(particularly of shuttle trains) and the requirements for power supply in the tunnel were reduced by pressure relief ducts (see [4]).

However, the resulting re-circulation of air reduced the air-exchange and increased the expected temperatures in the tunnel. In order to control the temperature and in order to compensate the effect of internal re-circulation of air, a dry-cooling system became introduced (cooling pipes, refrigeration plants), i.e. aerodynamic improvements for trains caused substantial investments for a cooling system. Today's experience shows that the cooling system is not necessary. Very likely, a holistic design approach with adapted design specifications for rolling stock, power supply and tunnel design, had avoided the requirement for the cooling system in the first place

7.3 Air Exchange in Tunnel

In warm climate and/or in heavy duty underground systems, the heat release from trains due to braking and air-conditioning of coaches needs to be removed. During normal mode of operation, excessive temperatures might be noted particularly at stations [6]. This is for the following reasons:

- Prior to the stop at the station, the resistor grids and bogies heat up due to breaking.
- Air-conditioning units release heat at a certain location for an extended period of time.
- The piston-effect of the train decays.

Similarly, during congested mode of operation with short-time stops of trains in tunnels, the resulting built-up of heat in the tunnels needs to be handled by forced longitudinal ventilation of the tunnel.

In underground systems with full-height platform screen doors, the trackway space and the platform region are essentially decoupled. As a consequence, the tunnels do not need to be cooled down to the temperature which is expected on the platform for comfort reasons.

Fig. 9 Heat extraction at station by overtrack exhaust (OTE, *top*) and under-platform exhaust (UPE, *bottom*) along the station trackway (*left* Station trackway along platform, *right* station model [5])

Heat removal from tunnel and stations is achieved and supported by:

- Shafts at stations and in tunnels utilizing the train-induced piston effect and natural draft
- Portal and cross-over design of twin-tube, double-track systems
- Overtrack exhaust (OTE) along the trackway at the stations (see Fig. 9)
- Underplatform exhaust (UPE) along the trackway at the stations (see Fig. 9)

Thus, balancing the measures to control the temperature in vehicles and tunnels is the challenge. The following examples show a variety of measures:

(A) Reduce energy consumption of vehicle:

- reduce tractive requirements by tunnel design
  - provide larger cross-section for less aerodynamic resistance
  - introduce shafts/openings for less aerodynamic resistance
  - implement improved vertical track alignment (slopes for supporting acceleration and
  - braking near stations)
- reduce traction and power, demands by vehicle design
  - design lighter trains and better streamlined trains
  - utilize more efficient drives and adapted signaling systems
  - use more efficient regenerative breaking, i.e. reduce amount of heat dissipated to tunnel air
  - use more efficient on-board services

(B) Reduce energy consumption of infrastructure:

- reduce the energy effort to remove heat
  - allow for natural ventilation
  - allow for efficient train-induced ventilation
  - use energy-efficient control strategies of tunnel ventilation and platform cooling

  - increase efficiency of heat removal from train by better adapted interface of UPE and OTE to trains at stations

- reduce energy consumption for vehicle (see A)

Vehicle and infrastructure designers the mutual impact of measures is often not quantifiable. Three examples may illustrate typical design question:

1. To which extend shafts should be increased in size and how should they be arranged in order to allow for less powerful HVAC systems on trains?
2. Where should exhaust air from air-conditioning units of trains be released?
3. To which extend warm air from condenser units from trains can be redirected to the extraction points of under-platform exhaust (UPE)? How can the UPE be optimized to extract more efficiently heat from trains?

The location of the train air-conditioning unit has an impact on the tunnel air temperature and the tunnel ventilation requirements. During a train stoppage, the heat will accumulate in the annular space along the train and the warmer less dense air will rise and collect along the tunnel ceiling. Ceiling mounted A/C units aggravate the problem because the heat discharged by the upstream A/C units cascades along the train in the upper regions of the annulus and can make the condenser intake temperature of the downstream A/C units about 7° C (12° F) higher than the average tunnel air temperature at the same location, even with the operation of the tunnel ventilation system (see [3]). In comparison, underneath train A/C units can reduce the airflow requirement to ventilate the tunnel during congested operation due to the stratification effect and, in turn, reduce the associated civil provision.

It would be beneficial to elaborate a better data basis of the relationships. This would allow to direct investments in the most reasonable manner. So far, the improved energy efficiency has mainly been achieved by advanced drive systems and components for traction through the acquisition of new rolling stock, but this improvement is off set increasingly by demanding requirements from the customer in terms of overall travel comfort, escalators, HVAC systems, etc.

8 Proposal for Future Work and Objectives of Design Optimization

Many parameters determine the aerodynamics and resulting climate of an underground system. Many factors can be modified to create a better passenger environment, improved performance and reduced operational costs. Therefore, a basis of decision-making is needed to improve an underground system.

A set of best-practice-recommendations and checklists would be beneficial to support designers. The basis of decision-making should be improved to allow for sustainable design. In order to advance, research should follow the following objectives:

- Create awareness of interrelationships
- Illustrate sensitivity and impact of design parameter and measures by case studies
- Show systems limitations
- Provide key cost data for vehicles and infrastructure as basis for life-cycle cost analysis and as basis for decision making
- Issue check-lists and present best-practice recommendations
- Motivate improvement of rail tunnels and metro systems design in the context of life-cycle costs and energy demand

Most likely, the proposal will not yield “a technical revolution”. However, spending more research in system integration may lead to a set of cost-efficient improvements and optimization. Design decisions would be based on more sound knowledge.

9 Conclusions

Aerodynamic conditions in underground rails systems and the related climate can be influenced by both, measures at rolling stock or at the infrastructure. Improvements of aero-thermal conditions require adaptation at the interface of rolling stock and infrastructure and balance of investments for rolling stock and infrastructure. The design and analysis of underground systems is a complex process, requiring thoughtful consideration of both technical advantages and limitations of various alternatives, as well as the cost implications and long-term sustainability. Key design parameters are often set without consideration of the aero-thermal aspects or consequences for ventilation of vehicles and tunnels. For improved design, information about the relationship between the major design parameters is needed. Cost data is required for design optimization. Best-practice recommendations are needed to allow improvements. Progress is expected not by one “big thing”, but as a result of hundreds of little things.

References

1. Busslinger, A., Nyfeler, S., Reinke, P.: Wirksamkeit von Druckentlastungsschächten beim Hochgeschwindigkeitsbahnverkehr in Tunneln, Seminar “Aerod. Anford. an Schienenfahrzeuge”, IVF Bahntechnik (2010)
2. Figura-Hardy, G.I.: Pressure relief -trends and benefits of incorporating airshafts into railway tunnels. In: 10th International Symposium Aerodynamics and Ventilation of Vehicle Tunnels, Boston (2000)
3. Gehrke, P., Stacey, C.: Rail tunnel temperature stratification and implications for train and tunnel ventilation design. In: Proceedings of the 11th International Symposium on Aerodynamics and Ventilation of Vehicle Tunnels, Luzern, Switzerland, July 2003
4. Henson, D.: Aerodynamics, ventilation and cooling. In: Engineering the Channel Tunnel, E & FN Spon and Chapman & Hall, London (1995)

5. Metropolitan Transportation Authority. In: E.G. Sander (ed.) Greening Mass Transit and Metro Regions: The Final Report of the Blue Ribbon Commission on Sustainability and the MTA, New York, USA, 2008
6. Pope, C.W., Newman, D.G., Henson, D.A.: The factors affecting draught relief and air temperature in an underground metro system. In: 10th International Symposium Aerodynamics and Ventilation of Vehicle Tunnels, Boston (2000)
7. Steinmann, N., Schär, R.: Traktionsenergiebedarf der Gotthard Basislinie—Traction energy demand on the Gotthard base line, Elektrische Bahnen—Elektrotechnik im Verkehrswesen, 7/2009
8. Vardy, A.E., Reinke, P.: Estimation of train resistance coefficients in tunnels from measurements during routine operation. In: Proceedings of the Institution of Mechanical Engineers, Part F: Journal of Rail and Rapid Transit, vol. 213, No. 2 (1992)

Part II
Train Aerodynamics: Cross-wind Effects

Numerical Investigation on the Embankment Configuration for Cross Wind Stability of ICE3

Irfan Ali, Arnd Rüter and Carlos-José Rodriguez Ahlert

Abstract Cross wind stability of the TSI HST reference vehicle ICE3 is numerically investigated for the six meter high embankment configuration. The focus of this numerical study is to investigate the sensitivity of the results on the embankment geometry and the position of the train on the embankment. This study is relevant as TSI HST does not explicitly specify the embankment configuration in the wind tunnel.

1 Introduction

Cross wind stability is a key topic for the homologation of trains and hence an important issue in railway design and operation. This is due to the increase in extreme wind events, coupled with low weight train concepts and increasing operating speeds [1]. Higher train speed results in the reduction of the effective yaw angle thereby reducing the cross wind effect. However, for high speed, lighter trains are also preferred as they have significant benefits in terms of energy consumption and other performances like acceleration, wheel-rail wear and damage.

Compared to Japan [2], the interests in Europe in cross wind stability of trains is a relatively new topic. Although in the early 20th century several wind induced accidents were reported [3], it was only after the entry of the high speed trains, that this issue became relevant [4]. A large European research project TRANSAERO [5] and the Franco-German DEUFRAKO cross wind program contributed to the research activity in Europe in this field. This research contributed to the section of the Technical Specification of Interoperability (TSI) dealing with the crosswind stability [6].

I. Ali (✉)
Project Management, Mobility Division, Siemens AG, Krefeld, Germany
e-mail: irfan.ali@siemens.com

A. Rüter · C.-J. Rodriguez Ahlert
Aerodynamics, Mobility Division, Siemens AG, Krefeld, Germany

A. Dillmann and A. Orellano (eds.), *The Aerodynamics of Heavy Vehicles III*,
Lecture Notes in Applied and Computational Mechanics 79,
DOI 10.1007/978-3-319-20122-1_5

Three different wind tunnel set-ups (true flat ground (TFG), single ballast and rail (STBR) and embankment) are defined by different European standards used to investigate the cross wind stability of a train [6, 10]. Among these the embankment configuration is the most critical scenario, delivering lowest allowable cross winds. The block flow field on reaching the embankment is accelerated resulting in higher local velocities than the STBR or TFG configuration [11]. Another aspect contributing to it is that the embankment configuration is exposed more often to the high velocity gusts. For all these reason it is of extreme importance that it is correctly represented in the wind tunnel to analyse a safety relevant issue.

With the increase in computational resources CFD has been becoming an important tool in numerical investigations. Flow structures contributing to the cross wind instability have been identified by CFD [7]. Several successful attempts have been made to obtain good comparison with wind tunnel results for both the ground and embankment scenarios [8, 9]. Still the application of industry standard CFD methods is considered inappropriate by the TSI group on cross winds.

In this work a numerical study on the embankment configuration for cross wind stability of ICE3 is performed. In the first part the simulation method is validated with the help of available wind tunnel data. Numerical simulations are performed for a yaw angle of 15° where the boundary conditions are kept similar to the wind tunnel. The Reynolds average Navier-Stokes (RANS) method is used in the commercial code Fluent to calculate the flow field. A grid sensitivity analysis is performed, based on which further grids are generated. In the second part the shape of the embankment leading edge and position of the train on the embankment is varied. The obtained results are discussed and are analysed highlighting some of the open issues in the TSI HST [12].

2 Experimental Analysis

The measurements used for these investigations have been carried out in June 2009 in the DNW Large-Low-Speed-Facility (LLF) wind tunnel in Emmeloord (Netherlands). The LLF of DNW is an atmospheric, single return wind tunnel with two exchangeable closed test section arrangements, but it can be operated in an open jet mode as well. The measurements for the train test were performed in a closed test section with a cross section of $8 \times 6\,\text{m}$. Due to its large test sections and high achievable flow velocities the LLF is well suitable for fast ground vehicles like high speed trains at lower blockage ratio. An internal 6-component balance was used for force measurements.

2.1 Nomenclature

In this paper the coordinate system and non-dimensional coefficients are used as recommended in the EN14067-1 [13]. Origin of the coordinate system is located

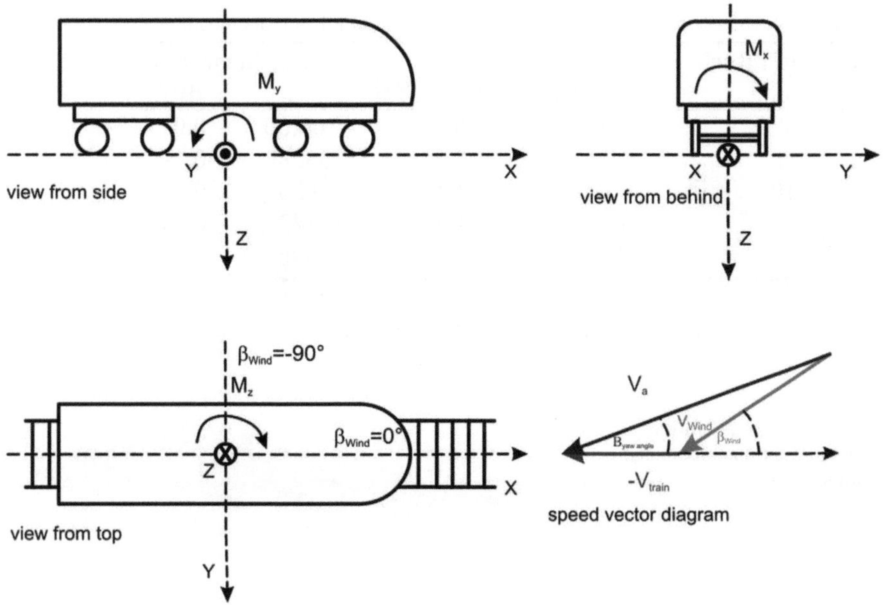

Fig. 1 Coordinate system and definition of velocity vectors

in the middle of the bogies, in the middle of the tracks on the top of rail (Fig. 1). The aerodynamic coefficients of drag, side and lift forces are represented by c_{fx}, c_{fy} and c_{fz} and the roll, pitch and yaw moment coefficients by c_{mx}, c_{my} and c_{mz}. The leewards rail rolling moment coefficient $c_{mx,lee}$ is obtained by the roll moment and lift force coefficient where $c_{mx,lee} = c_{mx} - c_{fz} \cdot b_A/d_0$ and is often used to quantify the overturning moment of trains. As a custom in train aerodynamics the reference length of $d_0 = 3$ m and reference area of $A_0 = 10\ m^2$ with respect to full scale is also used in this work. The width between the supports $(2 \cdot b_A)$ is equal to 1.5 m for standard UIC track gauge.

2.2 Benchmark Test

Wind tunnel tests which are used within the assessment process of high speed rolling stock shall include a benchmark test. A comparison of CSTB [14] and LLF measurement campaign for the leeward side of the embankment are shown in Table 1. For the relevant yaw angles in this paper (10° to 25°) minimum requirements of TSI [12] are fulfilled, since the $c_{mx,lee}$ aerodynamic coefficients confirms the quality within 20 % for the embankment configuration. The major part of the numerical investigations was done on a yaw angle of 15° in this paper.

Table 1 Benchmark test ICE3 (LLF)–ICE3 (CSTB), leeward side of embankment

Angle of attack deg	$c_{mx,lee}$ ICE3-LLF	$c_{mx,lee}$ ICE3-CSTB	$c_{mx,lee}$ (leeward) %	diff
0	-0.01	0.00	100.00	0.01
5	0.36	0.55	33.94	0.19
10	1.04	1.15	9.09	0.10
15	1.93	1.84	4.42	0.09
20	2.82	2.66	5.96	0.17
25	3.65	3.55	2.81	0.10
30	4.54	4.48	1.53	0.07

Fig. 2 Test configuration with ICE3, embankment and splitter plate

2.3 Test Setup

The wind tunnel measurements were performed on a 1:25 scaled ICE3 model on a 6 m embankment ground configuration, depicted in Fig. 2. The model was installed on a splitter plate located 200 mm above the wind tunnel floor in order to be outside of the wind tunnel boundary layer. This setup allows a fresh boundary layer on the splitter plate which is small enough compared to the model height as required by TSI HST.

The aerodynamic force and moment coefficients were measured for a Reynolds number of $Re = 0.75 \cdot 10^6$. Changing blockages induced by different yawing angles and slight temperature variations were compensated by Mach number adjustments. For the investigations a Mach number of approx. $Ma = 0.28$ was used, that is equivalent to an inlet flow velocity of $u_0 = 99$ m/s. The turbulent intensity $Tu' = \sqrt{u'^2/U^2}$ in the free stream flow was around about 0.1 %.

The blockage ratio is defined as the ratio of the projected area of the model, the splitter plate with the standard 6 m embankment, its support and the test section area and corresponds to 3 % for a yaw angle of 30°. Since it is less than 5 % this TSI HST requirement is also fulfilled.

3 Numerical Analysis

In order to reduce modelling and computational costs initially some simplifications on the geometry were carried out. A levitating splitter plate was considered, neglecting the supporting guides underneath it. Additionally the supporting struts between the

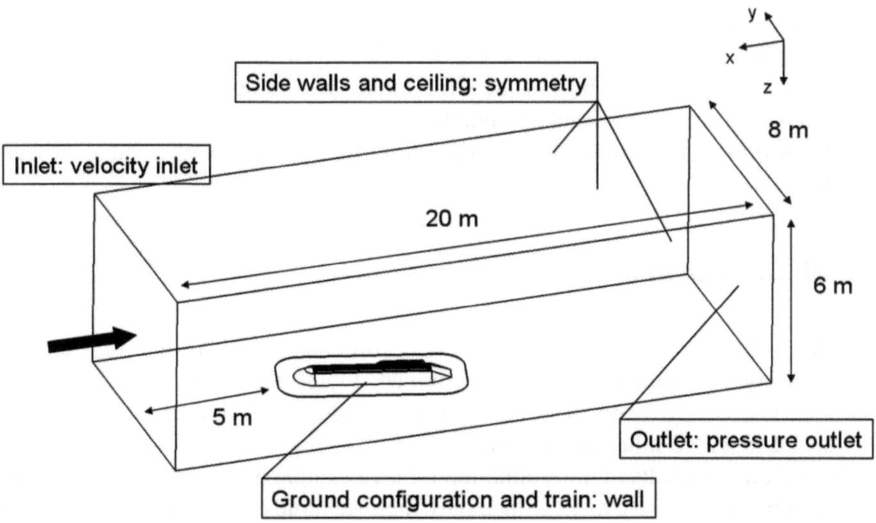

Fig. 3 Setup of the computational domain for the wind tunnel

Fig. 4 Surface mesh around the train

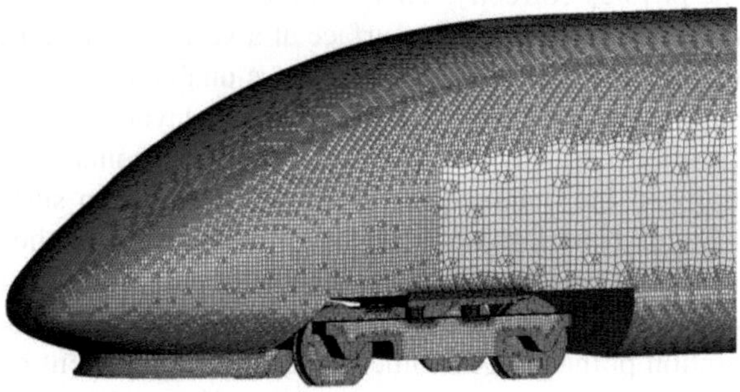

wake body and the ground plate were neglected. Finally some slight simplifications in the geometry of the bogies were performed. The computational domain resembles the wind tunnel geometry as far as the cross-sectional area is concerned, depicted in Fig. 3.

As a meshing algorithm an octree based meshing method was used, that produces boundary fitted, locally refined, conformal meshes with prism layers. Figure 4 shows a surface mesh around the train. The generated numerical grids were used for fluid simulations with the commercial CFD solver ANSYS Fluent. For the turbulence modelling RANS approach was used with the realizable k-epsilon model with enhanced wall treatment. The SIMPLE pressure-velocity coupling and a second order upwind spatial discretization were used. The boundary conditions used in the simulations are shown in Fig. 3 and are as follows:

(a) A block flow velocity inlet with a constant velocity $u_x = -99$ m/s, turbulent intensity of 0.1 % and turbulent length scale of 0.01m.
(b) The side walls and ceiling of the wind tunnel were assumed to be smooth and slip wall (symmetry boundary condition)
(c) At the outlet a pressure outlet boundary condition was used with a gauge pressure of 0.

3.1 Mesh Validation

In order to validate the grid quality different refinement levels of the train surface, boundary layer and bulk volume were examined. Table 2 shows the six different grids that were tested. A y^+ value less than 350 is achieved with 2 boundary layer; a value less than 150 with 5 boundary layers.

A look on the results of the simulation for a yaw angle of 15° in Fig. 5 shows that there is a high sensitivity to y^+. Since only high Reynolds flows were considered in these investigations, the aim was not to get a $y^+ \sim 1$. However it is clear that an insufficient amount of boundary layer as in mesh 1 and mesh 2 does not reproduce the physics correctly. This becomes clearer in Fig. 6. Here the pressure coefficient distribution on the train surface at a vertical cut 520 mm (1:25) from the train nose can be seen. Mesh 1 and mesh 2 are unable to capture the physical effects like flow separation. As a next step the boundary layers at the complete ground configuration (splitter plate and embankment) were additionally refined. In contrast to the sensitivity of the obtained results on y^+ at the train surface, a reduction of y^+ on the embankment does not have a significant effect on the resulting pressure distribution and on $c_{mx,lee}$. This is most probably due to the fact that the train surface has rounded edges and thus needs a better resolution of the boundary layer to capture the separation points. On the other hand the embankment has only sharp edges where the separation points are clearly defined. Hence there is no need to invest more cells in refining the boundary layers on the embankment as has been seen in this work.

The difference between the different refinement levels of the mesh on the train surface is visually hardly recognizable. The difference in $c_{mx,lee}$ in Table 2 is minimal and only a closer look on the pressure distribution in Fig. 6 at the upper right corner

Table 2 Setup and numerical results of the six tested grids for a yaw angle of 15°

ID	Mesh	$c_{mx,lee}$
1	Coarse train surface with $y^+ < 350$ (train)	1.635
2	Fine train surface with $y^+ < 350$ (train)	1.690
3	Coarse train surface with $y^+ < 150$ (train)	1.746
4	Fine train surface with $y^+ < 150$ (train)	1.744
5	Fine train surface with $y^+ < 150$ (train & embankment)	1.752
6	Fine train surface & refined volume with $y^+ < 150$ (train)	1.750

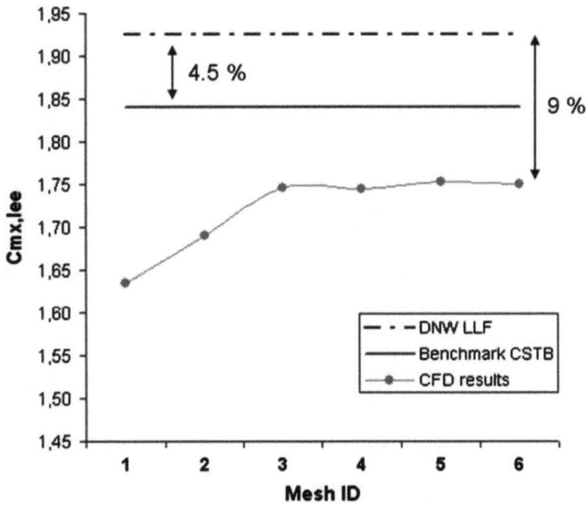

Fig. 5 Comparison of $c_{mx,lee}$ for the six tested grids for a yaw angle of 15°

shows that the grids with a coarse surface mesh have an inadequate resolution of the design features.

As a last grid validation step a local refinement of the volume was performed. At a yaw angle of 15° a stable vortex develops at the leeward side of the train that has a considerable effect on the force coefficients on the train surface. To get a proper resolution of this vortex the volume around it was locally refined in mesh 6. There is a slight difference in the results compared with a coarser mesh but considerable increase in computational costs (around 16 %) as can be seen in Table 3. Hence it was decided not to use mesh 6 for further investigations. All computations were done on an Intel(R) Xeon(R) 8 core computer with 32 GB RAM. The drag coefficient of the end car was used as the convergence criteria.

The best trade-off between grid quality and computational costs is achieved with mesh 4. The refinement levels of this mesh are used for all further simulations. Figure 5 shows that there is a difference in $c_{mx,lee}$ of approx. 9 %. The Fig. 7 shows that all force coefficients c_{fx}, c_{fy} and c_{fz} are underestimated. In order to improve the comparison with the experimental results some variations in numerical setup were performed as described in Sect. 3.2.

3.2 Variations in Numerical Setup

In this section different changes in the numerical setup were tested. On the one hand the goal was to achieve a better agreement between simulation results and wind tunnel measurements (Sects. 3.2.1 and 3.2.2) and on the other hand the effects of different train positions and different shapes of the embankment leading edge on the force and moment coefficients were investigated (Sects. 3.2.3 and 3.2.4).

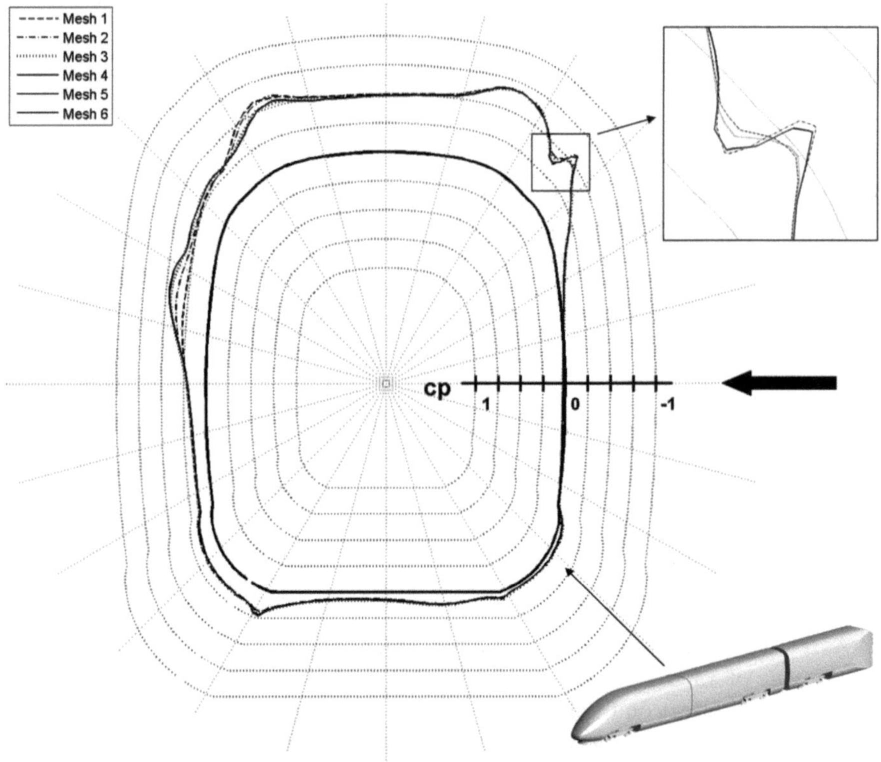

Fig. 6 Pressure coefficient distribution on the train surface on a vertical cut 520 mm (1:25) from the nose for the six tested grids for a yaw angle of 15°

Table 3 Computational costs

ID	# Cells	Computing time (h)
4	10685603	6.15
5	12675548	7.15

3.2.1 Consideration of Wind Tunnel Walls

In order to consider the wind tunnel effects like pressure loss or flow acceleration due to the increase of the boundary layer at the wind tunnel walls, the boundary conditions at the wind tunnel walls (bc wtw) in the simulation were changed to no slip condition (wall). The mesh at the wind tunnel walls was refined to reach a y^+ value less than 150. Looking onto the results in Fig. 7 it can be seen that there is no significant improvement in the prediction of $c_{mx,lee}$. There is only a difference of 1 % and the force coefficients are still underestimated. Taking the considerable increment of computational costs into account (approx. 3 million more cells and 2 more hours in computing time) it is decided against considering the effects of the

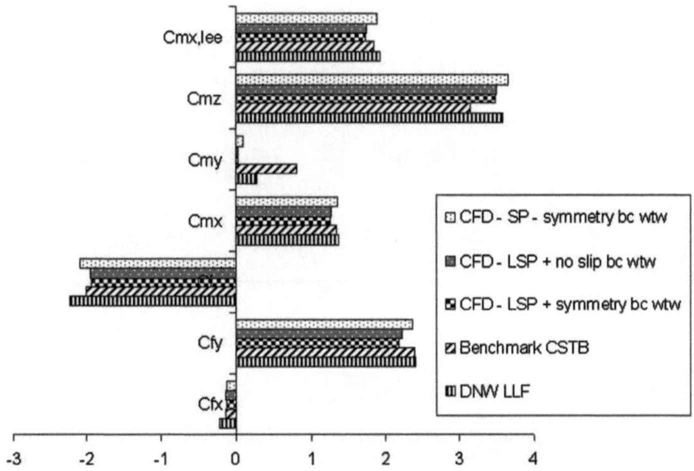

Fig. 7 Comparison of the force and moment coefficients for different configurations for a yaw angle of 15°

wind tunnel walls. Therefore the symmetry boundary condition was used for further investigations.

3.2.2 Changes in Splitter Plate Configuration

To reduce modelling and computational costs, initially the geometry of the splitter plate was simplified and a levitating splitter plate (LSP) was considered. Since the accordance between the measurements and the computation results is not good as shown in Fig. 7, in this chapter the exact geometry of the splitter plate with supporting guides (SP) is meshed and simulated to get a better representation of the wind tunnel setup.

Figure 7 shows that considering the supporting guides below the splitter plate leads to a very good accordance with the measurements. Now only a difference of 2 % in $c_{mx,lee}$ is achieved. Also the force coefficients show a better agreement.

The reason for a better comparison can be observed in Fig. 8, there the velocity in y-direction is plotted at a vertical slice of the flow volume at position x = 0 (a little before train nose). Even if the global blockage ratio of the ground configuration does not increase too much compared to a levitating splitter plate there is a significant local blockage that induces an overpressure below the splitter plate. Due to this overpressure the flow accelerates over the splitter plate. The higher flow velocity in y-direction on the top of the embankment leads to an increase in the force coefficients and therefore to an increase in $c_{mx,lee}$.

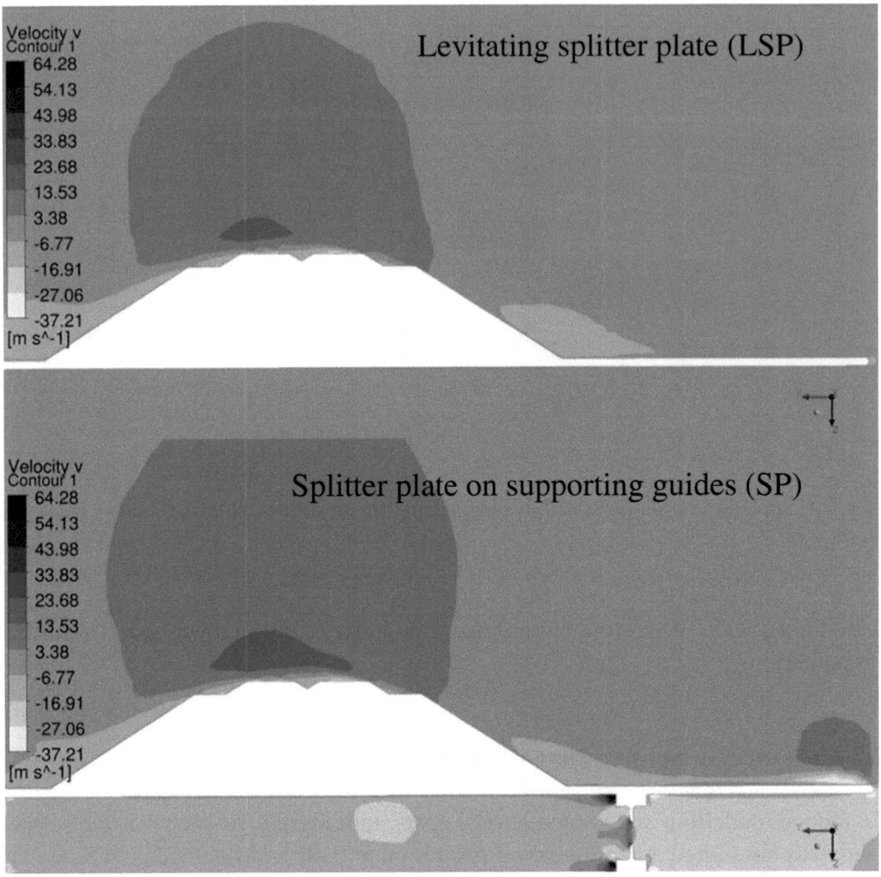

Fig. 8 Velocity distribution in y-direction on a slice at x = 0 parallel to the inlet surface for a yaw angle of 15°

Figure 9 shows that the accordance between experiments and simulation is also good for other yaw angles between 10° and 25°. A maximum of 2 % difference is achieved between experiments and simulations. In the scope of this work only small yaw angles have been tested. Past studies have shown that the difference in the results becomes larger with higher yaw angles because of greater flow separation areas with a transient character [14]. This makes it difficult to predict the results accurately with steady state RANS turbulence models. Since for small yaw angles a good agreement was achieved further investigations regarding the train position and the shape of the embankment leading edge were performed for a yaw angle of 15° with the exact geometry of the splitter plate tested in the wind tunnel.

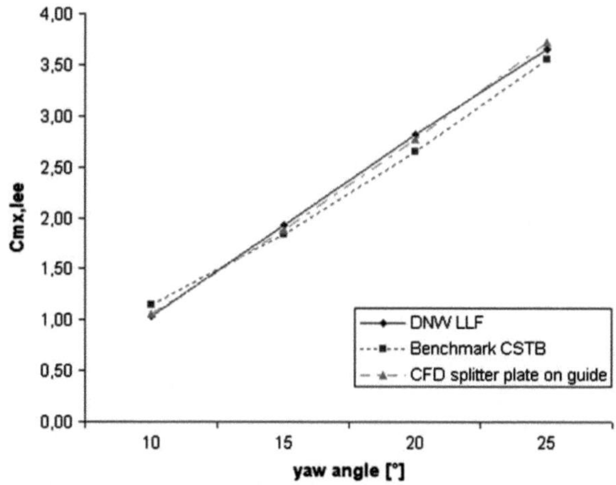

Fig. 9 Comparison of experimental and numerical results for different yaw angle

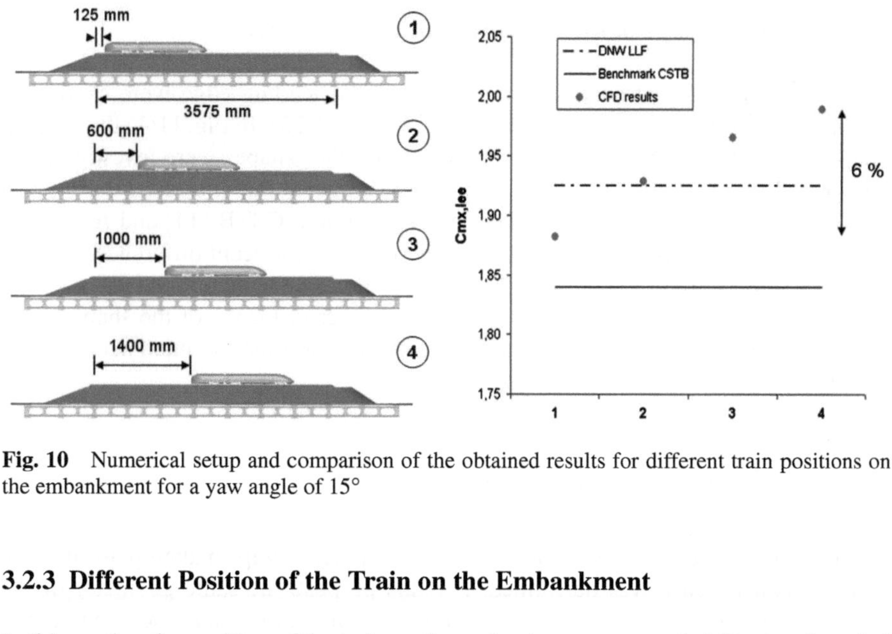

Fig. 10 Numerical setup and comparison of the obtained results for different train positions on the embankment for a yaw angle of 15°

3.2.3 Different Position of the Train on the Embankment

In this section the position of the train on the embankment was varied. During the wind tunnel tests the train was mounted such that the end of the wake body was 125 mm away from the end of the rails as can be seen in Fig. 10(1). The numerically simulated configurations can be taken from Fig. 10(2)–(4). The obtained results in Fig. 10 (right side) show that there is a significant difference of 6 % in $c_{mx,lee}$ between the two extreme positions. It can be seen that $c_{mx,lee}$ decreases with increasing distance to the front of the embankment.

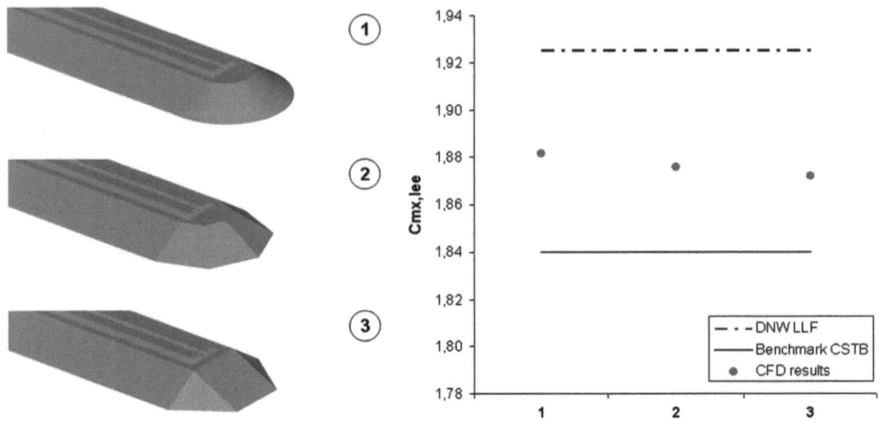

Fig. 11 Numerical setup and comparison of the obtained results for different shapes of the embankment leading edge for a yaw angle of 15°

3.2.4 Changed Shape of the Embankment Leading Edge

As a last step different shapes of the embankment leading edge were examined, keeping the position of the train constant at 125 mm (1:25). In Fig. 11(1) the original shape tested at the DNW LLF can be seen. Two other shapes were investigated as shown in Fig. 11(2) and (3). These shapes are similar to the ones used during past wind tunnel tests with an embankment configuration at CSTB [14] and Audi [15]. As can be observed in Fig. 11 (right side) there is no significant difference between the three shapes (max. 1 % between (1) and (3)). It has to be mentioned that the train position was kept constant as in Fig. 10(1). A bigger influence of the shape of the embankment leading edge should be expected with the train mounted nearer to the leading edge and at higher yaw angles.

3.3 Simulation of a More Realistic Case

Additionally to the different variations in the numerical setup an attempt to simulate a more realistic case was performed. For this purpose the same geometry of the ICE3 and the embankment was used with all simplifications as in the wind tunnel. The numerical setup is depicted in Fig. 12. The boundary conditions used in the simulation are as follows:

(a) A moving ground configuration was used as a boundary condition with a moving velocity equivalent to the train velocity.
(b) Two velocity inlets were set in order to have a yaw angle of 15°. At inlet (1) the velocity in x-direction was set equally to the train speed ($u_{train} = -95.63$ m/s) and the velocity in y-direction was set equally to the wind speed ($u_{wind} = 25.62$ m/s). The velocity in z-direction was set to 0.

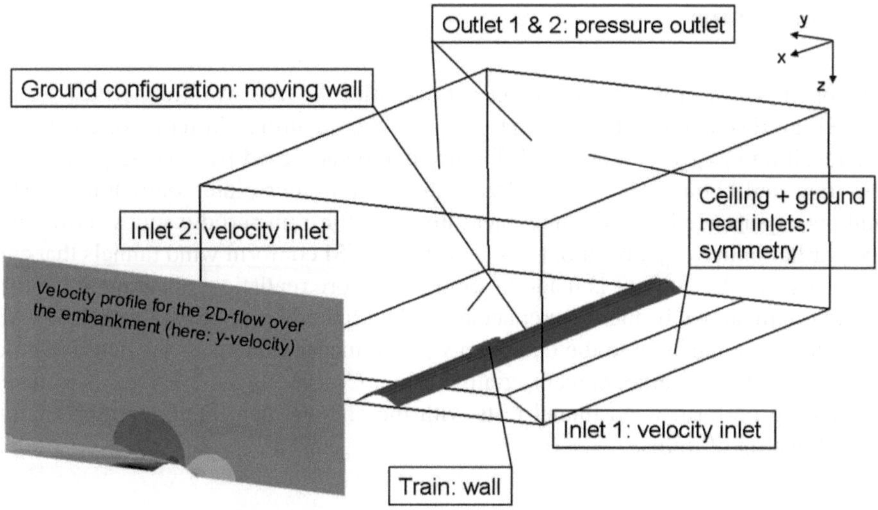

Fig. 12 Numerical setup for a simulation of a more realistic case

(c) At inlet (2) the velocity profile in Fig. 12 was used for the setting of the velocities in y- and z-direction. This profile was obtained by simulating the 2D-flow over the embankment with $u_{wind} = 25.62$ m/s. The velocity in x-direction was set to $u_{train} = -95.63$ m/s.
(d) Two pressure outlets were set with a gauge pressure of 0.
(e) For the ceiling and a small area of the ground near the inlets, symmetry was used as boundary condition.

The simulation of this more realistic case delivered completely different results than the wind tunnel test. Table 4 shows a considerable decrease of 46 % in $c_{mx,lee}$. First investigations showed that this is due to the fact that the flow at the leeward side of the embankment is very different compared to the flow around the wind tunnel configuration. It seems that the flow completely separates at the leeward side of the embankment. This phenomenon did not appear in the wind tunnel simulations where the flow reattaches after a few meters downstream.

This is a first attempt to get closer to reality. However the considerable decrease in $c_{mx,lee}$ shows that more investigations are needed in order to get more realistic boundary conditions for the embankment configuration.

Table 4 Numerical result for a more realistic case

	$c_{mx,lee}$
DNW LLF	1.925
CFD more realistic case	1.040

4 Conclusions

The open issues in the embankment configuration for the wind tunnel test set-up in TSI are discussed. Effects of these aspects are quantified in terms of change in the resulting $c_{mx,lee}$. About 6 % difference can be achieved by varying position of the train and about 7 % by considering different splitter plate configuration. The authors experience in wind tunnel measurements for train aerodynamics shows further variation in $c_{mx,lee}$ by about 5 % in different TSI conform wind tunnels that can be measured. An attempt is made to simulate a more realistic embankment configuration with shows further discrepancies in the obtained $c_{mx,lee}$ by about 46 %. All these factors contribute to the uncertainty in the measurements of the aerodynamic coefficients for the embankment configuration. Considering the above, it is questionable to perform wind tunnel tests with the embankment configuration for the homologation of trains.

References

1. Heine, C., Matschke, G.: Full scale tests on side wind effects on trains: evaluation of aerodynamic coefficients and efficiency of wind breaking devices. In: Proceedings of the Brite/Euram Project Symposium Transient Aerodynamics for Railway system Optimisation, Paris (2000)
2. Saito, M.: Japanese railway safety and the technology of the day. Jpn Railway Transp. Rev. **33**, 4–13 (2002)
3. Gawthorpe, R.G.: Wind effects on ground transportation. J. Wind Eng. Ind. Aerodyn. **52**, 73–92 (1994)
4. Cooper, R.K.: The effects of cross winds on trains. ASME J. Fluid Mech. **103**, 170–178 (1981)
5. Shulte-Werning, B., Grégoire, R., Matschke, G.: TRANSAERO—a European initiative on transient aerodynamics for railway system optimisation (2002)
6. Railway applications—aerodynamics—Part 6: requirements and test procedures for cross wind assessment, prEN 14067–6 (2007)
7. Khier, W., Breuer, M., Durst, F.: Flow structure around trains under side wind conditions: a numerical study. Comput. Fluids **29**, 179–195 (2000)
8. Diedrichs, B., Sima, M., Orellano, A., Tengstrand, H.: Crosswind stability of a high-speed train on a high embankment. Proc. I MechE Part F: J. Rail Rapid Transit **221** (2007)
9. Bocciolone, M., Cheli, F., Corradi, R., Muggiasca, S., Tomasini, G.: Crosswind action on rail vehicles: wind tunnel experimental analyses. J. Wind Eng. Ind. Aerodyn. **96**, 584–610 (2008)
10. RIL 80704: Aerodynamik/Seitenwind, Deutsche Bahn (2006)
11. Baker, C.J.: The determination of topographical exposure factors for railway embankments. J. Wind Eng. Ind. Aerodyn. **21**, 89–99 (1985)
12. Technical Specification for Interoperability, HS RST TSI (2008)
13. Railway applications—Aerodynamics—Part 1: Symbols and units, EN 14067–1 (2003)
14. Heine, C., Paradot, N., Rüter A.: Common DEUFRAKO Research on Cross Wind Effects on High Speed Railway Operation, T.R. Chapter 3: Vehicle Aerodynamics (2005)
15. Schober, M., Weise, M., Orellano, A., Deeg, P., Wetzel, W.: Wind tunnel investigation of an ICE 3 endcar on three standard ground scenarios. J. Wind Eng. Ind. Aerodyn. **98**, 345–352 (2010)

High-Speed Train Crosswind Analysis: CFD Study and Validation with Wind-Tunnel Tests

Carlo Catanzaro, Federico Cheli, Daniele Rocchi, Paolo Schito and Gisella Tomasini

Abstract The correct definition of the crosswind aerodynamic forces on a high speed train is important to judge the safety of the rolling stock. Different conditions in terms of scenario (DTBR, EMBK and presence of wind barriers) and in terms of train motion are compared in the paper considering the aerodynamic forces and moments that mostly contribute to the overturning risk of the train while it is running under crosswind. The results of CFD simulations are compared and validated with experimental wind tunnel tests. The numerical model allows to make a comparison between the aerodynamic forces computed on the train for different angles of incidence, simulating a still model and a moving train model. The still model reproduces what is tested in the wind tunnel, while the moving model simulates the real condition of a train running under crosswind. This work analyzes the aerodynamic behavior of a ETR-500 train under crosswind in different scenarios.

1 Introduction

Crosswind forces on trains are becoming more and more important since the increase of the train speed makes the relative air-train speed higher and leads to higher actions on the train. This problem is studied since many years and railways require a crosswind certification for the rolling stock [1]. The parameters that lead to the certification involve different aspects of the train (mass, geometry, aerodynamic characteristics), but considers also the infrastructure where the train is running (surrounding scenario, track characteristics, wind prediction) [2]. The crosswind performance of a train is described by the definition of the Characteristic Wind Curves (CWC), representing the maximum crosswind speed that a train may withstand under operating conditions [1, 3].

C. Catanzaro · F. Cheli · D. Rocchi (✉) · P. Schito · G. Tomasini
Politecnico di Milano, via La Masa 1, 20156 Milano, Italy
e-mail: daniele.rocchi@polimi.it

P. Schito
e-mail: paolo.schito@mail.polimi.it

A. Dillmann and A. Orellano (eds.), *The Aerodynamics of Heavy Vehicles III*,
Lecture Notes in Applied and Computational Mechanics 79,
DOI 10.1007/978-3-319-20122-1_6

Generally the definition of the aerodynamic behavior of the train is studied through wind tunnel tests on a scaled model, measuring the forces acting on the model at different wind yaw angles, defining the aerodynamic coefficients of the train [4]. This is a simplified model of the real operating conditions: in fact the wind tunnel tests are performed on a still model, where the crosswind is reproduced simply changing the yaw angle of the incoming wind. What happens in the real situation is that the relative wind speed, that produces the aerodynamic forces, is made by a contribution due to the wind blowing from a direction relative to the ground and by the train speed aligned to the track. In the wind tunnel the train is still with respect to the ground and the incoming wind speed represents the yaw incidence angle with respect to the train and to the ground. The major discrepancy between wind tunnel test and full scale operating conditions is the interaction between the geometry of the scenario and the crosswind flow. In fact the shape of the scenario the air encounters in the wind tunnel reproducing the relative wind incidence angle is different from the profile the absolute wind encounters while blowing across the line. In the wind tunnel tests on still models the wind is blowing on the scenario with a wrong yaw angle and speed that may affect the results. Wind tunnel tests that reproduce the train-ground relative motion are very expensive and difficult to set up, since there must be a moving ground or a moving vehicle. The setup of the moving ground is difficult for these kind of tests, since the ground is not flat as happens for trucks or cars aerodynamic tests, but consists on ballast and rails that cannot simply be reproduced with a moving carpet. A solution for experimental test is to perform tests on a moving train with a moving test rig. This kind of tests has already been performed at Politecnico di Milano Wind Tunnel by Bocciolone et al. [5], but also by Cooper [6] and Li et al. [7]. This kind of tests presents many technical difficulties, the acquisition time is very limited and requires several tests to acquire a good time averaging of the measured quantities and there is the need to depurate the results from the inertial forces. The complexity of such kind of tests, at the moment, prevented to define how large is the variation of the aerodynamic coefficients of the train if considering the motion of the vehicle.

In this paper an attempt to use CFD simulations to investigate this topic is proposed, analyzing both the influence of the scenario and the train motion on the aerodynamic coefficient. CFD is a tool that can analyze the influence of the train motion on the aerodynamic coefficients by reproducing the flow around the still and moving trains [8] and [9]. Most CFD calculations on crosswind on a moving train have been performed by solving the time dependent flow equations and moving the train on the track: this kind of model presents problems analogous to the wind tunnel tests: the analysis can be performed only for a limited time, depending on the length of the domain of the numerical model. This paper will discuss the results of a moving train using a steady calculation, assuming the train moving at constant speed and an infinite domain, imposing boundary conditions to reproduce the absolute wind on the steady ground and reproducing the relative motion of the train on the track.

2 CFD Model

The numerical model that analyzes the crosswind effect reproduces the 1 : 15 scaled ETR500 train model running on a Double Track with Ballast and Rail (DTBR) scenario in the same scale used in the Politecnico di Milano Wind Tunnel by Cheli et al. [10]. Changing on this scenario some geometrical details, wind break fences and embankment effect were studied. In order to limit the differences from one scenario to the other as well as to minimize the meshing effort for each different case, the domain has been divided in smaller volumes, containing the different geometrical details of the model.

The domain has a rectangular shape as can be seen in Fig. 1. Since the accuracy of the grid is mainly given by the cell's topology and dimension, tetrahedra, prisms and hexahedra are suitably used for each volume as can be seen in Fig. 2. Tetrahedral cells have been used where a 3D refinement is needed, while hexaedral elements have been

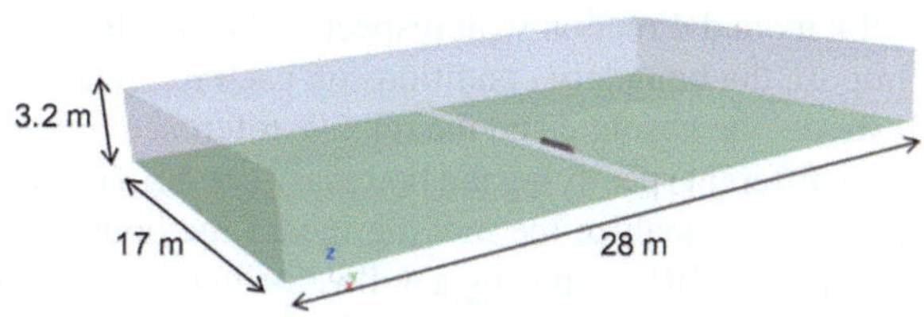

Fig. 1 Domain of the numerical simulation. In the figure are indicated the dimensions of the domain for the 1 : 15 scaled model of the ETR500 train

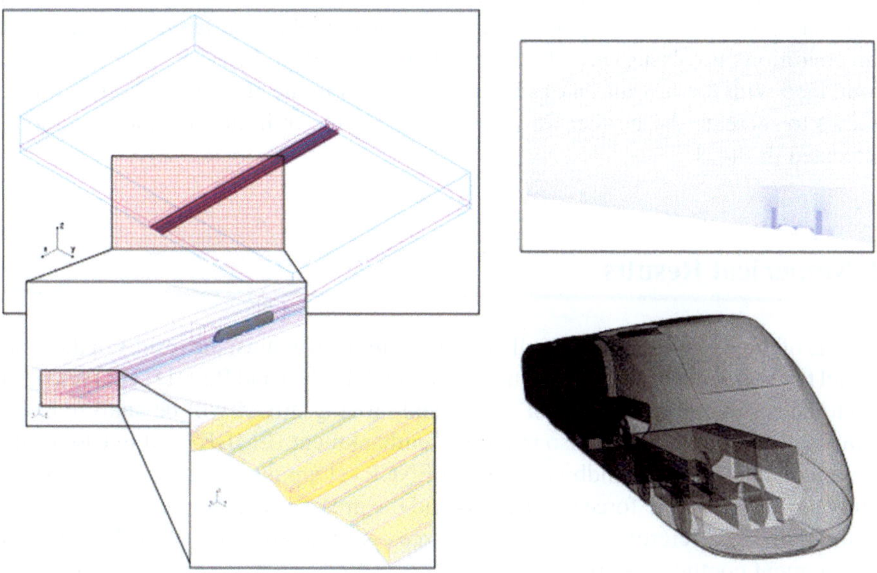

Fig. 2 Mesh detail around the train and in the domain

used elsewhere. The mesh has no boundary layer on the train surface and the non dimensional wall distance $y^+ > 10$. On this grid the steady state Reynolds Averaged Navier Stokes (RANS) flow equations have been solved, using the $k - \varepsilon$ turbulence closure model with standard wall functions. The management of the different wind yaw angles and control of the movement of the train has been performed by imposing proper boundary conditions on the grid and on the boundary of the domain, without any change to the mesh, except when some geometrical details have to be added to the scenario. Each simulation has the correct combination of boundary conditions to reproduce the different yaw angles, consisting in fixing the velocity and the pressure at the vertical faces of the domain. The upper face has a fixed null normal component of the fluid velocity. The other parts of the scenario are modeled with a null velocity of the fluid with respect to the face itself. Simulations on still train are performed using a single still reference frame. When the train is moving the grid has the same velocity magnitude and direction of the train itself defining a new moving relative reference system. From this statement it is possible to note that have been defined two different reference frames: one is the absolute one and one is a moving reference frame (MRF), with a constant velocity and direction with respect to the absolute one. At this point it is possible to impose the boundary conditions on each reference frame: when we consider a steady model the absolute and the relative reference frame are exactly the same, and makes no different specifying the boundary conditions on one or the other reference frame. When we analyze the moving model the boundary conditions on the train are defined in the MRF, imposing a null velocity on the surface of the train, while all other boundary conditions (the track, the terrain and the inlet and outlet) are imposed on the steady surfaces in the absolute reference frame (ARF). At this point it is possible to say that both simulations, the one with the moving model and the one with the steady train, use the same definition of the boundary conditions: the boundary conditions for the train are imposed on the relative reference frame, while the conditions involving directly the wind are defined in the ARF. This procedure, combined with the resolution of the flow equations in the absolute reference frame, allows to recreate the correct flow yaw angle for the train and for the scenario as indicated in Fig. 3.

3 Numerical Results

Results of the CFD simulations will be at first validated against the results of the wind tunnel tests on a still model [10] on a Double Track Ballast and Rail (DTBR) scenario. Then the results of the numerical steady and moving model will be analyzed. The same analysis is conducted also in case of embankment (EMBK) and in case of the presence of a 3 m high windbreak fence (FENCE) along the track. The analysis will study the aerodynamic force coefficients and the pressure coefficients on the surface of the train. The reference used for the forces is reported in Fig. 4 [11]. The force and moment coefficients are defined as:

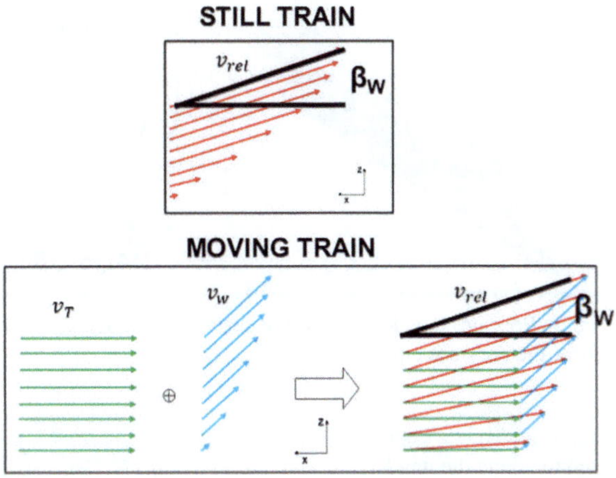

Fig. 3 Relative wind velocity profiles incoming to the still and the moving model of the train: in the steady case the wind profile has a null wind speed at the bottom, while in the moving case the relative wind velocity profile has a lower speed equal to the speed of the train and opposite direction, given by the combination of the train block profile and the incoming crosswind

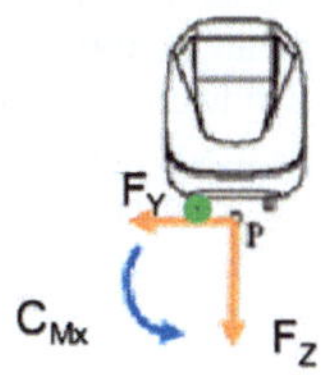

Fig. 4 Reference system used to report the forces acting on the ETR500 model

$$c_{Fi} = \frac{F_i}{\frac{1}{2}\rho V^2 A} \tag{1}$$

$$c_{Mi} = \frac{M_i}{\frac{1}{2}\rho V^2 Ah} \tag{2}$$

where F_i and M_i are the considered forces or moments, ρ the air density, V the relative air speed, A the reference area and h the reference height. In an analogous way is defined the pressure coefficient:

$$c_p = \frac{p - p_{ref}}{\frac{1}{2}\rho V^2} \tag{3}$$

where p_{ref} is the reference pressure and p the local static pressure. The wind tunnel model is equipped with 156 pressure taps distributed on transverse sections over the train surface as reported in Fig. 5. The reference for the wind incidence angle is reported in Fig. 6.

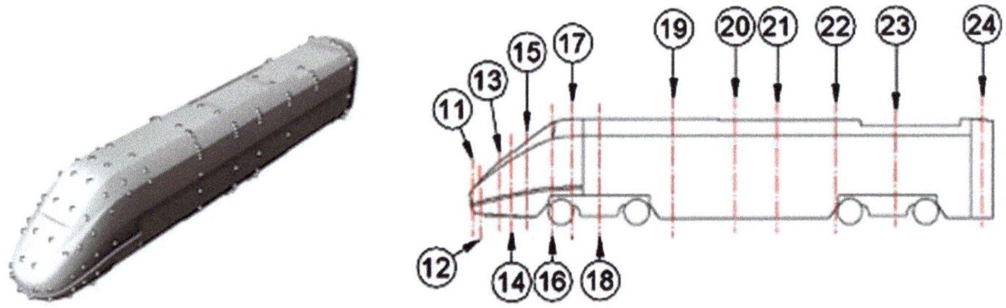

Fig. 5 Pressure taps located on the surface of the ETR500 train model and name of the sections where the pressure taps are located

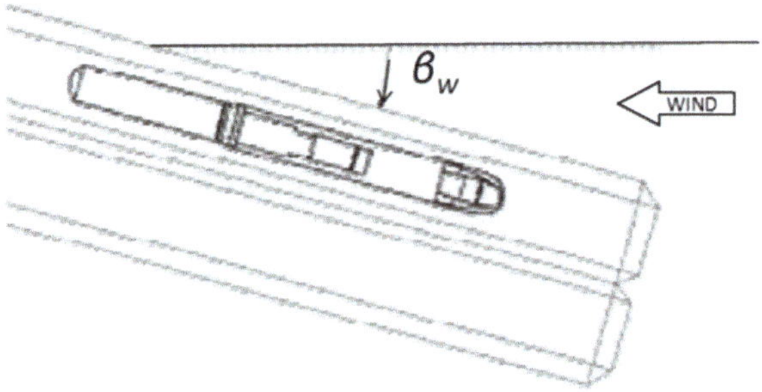

Fig. 6 Reference system used to define the wind incidence angle on the model: positive β angles mean the train is on the windward track, negative values mean the train is on the leeward track

3.1 Numerical Model Validation

The first step of this work is the validation of the numerical model, in order to show that the model is able to reproduce the real flow. The validation has been conducted on the DTBR scenario reported in Fig. 7. The numerical simulation is conducted on a domain that is not representative of the experimental setup, but, since the blockage of the wind tunnel test section is very low, the difference between the results reproducing the experimental tests and the new domain are negligible. The comparison of wind tunnel data and numerical results can be seen in Fig. 8, where it is possible to see that numerical simulations can reproduce with a quite good accuracy the global forces acting on the train in a crosswind condition. Lateral force and rolling moment are captured with a good accuracy up to yaw angles $\beta = 30°$,

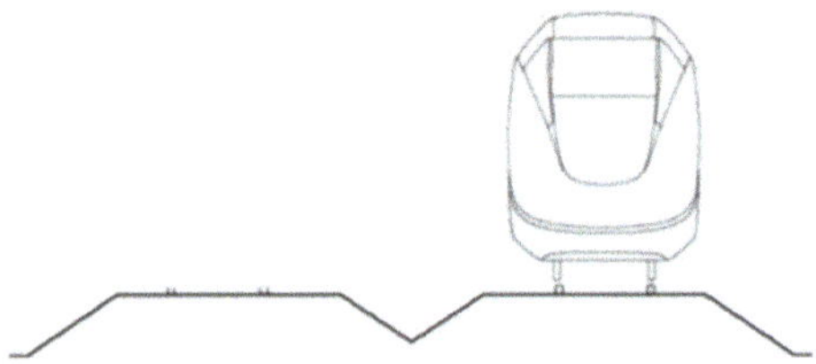

Fig. 7 Sketch of the Double Track Ballast and Rail (DTBR) scenario

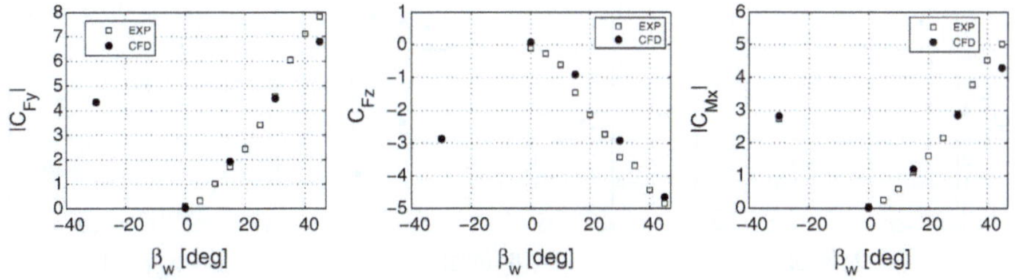

Fig. 8 Force coefficients as a function of the wind incidence angle for the DTBR scenario. Wind tunnel and CFD simulations are reported

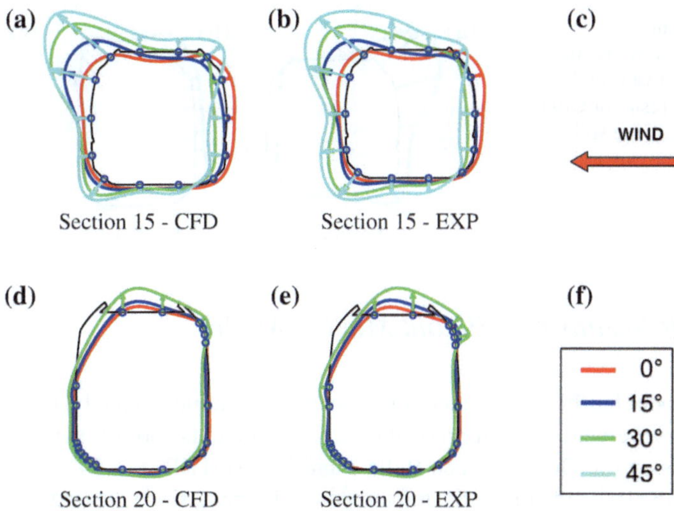

Fig. 9 Pressure distributions on two sections of the train at different wind incidence angles. The result of the wind tunnel tests and the CFD simulations are compared

but also the vertical component of the force, that usually is the most difficult to achieve in both, experimental and numerical tests, has a good correspondence. In the following will be considered a yaw angle range of $\beta = \pm 30°$, that represents the yaw angle that is typically experienced by a high speed train in operating conditions. The discrepancy at $\beta = 45°$ may be related to a decrease of the mesh properness for skewed yaw angles since the same grid is used for simulating all yaw angles. The pressure distribution on some sections of the train is reported in Fig. 9 in order to investigate in deep the analogies and the differences between the experimental and the numerical results.

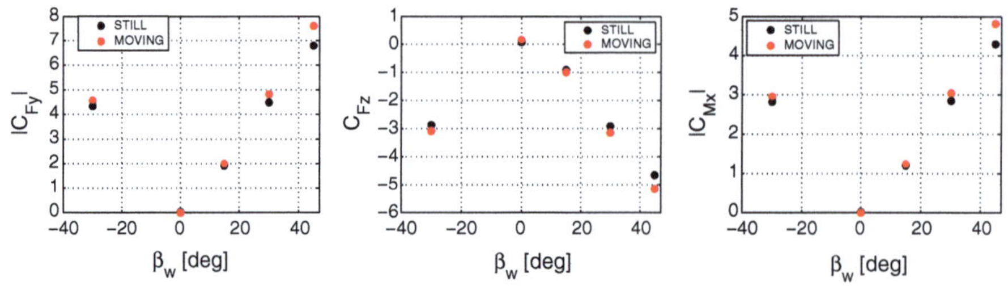

Fig. 10 Force coefficients as a function of the wind incidence angle for the DTBR scenario. CFD simulations with the still train and the moving train are reported

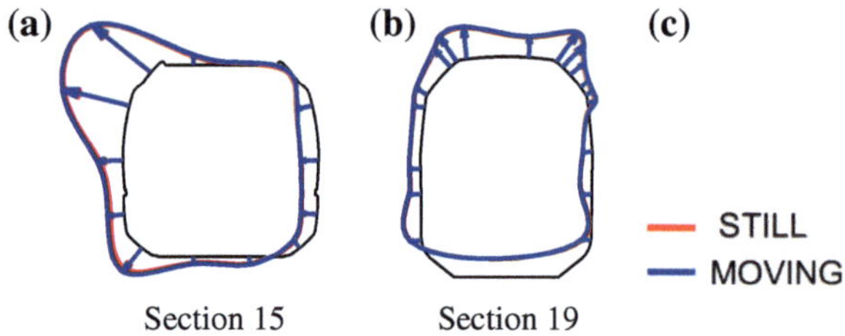

Fig. 11 Pressure distributions on two sections of the train at a yaw angle $\beta = 30°$. The result of CFD simulations with the still train and the moving train are compared

3.2 DTBR Scenario: Still and Moving Model

Once the model has been validated it is possible to conduct an analysis on the influence of the movement of the train on the force coefficients and pressure coefficients. The analysis on the Double Track Ballast and Rail (DTBR, Fig. 7) scenario can be assumed as the less sensible to the movement of the train, since the undisturbed wind is not very much affected by the geometry of the scenario, consisting only in the ballast and rail. The differences in terms of force coefficients, reported in Fig. 10 are relatively small with and increasing trend with the yaw angle. All the force coefficients show the same trend, with the lowest differences at 0°. Analyzing the distribution of pressure coefficient on the sections of the train, as reported in Fig. 11, it is possible to see that the motion of the train induces very small differences in the pressure distribution, mainly located in the region between the train and the track and on the leeward lower part. In all cases the pressure coefficient on the leeward side of the nose of the train presents lightly higher values. It is possible to say that the main difference on the forces of the train are given by the pressure distribution on the nose while on the body of the train there is not a visible difference on pressure distribution.

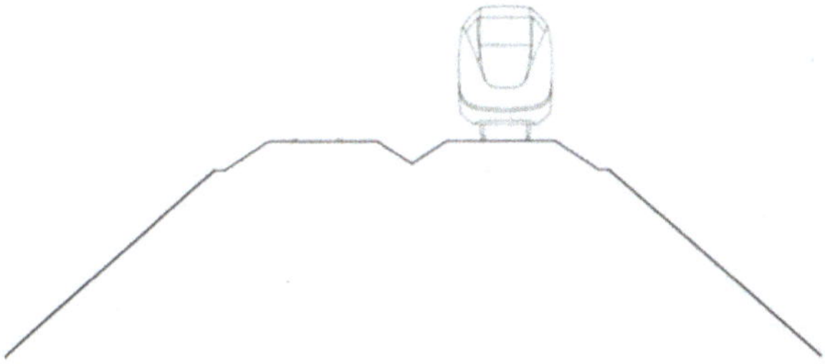

Fig. 12 Sketch of the embankment (EMBK) scenario

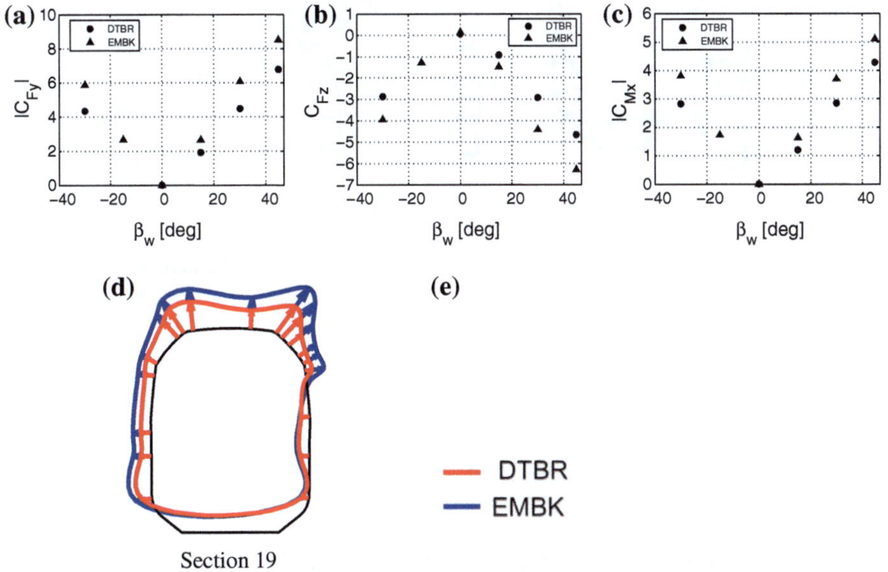

Fig. 13 Force coefficients and pressure distributions on a section of the train at different wind incidence angles for the EMBK scenario. The result of CFD simulations with the still train and the moving train are compared. The pressure distributions are reported for a wind incidence angle of 30

3.3 DTBR Versus EMBK Scenario

The comparison between DTBR and EMBK scenario is done using the same mesh around the train and the track, realizing the standard 6 m embankment reported in Fig. 12 adding the scenario underneath the ballast position. The results in terms of forces (see Fig. 13) on the first vehicle is higher for the EMBK case than in the case of DTBR case, showing a constant difference on the lateral and the rolling moment, but with a higher influence on the vertical force: the effect of the embankment scenario can be compared to a speedup of the relative wind incoming to the rolling stock. This can be seen also observing the different pressure distributions on a train sections: the

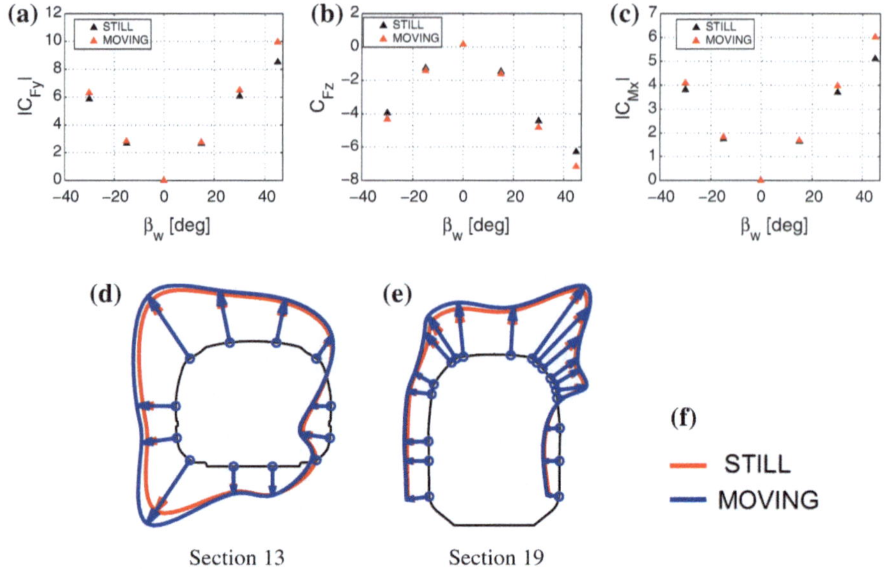

Fig. 14 Force coefficients and pressure distributions on a section of the train at different wind incidence angles for the EMBK scenario. The result of CFD simulations with the still train and the moving train are compared. The pressure distributions are reported for a wind incidence angle of 30

pressure distribution show a similar shape in the two cases, but changes in magnitude, as can be expected with a higher wind incidence magnitude.

3.4 EMBK Scenario: Still and Moving Model

The simulations that represent the motion of the train show a difference with respect to the still case of the same order of magnitude of the DTBR case (see Fig. 14). The difference still increases with the wind incidence angle. This means that the train moving on an embankment has an increment of forces with respect to the still case in the order of magnitude that has been already seen for the DTBR case. In this case the difference in terms of pressure distribution is mainly visible in the front part of the first vehicle, especially on the lower leeward side of the car.

3.5 DTBR Scenario: Without and with Windbreak Fences

Wind tunnel tests performed on the DTBR scenario are available with and without windbreak fences [10, 12, 13] along the track as it is reported in Fig. 15, and it

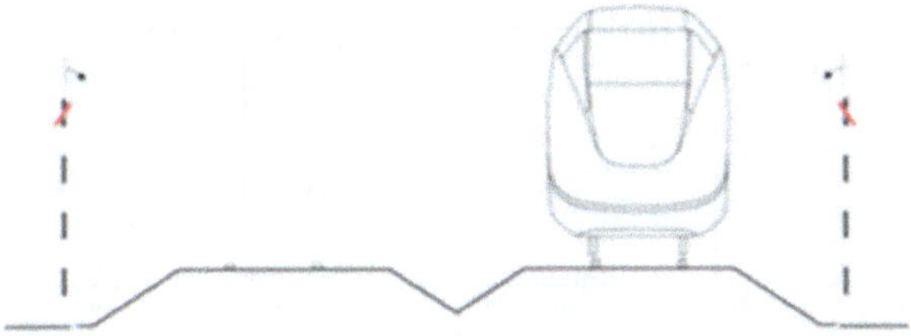

Fig. 15 Sketch of the DTBR scenario with 3 m high windbreak fences (FENCE) scenario

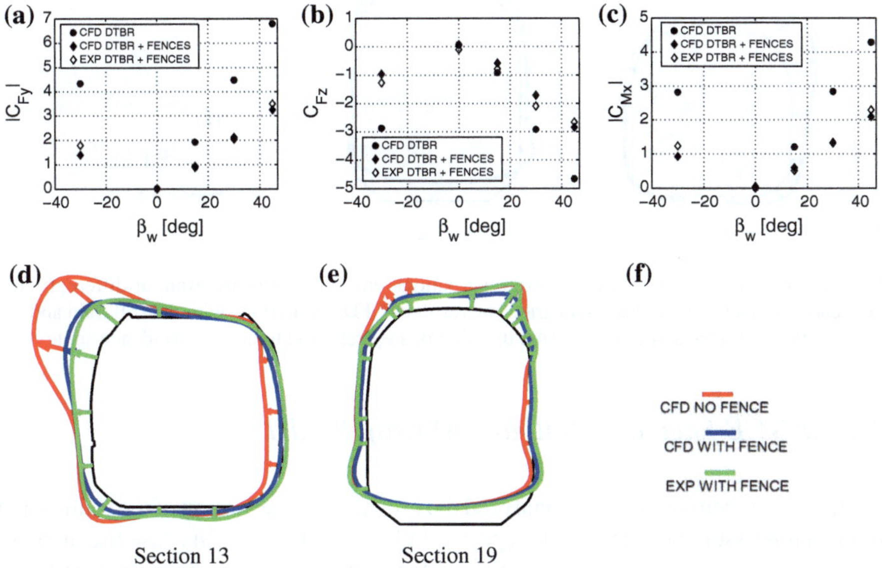

Fig. 16 Force coefficients and pressure distributions on a section of the train at different wind incidence angles for the EMBK scenario. The result of CFD simulations with the still train and the moving train are compared. The pressure distributions are reported for a wind incidence angle of 30

is possible to check the numerical results with the experimental data. In Fig. 16 is reported the force and moments coefficients without and with the windbreak fences, in terms of experimental and numerical results. The effect of the fences is clear, decreasing the loads on the train of about 50 %. The agreement between numerical and experimental data in the case of fence is good and the pressure distribution around the train sections show the decrease of pressure load especially on the windward face of the train, sheltered by the barrier, but also reducing the peak on the upper and lower leeward face.

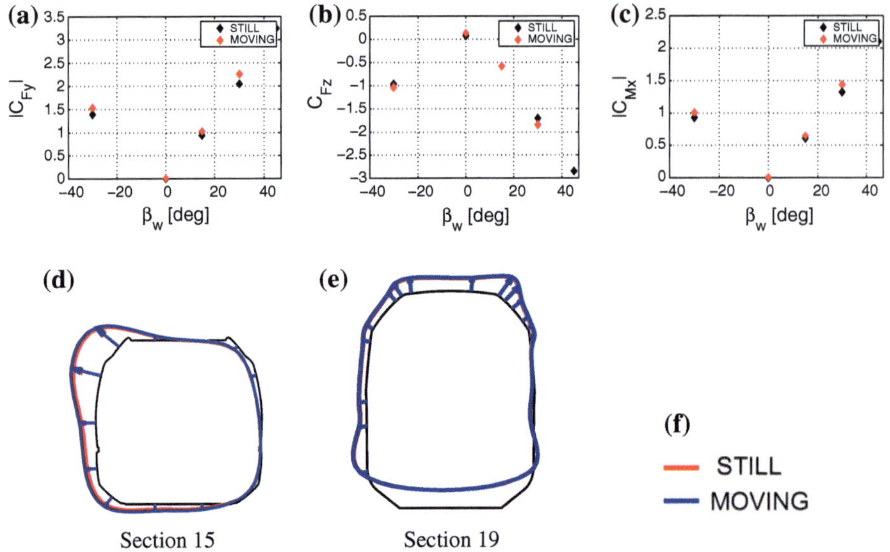

Fig. 17 Force coefficients and pressure distributions on a section of the train at different wind incidence angles for the EMBK scenario. The result of CFD simulations with the still train and the moving train are compared. The pressure distributions are reported for a wind incidence angle of 30

3.6 FENCE Scenario: Still and Moving Model

For the fence scenario has been analyzed the influence of the motion of the train on the force and pressure coefficients as reported in Fig. 17. Also for this case the increase is in the same order of magnitude of the other scenario that has been analyzed, and the pressure distribution locates the zones of major difference in the two cases on the leeward side of the nose of the train. Due to the smaller value of the pressures on the upper side of the train the differences are less evident but still present as for the other cases.

4 Conclusion

A computational fluid dynamic analysis has been conducted on a high speed train in crosswind condition with different scenario configurations and considering still and moving vehicles. The analysis regards the influence of the movement of the train with respect to the track and the infrastructure. The movement of the train has been reproduced through the imposition of the boundary conditions on the grid and the flow equations have been solved in the steady formulation. The influence of the movement of the train induces on the aerodynamic coefficients lower variations than the change of scenario. The influence on the force and moment coefficients of the

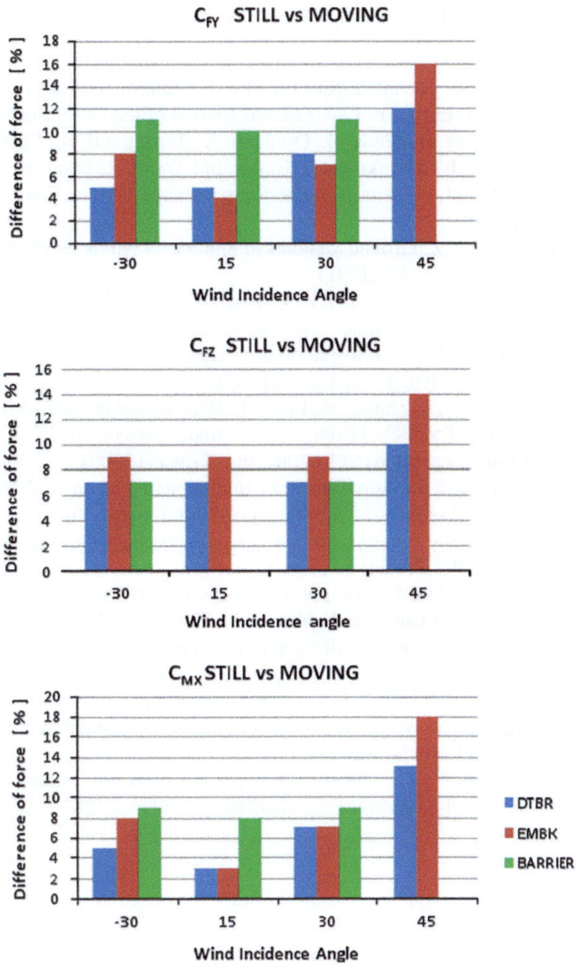

Fig. 18 Moving train over still train force coefficient ratio. It is possible to see that the forces on the moving train are always higher than in the still condition

movement of the train seems independent from the scenario, and is less than 10 % for yaw angles lower than $\beta = 30°$, that are the typical range experienced by high speed trains as can be seen in Fig. 18. The differences between the steady and the moving results seem higher when the scenario has a great influence on the flow incoming on the track, and are more risky for the embankment and the windbreak fence configuration.

References

1. EC, TSI, Technical Specification for Interoperability of the trans-European high speed rail system. European Law, Official Journal of the European Communities (2006)
2. Dietrichs, B., Sima, M., Orellano, A., Tengstrand, H.: Crosswind stability of a high speed train on a high embankment. J. Rail Rapid Transit **221**(Part F), 205–225 (2006)
3. Tomasini, G.: Analisi numerico-sperimentale delle forze aerodinamiche sui veicoli ferroviari (2005)
4. Baker, C.J.: Ground vehicles in high cross winds part I. Steady aerodynamic forces. J. Fluids Struct. **5**, 69–90 (1991)
5. Bocciolone, M., Cheli, F., Corradi, R., Muggiasca, S., Tomasini, G.: Crosswind action on rail vehicles: wind tunnel experimental analyses. J. Wind Eng. Ind. Aerodyn. **96**, 584–610 (2008)
6. Cooper, R.K.: Atmospheric turbulence with respect to moving ground vehicles. J. Wind Eng. Ind. Aerodyn. **17**, 215–238 (1985)
7. Li, Y., Hu, P., Zhang, M., Liao, H.: Wind tunnel test with moving vehicle model for aerodynamic forces of vehicle-bridge systems under crosswind. In: APCWE 7 The Seventh Asia-Pacific Conference on Wind Engineering, Taiwan (2009)
8. Dietrichs, B.: Aerodynamic calculations of crosswind stability af a high speed train using control volumes of arbitrary polyhedral shape. In: BBAA VI International Colloquium on Bluff Body Aerodynamics and Applications, Milano (2008)
9. Khier, W., Breuer, M., Durst, F.: Flow structure around trains under side wind conditions: a numerical study. Comput. Fluids **29**, 179–195 (2000)
10. Cheli, F., Diana, G., Schito, P., Tomasini, G., Volpe, R.: Sulleffetto di barriere antivento sui veicoli ferroviari: analisi sperimentali in galleria del vento. In: IN-VENTO 2010, Spoleto, Italy (2010)
11. CEN, pr EN 14067–1: Railway Applications—Aerodynamics, Part I: Symbols and Units (2002)
12. Vigano, A., Volpe, R.: Analisi numerico-sperimentale dell'effetto di barriere frangivento sui veicoli ferroviari (2009)
13. Catanzaro, C.: Modelli CFD per l'analisi dell'effetto di barriere frangivento sui veicoli ferroviari—confronti con risultati sperimentali (2010)

Numerical and Experimental Investigations of the Flow Around a High-Speed Train on an Embankment Under Side Wind Conditions

Dan Zhou, Hong-qi Tian, Mark Thompson and John Sheridan

Abstract In this paper the aerodynamic forces and moments on a high-speed train in a cross wind were obtained both numerically and experimentally. The train was positioned at the top of a 2 m high embankment. Experimental data was obtained using the Monash wind tunnel for a 1:20 scale model. These measurements were made over a range of yaw angles spanning 0°–90°. The numerical computations were performed using the commercial code *FLUENT*. These computations focused on the structure of the flow field around the high-speed train at different yaw angles. The model was discretized using a hybrid grid, with a total cell count of 1.02×10^7. The predictions show a good match for the calculated aerodynamic force and moment coefficients with experimental measurements. Aerodynamic forces were determined for two cases: with the train on either the windward or the leeward track on top of the embankment. The side force is found to be greater for the leeward case, while the lift force for the leeward case is less than that for the windward case.

1 Introduction

There have been many wind-induced train accidents in China, a large proportion of which occurred on the Lan-Xin Railway. Since 1960 there have been at least 35 railway accidents involving the derailing or overturning of passenger trains on the

D. Zhou (✉) · H. Tian
Key Laboratory of Train Safety, Central South University, Changsha, Hunan 410075, China
e-mail: qlzqjd@mail.csu.edu.cn

H. Tian
e-mail: thq@mail.csu.edu.cn

M. Thompson · J. Sheridan
Fluids Laboratory for Aeronautical and Industrial Research (FLAIR), Department of Mechanical and Aerospace Engineering, Monash University, Clayton 3800, Australia
e-mail: mark.thompson@monash.edu

J. Sheridan
e-mail: john.sheridan@monash.edu

A. Dillmann and A. Orellano (eds.), *The Aerodynamics of Heavy Vehicles III*,
Lecture Notes in Applied and Computational Mechanics 79,
DOI 10.1007/978-3-319-20122-1_7

Lan-Xin line caused by cross winds: examples of the resulting damage are shown in Fig. 1. The Lan-Xin Railway is located in the northwest part of China, running between Lanzhou and Wulumuqi. Due to the special geographical position and climate conditions on this line, strong prevailing winds from Siberia often cause sections of the railway running west-east to be exposed to dangerous cross winds. Figure 2 shows the location of particularly windy track sections and indicates the general prevailing wind direction. There are two major wind sections, one is the Lan-Xin railway wind region between GrassLand and Liaodun, and the other is the Lanjiang line wind region from Yuergou to Tulufan. The lengths of railway in question are approximately 100 km for both sections. Historical wind data shows that the wind is characteristically of high-speed and long-lasting, with large seasonal velocity fluctuations. According to weather records, days with strong winds (wind speeds in excess of 20.7 m/s) exceed 100 days/year. Peak wind speeds are 64 and 60 m/s for the Lanjiang and Lan-Xie sections, respectively.

Fig. 1 Train overturned (*left*) and badly damaged (*right*) on the Lan-Xin railway

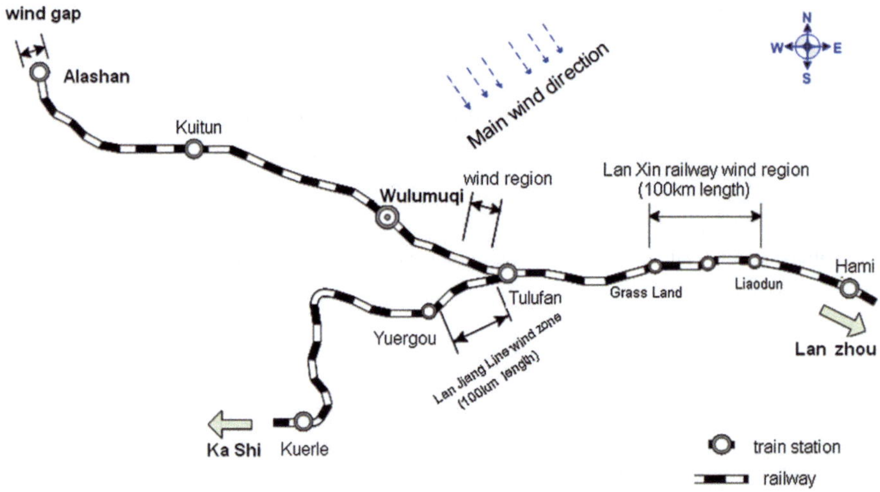

Fig. 2 Locations and direction of the main high-speed wind zones for the Lan-Xin railway

Extreme cross winds on traveling trains can affect traveling safety. Trains running in cross winds experience additional aerodynamic forces and moments, which tend to roll the train on the tracks [1, 2]. In some particularly windy circumstances, such as on large bridges, embankments, in high mountains passes etc., aerodynamic forces increase significantly, which may result in a greater likelihood of trains derailing or overturning [3–5].

The effects of strong cross winds on vehicle handling and controllability are important factors to be considered in ensuring safety in rail transportation. The first, and most important, step in order to establish safety criteria is to characterize the aerodynamics of rail vehicles. For this reason many research projects have examined the effects of cross-winds on rail vehicles since the 1960s [6–8]. In general, there are two ways to research the aerodynamic characteristics of trains under strong cross winds, one is to use scale-model wind-tunnel testing, the other is to use computational fluid dynamics (CFD) to simulate the flow.

For wind-tunnel research, aerodynamic tests on vehicles are generally carried out through a static approach, which consists of reproducing the wind–train interaction by means of a static train model exposed to a wind-tunnel flow. The train speed, the wind speed and the angle between the wind direction and the train longitudinal axis define the yaw angle between the train and the relative wind. With the rapid development of CFD, more researchers have now investigated the flow structures numerically to obtain a better understanding of the flow behavior under cross-wind conditions. Most of these studies are based on forms of the Reynolds-averaged Navier-Stokes equations. Khier et al. [9] numerically investigated the flow around a simplified train geometry at high Reynolds number under different yaw conditions. They solved the three-dimensional Reynolds-averaged Navier-Stokes equations, based on the $k - \varepsilon$ turbulence model, in finite-volume form, employing a multi-block structured grid. Paradot et al. [10] investigated the cross-wind sensitivity of the TGV Duplex. They compared experimental and numerical results for the side and lift forces, and lee rail rolling moments acting on the TGV. The results showed good agreement between the two sets of results for the TGV at relative wind flow angles of 30° or less. Beyond this, the differences between the results are more pronounced with a maximum difference occurring for a flow angle of approximately 60°. Chiu [11] employed two- and three-dimensional source/vortex panel techniques to predict the pressure distribution on an idealized train model. In these computations the base pressure was required in advance, but given that information, the results were in good agreement with experiments. Since that time more sophisticated numerical models have been used, for example Hemida [12, 13] adopted the large-eddy simulation (LES) technique, with a standard Smagorinsky sub-grid scale (SGS) model with the Smagorinsky constant set to $C_s = 0.1$ to compute the side-wind flow around stationary generic train models.

2 Method

2.1 Numerical Computational Method

Results of the present study have been obtained from the commercial flow simulation package FLUENT. The underlying algorithm used is based on the finite-volume method with a second-order accurate central-difference formulation used to discretise the diffusion terms, while convection terms were discretized using the QUICK scheme. Simulations were performed considering the flow as incompressible and using a collocated grid solver. The velocity-pressure coupling was performed with a SIMPLEC method and a realization of the $k-\varepsilon$ turbulence model was adopted using a wall-function model for the regions adjacent to the solid boundaries.

The calculation and experimental train models consisted of a leading car, middle car and end car, as shown on Fig. 3. The total length of the train being modelled was $L = 68.3$ m and its height $D = 3.84$ m. Figure 4 shows the cross-sectional geometry of the embankment and superstructure. The height of the embankment is 2 m with slope gradient of 1.5 and the height of the superstructure is 0.4 m. The train's aerodynamic characteristics were studied with the train on either the windward and or leeward track at top of the embankment.

For the numerical predictions, the flow domain was discretized using a *hybrid* grid. The train surface was discretized by a structured grid and the bogies of the train were discretized by an unstructured grid Fig. 5 shows the train's surface grid distribution and Fig. 6 shows the surface grid of the bogies. In order to resolve the boundary

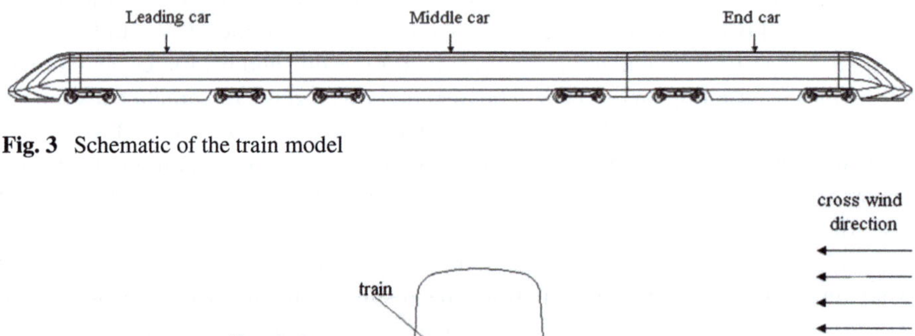

Fig. 3 Schematic of the train model

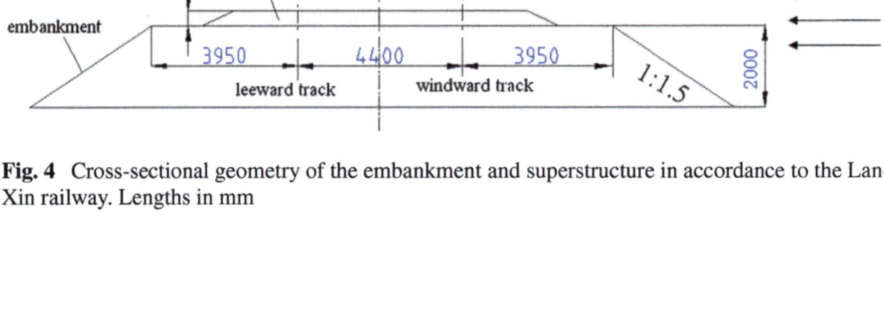

Fig. 4 Cross-sectional geometry of the embankment and superstructure in accordance to the Lan-Xin railway. Lengths in mm

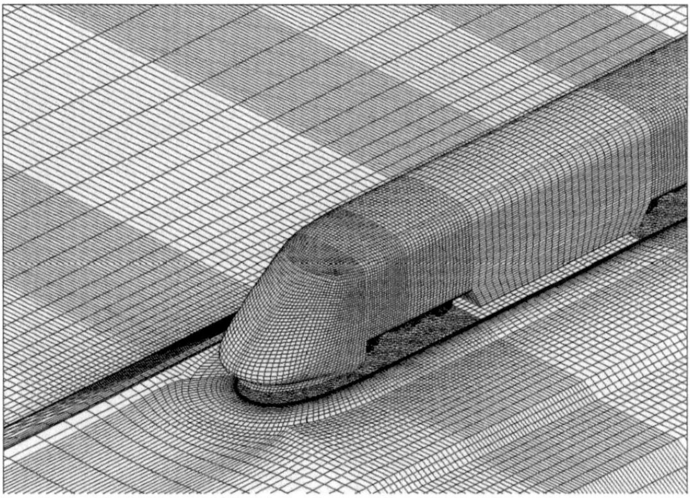

Fig. 5 Distribution of grid on the surface of the train

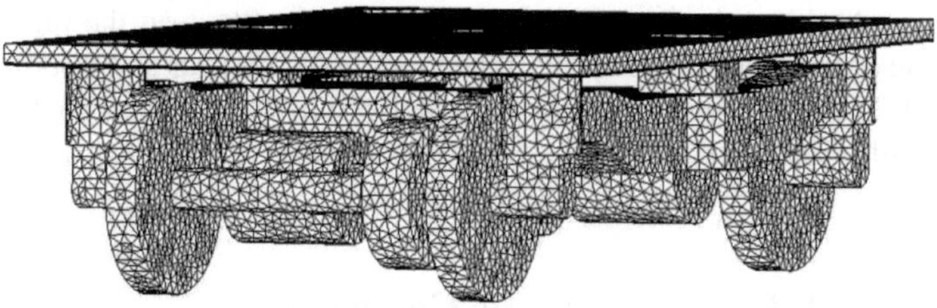

Fig. 6 Distribution of the grid on the bogie surface

layers, the first cells adjacent to the walls of the train, ground and embankment are adjusted to meet to wall function requirement that $30 < y^+ < 150$. The total number of cells in the calculation model was 1.02×10^7, reflecting the complex geometry of the train embankment combination.

2.2 Description of the Wind Tunnel Tests

Wind tunnel tests were carried out in the 1.6MW Monash University wind tunnel using an experimental set-up as shown in Fig. 7. The tunnel's working cross-sectional area has dimensions 12 m × 5 m. The maximum wind flow velocity in the test section is 50 m/s. For this study, the test wind velocity was set at 40 m/s. The mock-up scale model of the train is at 1/20th scale and consisted of three carriages: a leading car, a middle car and a tail car. The height of the embankment in the wind tunnel is 100 mm and tests were run at a Reynolds number of $\text{Re} = 5.3 \times 10^5$ based on the

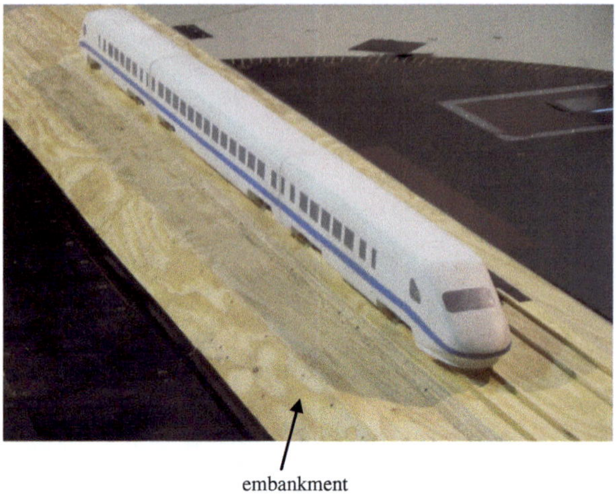

Fig. 7 Photograph showing the train model positioned on embankment placed on the Monash wind-tunnel turntable

free-stream velocity and the height of the train. The aerodynamic forces and moments were measured with a six-component internal balance fitted on the turntable. The aerodynamic performance of the high speed train's leading car and middle car were investigated. The model was tested at yaw angles from 0° (train axis) to 90° (normal direction) with respect to the relative wind direction.

3 Definition of the Aerodynamic Coefficients

Side winds mainly cause increased aerodynamic forces and moments. Under cross-wind conditions, the aerodynamic forces and moments, especially the side force F_s, lift force F_l and overturning moment M_r, will exhibit a significant increase. To compare the aerodynamic forces produced by different train shapes regardless of their size or driving speed we need a simple basis for the comparison. This is most conveniently provided by dimensionless aerodynamic coefficients. The lift coefficient C_l, the side force coefficient C_s and the overturning moment coefficient C_m are defined as:

$$C_s = \frac{F_s}{1/2\rho(V_t^2 + V_w^2)S_{ref}}$$

$$C_l = \frac{F_l}{1/2\rho(V_t^2 + V_w^2)S_{ref}}$$

$$C_m = \frac{F_s}{1/2\rho(V_t^2 + V_w^2)S_{ref}B}$$

where S_{ref} is the reference area, which here is taken as the cross sectional area of the train $S_{ref} = 12\,\mathrm{m}^2$, B is the length of car, V_t is the speed of high speed train and V_w is the wind speed.

4 Results and Discussion

4.1 Comparison with Wind Tunnel Tests

The simulation results were compared with the experimental test results for the leading car and middle car, as shown in Figs. 8 and 9. From the figure, the trends in the variations for the side and lift force coefficients and the overturning moment coefficient with yaw angle are shown to be almost the same for the experimental and numerical simulation results. The maximum side force coefficient on the leading car occurs at a yaw angle of 50°. The side force coefficient of the middle car shows a monotonically increasing trend from 0° to 60° and then remains approximately constant between 60° and 90°. For the overturning moment coefficient, the effect of yaw angle is almost the same as that seen for the side force. The lift coefficients of both the leading car and the middle cars increase with increasing yaw angle; however,

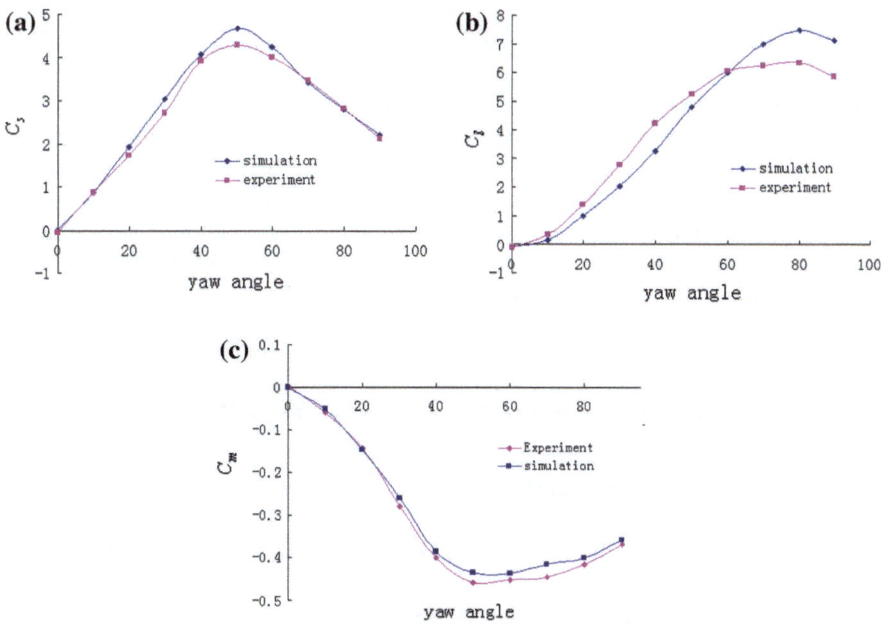

Fig. 8 Aerodynamic force and moment coefficients for the leading car—comparing experiment and simulation. **a** Side force coefficient. **b** Lift force coefficient. **c** Overturning moment coefficient

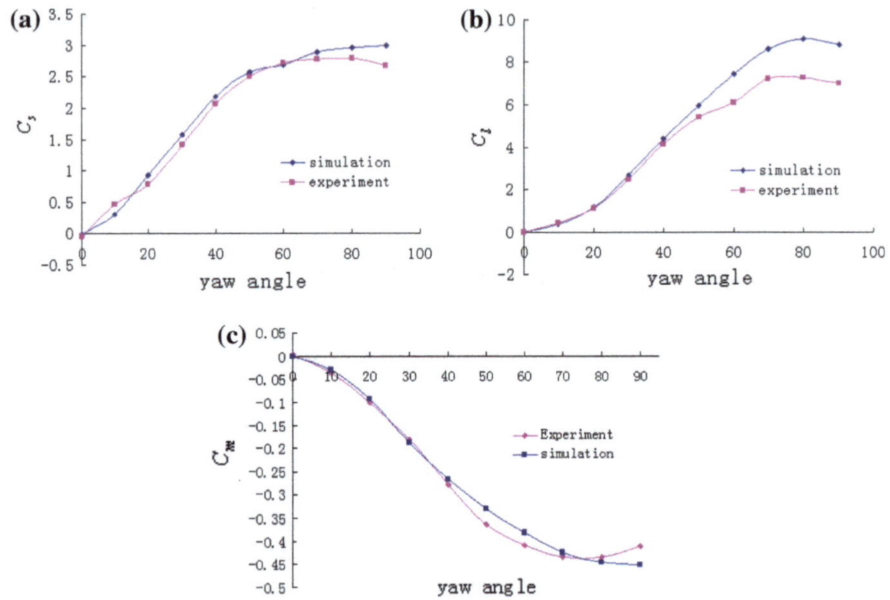

Fig. 9 Aerodynamic force and moment coefficients for the middle car—comparing experiment and simulation. **a** Side force coefficient. **b** Lift force coefficient. **c** Overturning moment coefficient

the lift coefficients of the cars decrease as the yaw angle approaches 90°. The middle car experiences smaller lateral forces than the leading car because, unlike the leading car, it does not expose a frontal surface to the wind.

For the side force coefficient and overturning moment coefficient it can be seen from the figures that there is relatively good agreement between the measured and computed results for both the leading and middle cars, the discrepancies being less than 10 %. For the lift force, relatively good agreement was obtained at yaw angles between 0° and 50° but the calculated results deviate by about 20 % from those of the experiments at high yaw angles.

4.2 Aerodynamic Characteristics Under a Cross Wind

4.2.1 Flow Structure Around the Train

The recirculating flow in the wake on the leeward side of the train has a large effect on the pressure distribution on the train, and hence on the aerodynamic forces under cross-wind conditions. The following section discusses the different flow structures at $\beta = 20°$ and $\beta = 50°$.

Figure 10 shows that strongly swirling flow stretches in the wake all the way from the nose tip to the end car for the case with $\beta = 20°$. Figure 11 shows the

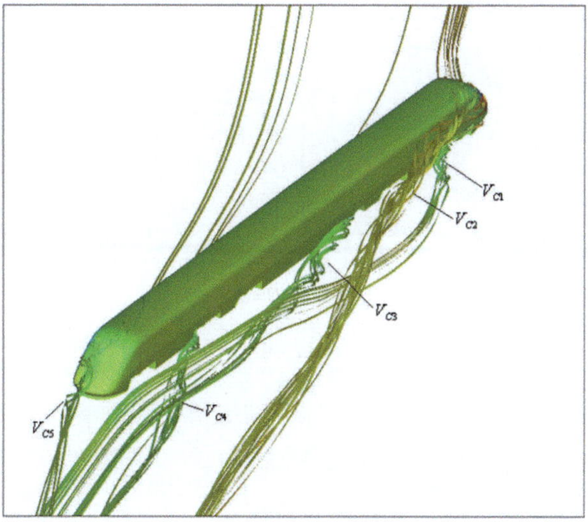

Fig. 10 Wake flow structures for $\beta = 20°$

cross-sectional flow structure depicted using two-dimensional streamlines at different locations on the x-axis. (Small yaw angles mean the speed of train is larger than the speed of wind.) This can be associated with the generation of strong axial flow in the recirculating flow on the leeward side of the train. A separation from the roof and the detachment of the flow from the leeward face begins to occur at the axial location $X/D = 1.7$. Further along the train, a strong recirculating cross-flow persists.

A strong swirling vortical structure, V_{C1}, emerges on the leeward side of the upper edge of the leading nose at the location of $X/D = 1.7$. It is attached to the upper half of the leeward side face of the train along the length of the leading car until it separates. Another recognizable trailing vortical flow structure, V_{C2}, originates on the first bogie of the leading car at the location of $X/D = 3$ and grows quickly in size. After a short distance in the direction of the length of the train, the vortex detaches from the train surface and changes its direction toward the wake. The upper vortex along the leeward side has a strong interaction with the lower vortices, which are of opposite sign. Similarly, other strong vortical flows, V_{C3} and V_{C4}, emerge from the lee-side underbody edge near the bogies of the middle and end cars. There is also a large-scale vortex, V_{C5}, near the vehicle wake. These results show dominant wake structures that are similar to those found in the experimental investigations by Copley [14]. Figure 12 shows a schematic representation from Copley [14] of the flow around a generic train under side-wind conditions. Copley investigated the wake shedding as a progressive pair of vortices shed from the upper and lower edges of the leeward side face at the same position.

Figure 13 shows that the trailing vortical structures stretch in the wake flow all the way from the nose tip to the end car even for the case with $\beta = 50°$. Figure 14 shows the flow structure for this case in terms of two-dimensional streamlines at different locations on the x-axis. An increase in the yaw angle results in an advance of the formation position and the breakdown of the vortex. The leeward wake at 50°

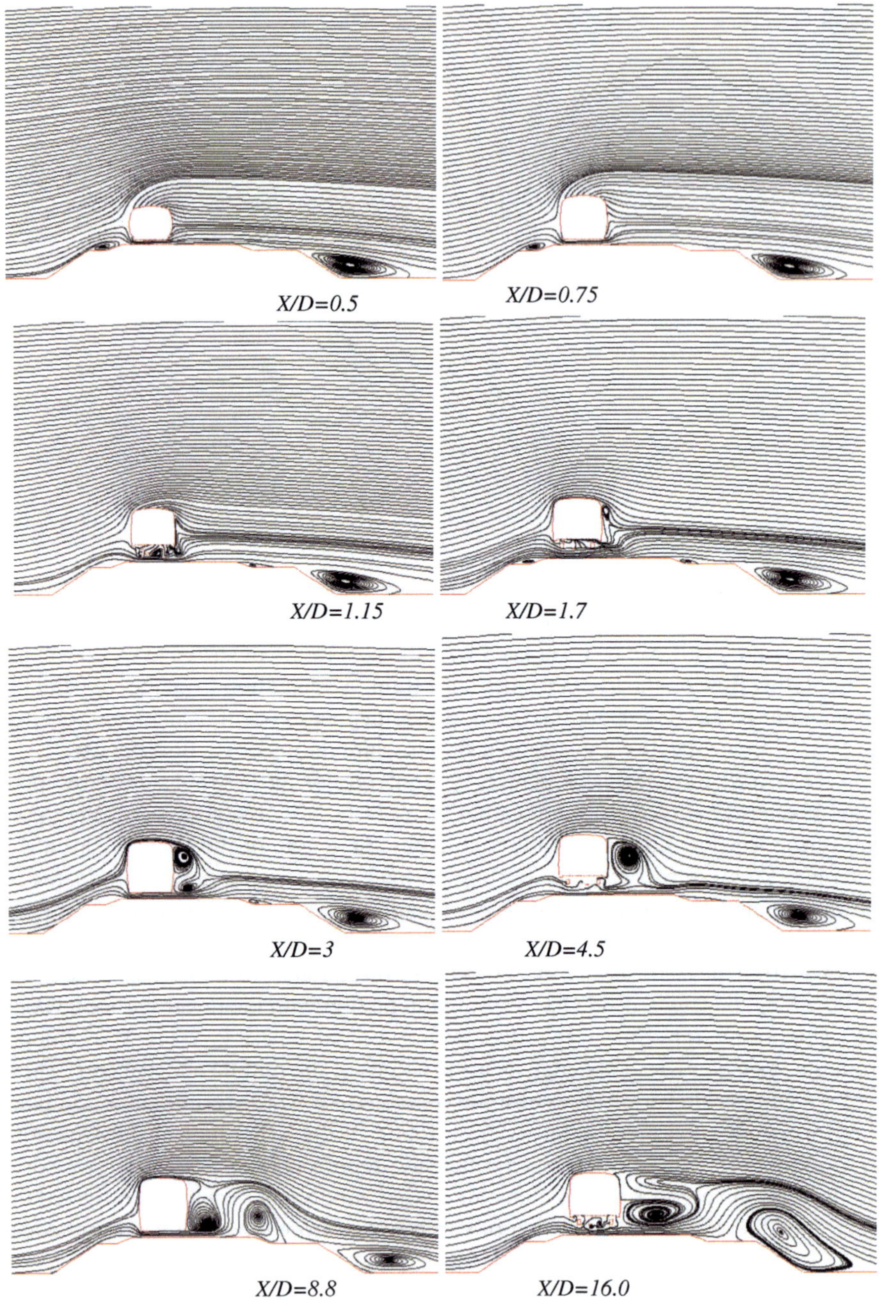

Fig. 11 2-D streamlines for different locations on the x-axis for $\beta = 20°$

Fig. 12 Sketch of the vortical flow pattern in the wake of a train in a cross-wind; taken from Copley (1987)

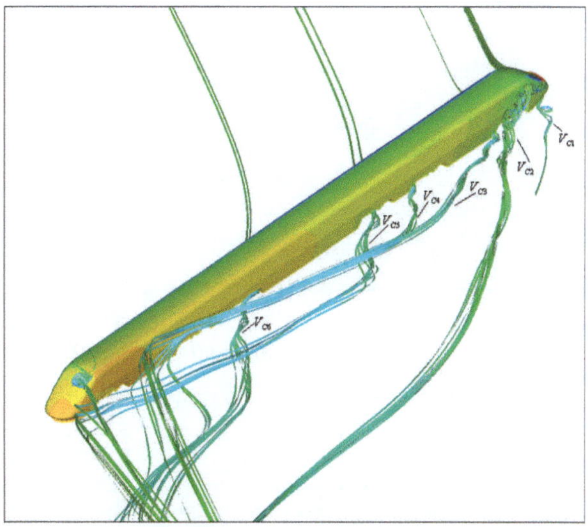

Fig. 13 Flow tracers showing the complex downwind wake structure for $\beta = 50°$

is more complex than it was at small yaw angles. A vortex V_{C1} is generated on the leeward-side of the upper edge of the leading nose at a location of $X/D = 0.75$. It forms earlier than it does for a yaw angle of 20°. Another vortex V_{C2} shed from the underside and roof of the train, rolls up to form a large trailing vortex, which moves away from the train. Further flow tracers, V_{C3}, V_{C4}, V_{C5}, V_{C6}, released into the flow on the windward side appear to become trapped in the recirculating flow on the leeward side of the train, as shown in Fig. 13. They are released into the wake towards the end of the train, effectively joining up with the previously separated trailing vortical structure V_{C2} to form a complex swirling wake, which advects in the wind direction.

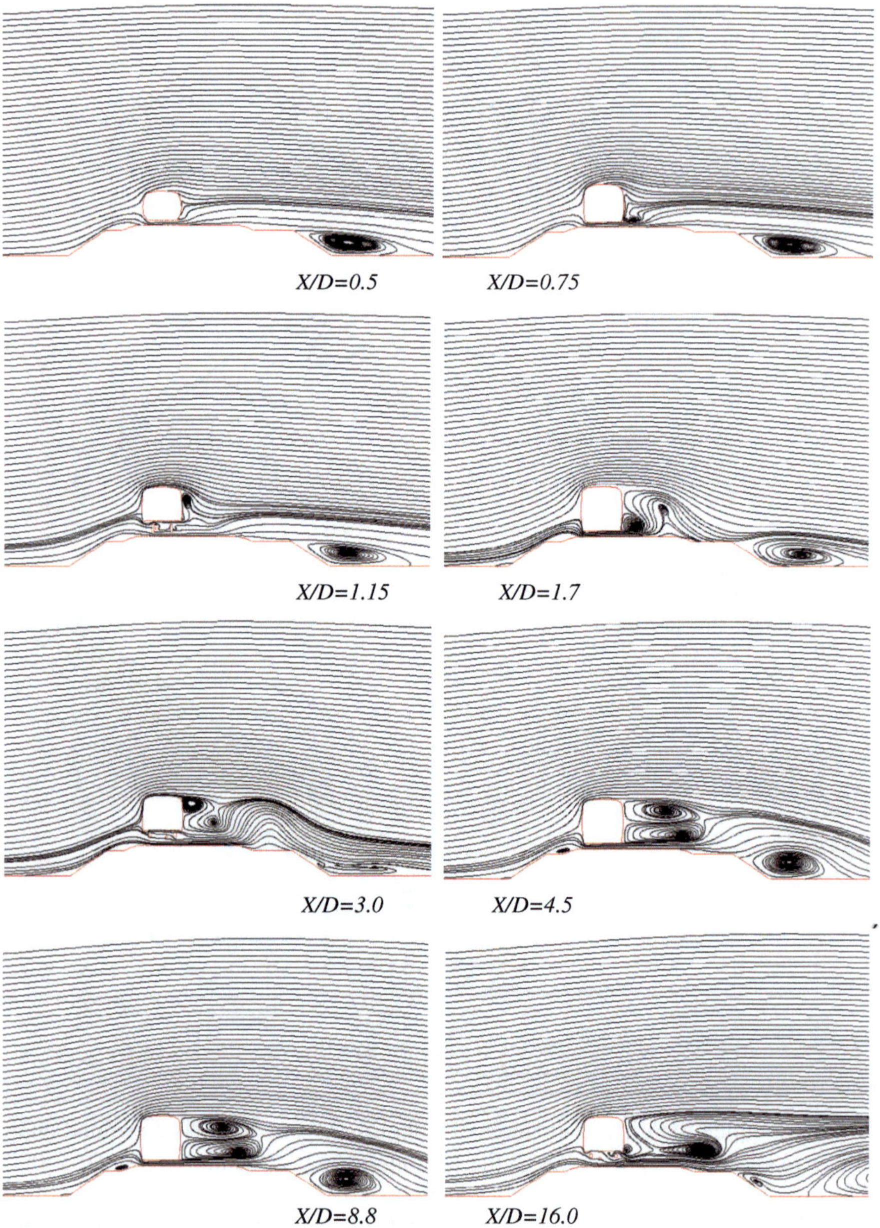

Fig. 14 2-D streamlines for different locations on the x-axis for $\beta = 50°$

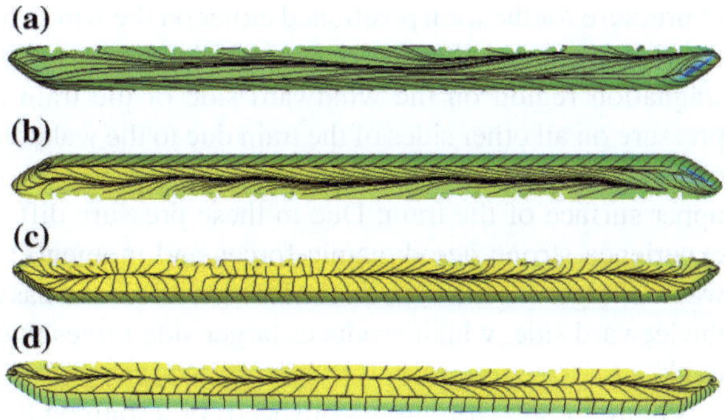

Fig. 15 Surface streamline patterns on the leeward side for different yaw angles. The flow direction is from *right to left*. **a** $\beta = 20°$. **b** $\beta = 50°$. **c** $\beta = 80°$. **d** $\beta = 90°$

4.2.2 Surface Flow Patterns

The variation of the flow structure with the yaw angle can be investigated by observing the surface streamline patterns on the leeward side of the train. Figure 15 shows the complicated flow patterns on the lee-side face caused by the attachment and detachment of the flow. A main separation line on the leeward upper edge can be seen along the train. The leeward face is completely covered by the separated flow, consistent with the two-dimensional streamline patterns shown in Fig. 14.

The complex separation-attachment pattern on the leeward side is caused by the detachment of line vortices, or the vorticity-rich separating boundary layers, from the upper and lower edges of the train surface. At yaw angles less that 50° there is a significant axial flow inside the generally recirculating flow formed from these separating boundary layers (see Figs. 11 and 14, and discussion above). This axial component tends to carry the initially strongly three-dimensional flow formed at the nose downstream axially along the train. Since the axial velocity component is stronger at smaller yaw angles, the three-dimensional effects of the nose can thus penetrate further away from the nose at yaw angles less than 50°. The lee side surface streamline pattern is more complicated at small yaw angles than at large yaw angles. Indeed, within the yaw angle range from 80° to 90°, the streamline pattern away from the nose does not vary observably. Therefore, it is suggested that the effects of the cross-wind velocity and train velocity might be relatively independent at locations far from the nose for these high yaw cases. As the yaw angle approaches $\beta = 90°$, the flow clearly becomes symmetric. Downstream of the front nose a clear separation line originates from the nose vortex, extends along the leeward upper edge and finally terminates symmetrically at the rear nose vortex.

4.3 Effects of the Windward and Leeward Track

4.3.1 Flow Structure

A sample of the computational results for $\beta = 50°$ for the train running on the windward or leeward track is provided in Figs. 16 and 17. Figure 16 shows contours of pressure for the train positioned either on the windward or leeward track. From this figure, we can see that under strong side-wind conditions, there is a high pressure stagnation region on the windward side of the train and embankment, and a low pressure on all other sides of the train due to the wake flow. In addition, the air moves over the train roof at high velocity, making a region of even lower pressure on the upper surface of the train. Due to these pressure differences, the high-speed trains experience strong aerodynamic forces and moments. The images indicate that the wake is larger for the train on the leeward track and has a relatively lower pressure on the leeward side, which produces larger side forces than for a train on the windward track.

Figure 17 shows contours of speed for a train on the windward or leeward track. The speed contours clearly show that the air is accelerated as it passes over embankment and around the train. There is a low speed wake directly behind the embankment and train. The differences between the speed and pressure contours surrounding a train traveling on the windward or leeward track is largely due to the slope of the embankment. Therefore, the effective wind speed is larger acting on the train on windward track and this has the effect of increasing the lift load.

4.3.2 Aerodynamic Forces

The aerodynamic force components for a train positioned on the windward and leeward tracks are shown in Figs. 18 and 19. It can be seen that, for small yaw angles, side force coefficients of the train in the leeward lane are slightly greater than those for the train in windward lane. According to Bocciolone [15], beyond the transition angle (45° for the ETR 480 train), the vehicle behaves as a bluff body; the phenomenon becomes bi-dimensional and it is governed by the amplitude of the wake structure. Under these conditions, the main parameter which influences the train aerodynamic forces is wake size; the leeward lane position modifies the amplitude of the wake and, as a consequence, the aerodynamic forces. As the yaw angle approaches 50°, the side force coefficients of the train in the leeward lane are obviously greater than those for the train on the windward lane. For the leading car, the side force on the train in the leeward lane is approximately 12 % greater than that for a train in the windward lane at $\beta = 70°$. For the middle car, the side force on the train in the leeward lane is approximately 15 % greater at $\beta = 60°$.

Lift force coefficients on the train in the windward lane are greater than those for a train in the leeward lane. This appears to be due to the different wind speed at the top of the train when positioned on the different tracks. It is obvious that the embankment configuration causes an acceleration of the flow and a corresponding overpressure on the train running in the windward lane. For the leading car, the lift force on the train in the leeward lane is approximately 30 % greater than for the train in the windward lane. For the middle car, the side force is approximately 35 % greater for the leeward lane.

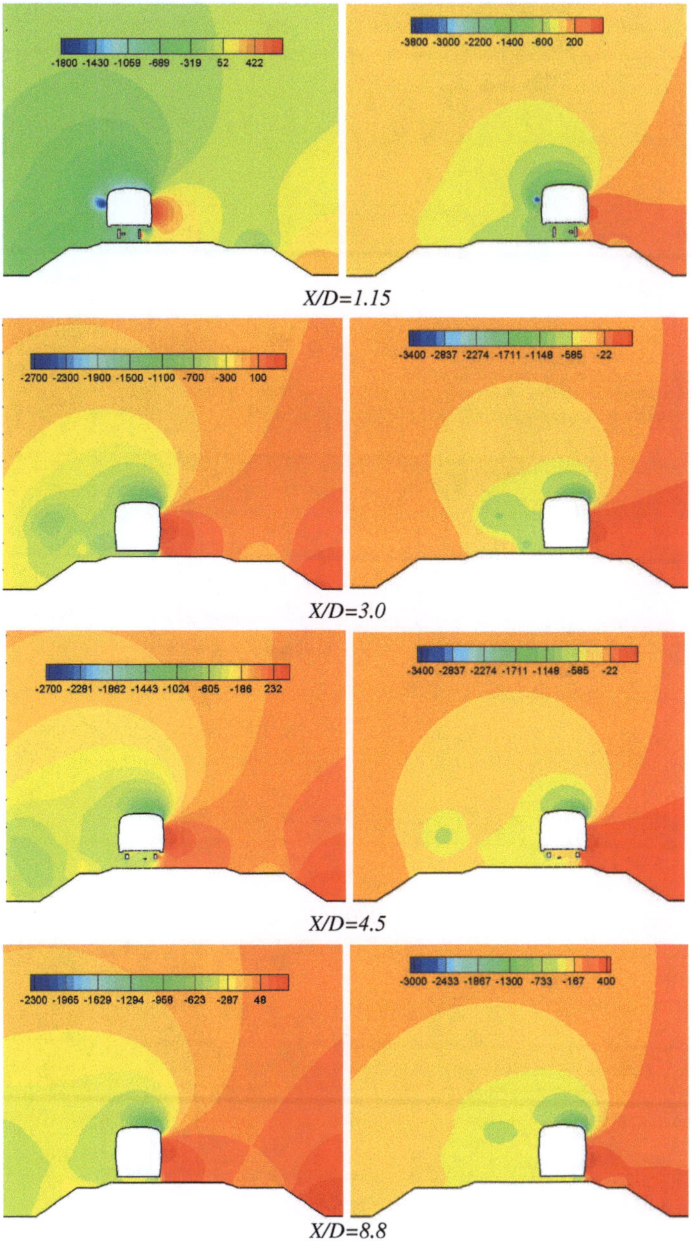

Fig. 16 Contours of pressure from the simulation results corresponding to leeward (*left*) and windward (*right*) for $\beta = 50°$. Crossflow is *right to left*

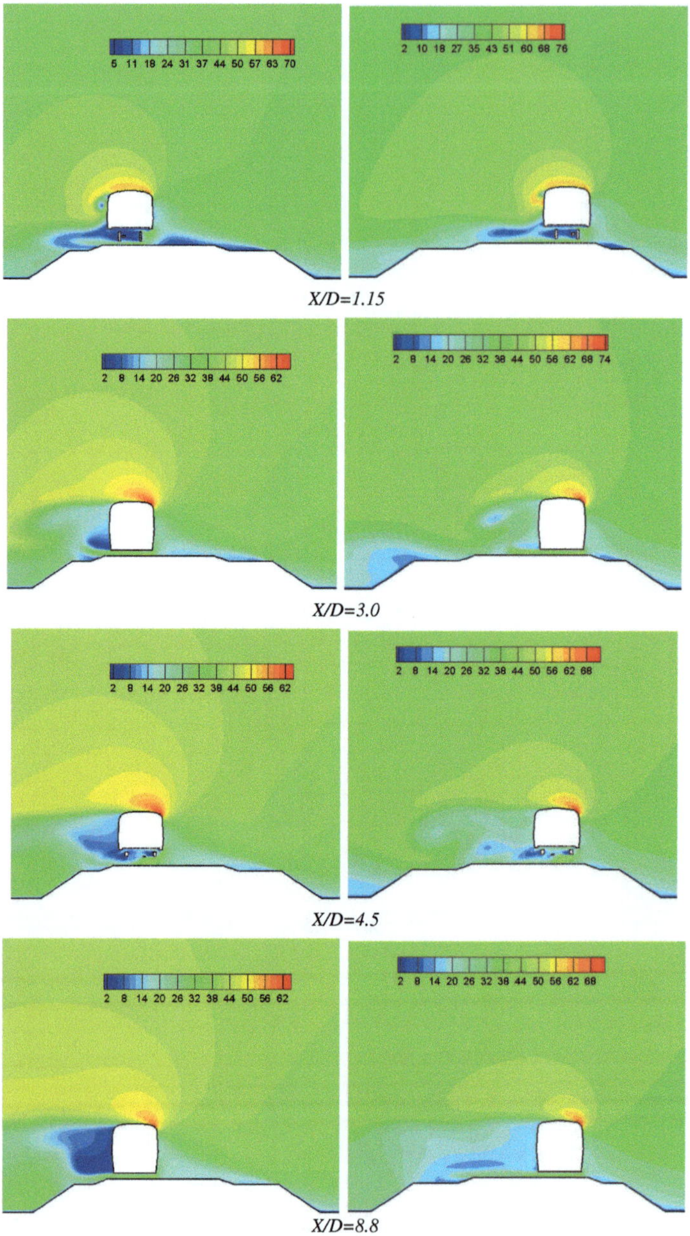

Fig. 17 Contours of velocity from the simulations corresponding to the leeward (*left*) and windward (*right*) case with for $\beta = 50°$

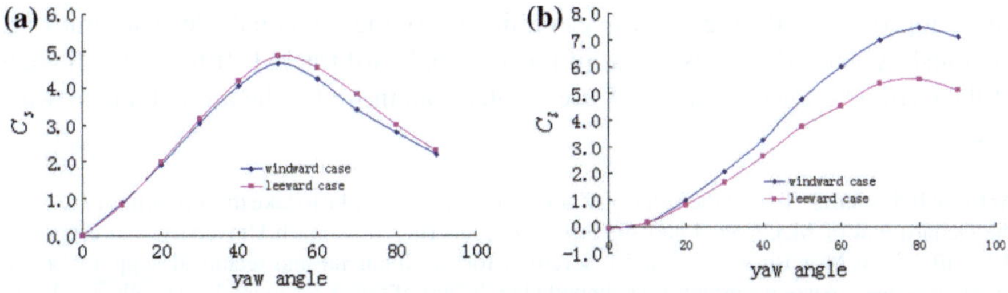

Fig. 18 Aerodynamic force coefficients for the leading car—comparison between windward and leeward lanes. **a** Side force coefficient. **b** Lift force coefficient

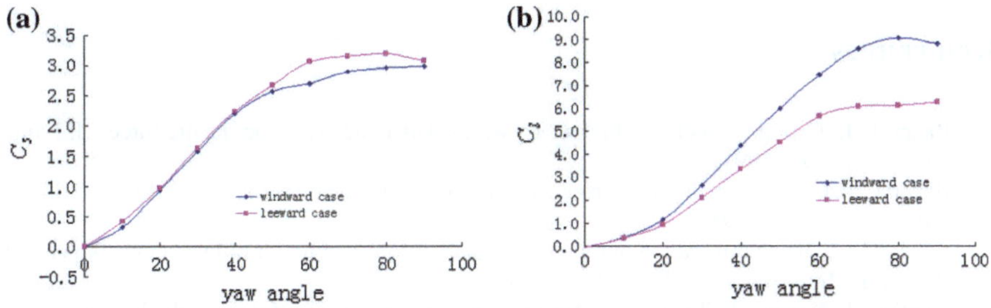

Fig. 19 Aerodynamic force coefficients for the middle car—comparison between windward and leeward lanes. **a** Side force coefficient. **b** Lift force coefficient

5 Conclusions

In this paper, the aerodynamic forces and moments acting on a train running on an embankment in a cross wind were experimentally and numerically studied. In addition, the flow structure around the train and embankment were explored using CFD. The numerical predictions are in good overall agreement with experimentally-measured force and moments coefficients. For the side force the discrepancies are less than 10 %. For the lift force, relatively good agreement was obtained at yaw angles from 0° to 50°. However, the numerical predictions deviate by about 20 % from experimental results at high yaw angles.

The results show that the maximum of the side force coefficient on the leading car is at about 50°. The side force coefficient on the middle car shows a monotonic increasing trend from 0 to 60° but is almost constant between 60° and 90°. For the overturning moment coefficient, the trend with yaw angle is the same as for the side force. The lift coefficients of both the leading car and the middle car increase as the yaw angle increases. However, for larger angles the lift coefficients of the cars decrease as the yaw angle approaches 90°. The middle car is subjected to smaller lateral forces than the leading car.

For small yaw angles, the side force coefficients when the train is on the leeward track are slightly greater than those for the train on the windward track. As the yaw angle approaches 50°, the side force coefficients of the train on the leeward track are obviously greater than those of train for the windward track. Lift force coefficients of the train on the windward track are greater than those for the train on the leeward track.

Acknowledgments Two of the authors (Zhou and Tian) would like to take this opportunity to thank the Department of Mechanical and Aerospace Engineering at Monash University, and especially the staff of the Monash wind tunnel laboratory, for the financial and technical support for this research project. Also, we gratefully acknowledge the use of the experimental data collected Xiong Xiao-Hui.

References

1. Baker, C.J.: Ground vehicles in high cross winds part I: steady aerodynamic forces. J. Fluids Struct. **5**, 69–90 (1991)
2. Baker, C.J.: Ground vehicles in high cross winds part II: unsteady aerodynamic forces. J. Fluids Struct. **5**, 91–111 (1991)
3. Baker, C.J.: The determination of topographical exposure factors for railway embankments. J. Wind Eng. Ind. Aerodyn. **21**, 89–99 (1985)
4. Suzuki, M., Katsuji, T., Tatsuo, M.: Aerodynamic characteristics of trains/vehicles under cross winds. J. Wind Eng. Ind. Aerodyn. **91**, 209–218 (2003)
5. Diedrichs, B., Simal, M., Orellano, A.: Crosswind stability of a high-speed train on a high embankment. J. Rail Rapid Transit. **221**, 205–225 (2007)
6. Chiu, T.W., Suire, L.C.: An experimental study of the flow over a train in a crosswind at large yaw angles up to 90. J. Wind Eng. Ind. Aerodyn. **45**, 47–74 (1992)
7. Gawthorpe, R.G.: Wind effects on ground transportation. J. Wind Eng. Ind. Aerodyn. **52**, 73–92 (1994)
8. Coleman, S.A., Baker, C.J.: An experimental study of the aerodynamic behaviour of high-sided lorries in cross winds. J. Wind Eng. Ind. Aerodyn. **53**, 401–429 (1994)
9. Khier, W., Breuer, M., Durst, F.: Flow structure around trains under side wind conditions: a numerical study. Comput. Fluids **29**, 179–195 (2000)
10. Paradot, N., Angel, B., Gautier, P.E., et al.: Numerical simulation and experimental investigation of the side loading on a high speed train. Eng. Turbul. Model. Exp. **6**, 657–666 (2005)
11. Chiu, T.W.: A two-dimensional second-order vortex panel method for the flow in a cross-wind over a train and other two-dimensional bluff bodies. J. Wind Eng. Ind. Aerodyn. **37**, 43–64 (1991)
12. Hemida, H.: Large-eddy Simulation of the Flow Around Simplified High-speed Trains Under Side Wind Conditions. Chalmers University, Dessertation (2006)
13. Hemida, H., Krajnovic, S.: LES study of the influence of the nose shape and yaw angles on flow structures around trains. J. Wind Eng. Ind. Aerodyn. **98**, 34–46 (2010)
14. Copley, J.M.: The three-dimensional flow around railway trains. J. Wind Eng. Ind. Aerodyn. **26**, 21–52 (1987)
15. Bocciolone, M., Cheli, F., Corradi, R.: Crosswind action on rail vehicles: wind tunnel experimental analyses. J. Wind Eng. Ind. Aerodyn. **95**, 584–610 (2008)

Part III
Train Aerodynamics: Experimental Techniques

The Influence of Wind Tunnel Grid Turbulence on Aerodynamic Coefficients of Trains

C.N. Nayeri, C. Strangfeld, C. Zellmann, M. Schober, A. Tietze and C.O. Paschereit

Abstract In the design process of trains wind tunnel tests are indispensable in order to assess the cross-wind sensitivity. The aerodynamic forces and moments acting on the leading cars or trailer cars can be measured and aerodynamic coefficients determined. Usually the tests are carried out at low turbulence conditions ("smooth flow"). During recent wind tunnel tests on new high speed train models in the Hermann-Föttinger Institute Berlin undesired Reynolds number effects were observed. This paper describes an approach to reduce these effects by changing the turbulence conditions of the flow. Wind tunnel measurement on a high speed train model in scale 1:25 are carried out with different turbulence conditions: increased free stream turbulence and a tripped boundary layer on the model. The free stream turbulence level is varied from 0.5 to 8 % by using grids with different geometries up stream of the test section. The experiments showed that the Reynolds number dependency on the saftey relevant rolling moment can be well reduced by a simple rectangular grid installed in the contraction of the wind tunnel.

1 Introduction

Cross wind sensitivity of trains is a safety relevant topic covered in European legislation within the framework of the Technical Specification for Interoperability (TSI). Train manufacturers are therefore required to meet the safety criteria defined therein.

C.N. Nayeri (✉) · C. Strangfeld · C. Zellmann · C.O. Paschereit
Hermann-Föttinger Institute (TU-Berlin), Müller-Breslau-Str.8, 10623 Berlin, Germany
christian.nayeri@tu-berlin.de

C. Strangfeld
strangfeld@tu-berlin.de

C. Zellmann
christoph@zellmann.de

M. Schober · A. Tietze · C.O. Paschereit
Bombardier Transportation, Am Rathenaupark, 16761 Hennigsdorf, Germany
martin.schober@de.transport.bombardier.com

A. Tietze
andreas.tietze@de.transport.bombardier.com

A. Dillmann and A. Orellano (eds.), *The Aerodynamics of Heavy Vehicles III*,
Lecture Notes in Applied and Computational Mechanics 79,
DOI 10.1007/978-3-319-20122-1_8

However, in order to achieve maximum performance and to increase operating efficiency, train manufacturers construct increasingly lighter trains. With decreasing weight, the impact of the aerodynamic forces increases, particularly for high-speed trains which are exposed to strong cross-winds. Nowadays, wind tunnel experiments and numerical studies (CFD) are the standard tools to estimate a trains sensitivity to such aerodynamic strains. This paper focuses on the rolling moment generated by cross winds about the lee ward rail described by the coefficient

$$c_{Mx,lee} = c_{Mx} - \frac{s}{2l} \cdot c_z \quad (1)$$

where s = 1.5 m defines the nominal lateral distance between the contact points of a wheel-set for a standard gauge track of 1435 mm and l is a fixed reference length of 3 m (values for full scale). c_z is the lift force coefficient and $c_{M}x$ the rolling moment.The modeling of cross wind acting on a train in a wind tunnel comprises different difficulties namely the stationary ground and the necessary high Reynolds numbers. The Reynolds number of modern high speed trains based on train width $b = 3$ m and a cruising speed of 350 km/h is approximately $Re = 2 \times 10^7$ which imposes a challenge for most conventional wind tunnels.

In the wind tunnel of the Hermann-Föttinger Institute experiments were carried out in the past at low Reynolds numbers below 300,000 with satisfactory predictive quality. However, some recently designed trains shape exhibit a strong Reynolds number dependence of the rolling moment coefficient $c_{Mx,lee}$ such that the design optimization process cannot be carried out in this wind tunnel regarding this safety relevant criteria. Figure 1 shows the values for $c_{Mx,lee}$ and the deviations $\Delta c_{Mx,lee}$ for Re = 160,000, 250,000 and 320,000 as a function of the yaw angle β obtained in the wind tunnel. Large deviations of up to 20 % are recognizable in the range of $\beta = 30°$ to 70° 1. In this yaw angle range strong vortices on the leeward side of the train develop. These vortices are mainly responsible for the low-pressure region on the leeward-side of the train body. The positions and strengths of these vortices in the wake have a large effect on the pressure distribution on the train, which explains its pronounced influence on the overall forces and moments, in particular on the rolling moment [1, 2]. For yaw angles larger than approximately 40°, these vortices can become unstable and exhibit a transient behavior. Instability leads eventually to a vortex breakdown for yaw angles exceeding 50° and for smaller angles the flow basically stays attached [3].

1.1 Influence of Reynolds Number

The Reynolds number plays generally an important role for the determination of aerodynamic coefficients of vehicles. In many cases Reynolds number independence of the coefficients is achieved at relatively low Reynolds numbers. This was shown

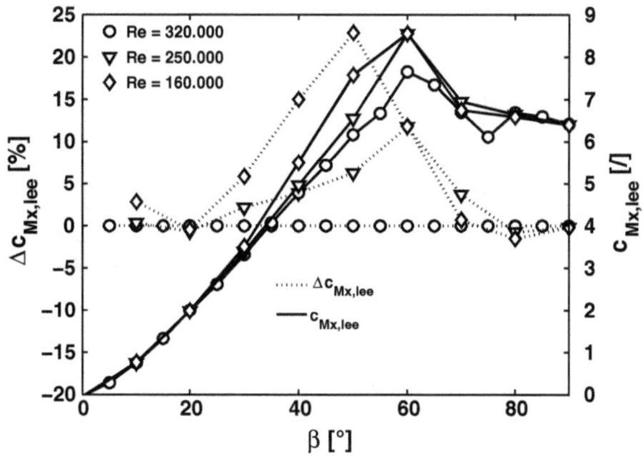

Fig. 1 Influence of Reynolds number on the rolling moment $c_{Mx,lee}$ (*baseline* case)

by Orellano and Schober [4] who reported very little Reynolds number effects in wind tunnel experiments on the important aerodynamic coefficients such as rolling moment, side force and lift force in the range of $Re = 0.6 - 1.4 \cdot 10^6$. However, in the present case lower Reynolds numbers are used and as mentioned above the measurements of the rolling moment show a Reynolds dependency.

This paper presents results of a campaign with the goal to reduce the Reynolds number dependency of the rolling moment coefficient in the large wind tunnel of the Hermann-Föttinger Institute which was shown in Fig. 1 for $Re \leq 320,000$. This would enable inexpensive and fast design studies on the geometry of high speed train in the scale of 1:25. The origin of the Reynolds number dependency is assumed to be the unstable and transient behavior of the leeward vortex mention above. This vortex is fed by the separating boundary layer shedding from the train roof forming a free shear layer and finally rolling up to the vortex. Thus, avoiding an unsteady, transitional behavior of the boundary layer on the roof for the Reynolds number regime considered is expected to reduce the Reynolds number dependency. This is typically achieved by forcing the boundary layer to a turbulent state by either increasing the free stream turbulence or tripping the boundary layer.

2 Experimental Set-Up

The campaign was performed in the closed loop 300 kW-powered wind tunnel at the Hermann-Föttinger Institute and the set-up is shown in Fig. 2. In the test section of 2 m x 1.4 m x 10 m (W x H x L) a maximum velocity of 42 m/s can be achieved. The corresponding Reynolds number is 320,000 based on the model width. The contraction ratio is 6.25 and the standard free stream turbulence level (*FST*) below 0.25 % (with installed filter mats). An external six component balance was used to measure the forces and moments. To reduce the influence of the boundary layer on the wind tunnel floor a splitter plate with an elliptical leading edge was used. The

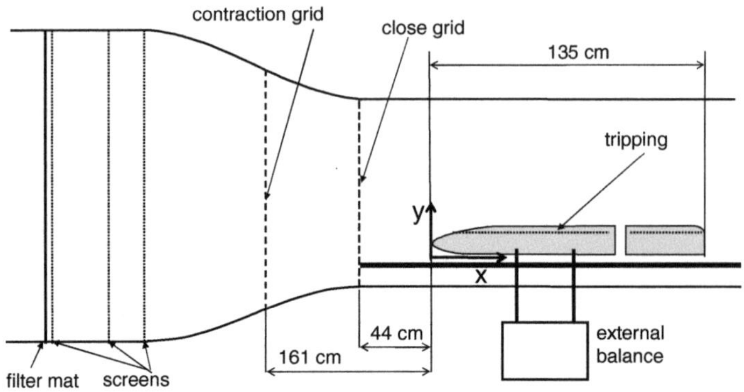

Fig. 2 The wind tunnel set-up (not to scale)

simplest ground scenario, i.e. flat ground, was applied. The investigated Reynolds number range was 140,000 to 340,000 based on the model width. The aerodynamic coefficients were determined for yaw angles β varying from 0° to 90°. The blockage ranged from 2.7 to 11.5 % for these yaw angles. The free stream mean velocity was measured with a pitot tube placed above the train in the test section.

The aluminum train model was scaled 1:25 and comprised a leading car, an inter car gab and a down stream dummy. The total length was 135 cm. A turn table in the splitter plate and in the wind tunnel floor allowed to adjust any yaw angle. The axis of rotation was about 1.25 m down stream of the leading edge of the splitter plate.

Various methods were tested in order to reduce the Reynolds number dependency: tripping, a fine turbulence grid and coarse turbulence grid. Two different positions of the coarse grid were studied i.e. close and distant (Fig. 2). The close position is immediately down stream of the contraction whereas the distant one is placed within the contraction. The wind tunnel is equipped with three turbulence screens in the settling chamber and a filter mat. For the experiments the filter mat was removed which increased the free stream turbulence level to about 0.5 %. As the first method the tripping of the boundary layer on the model was used. The tripping was achieved by a 3-D-turbulator (zig-zag) adhesive tape with 0.4 mm thickness (Fig. 3 left). The grids were made of wood and span across the entire cross section of the test section or contraction (Fig. 3 right).

In addition to the measurement of the forces with the external six component balance, a normal hot wire probe was used to determine mean axial velocity profiles $\overline{U}(y)$ and profiles of the axial turbulence level $\sqrt{\overline{u'^2}}(y)/\overline{U}(y)$. In the remainder of this paper *FST* represents $\sqrt{\overline{u'^2}}(y)/\overline{U}(y)$.

2.1 Design Criteria for the Turbulence Grids

Grids have been known as a standard device for either the generation of turbulence or for decreasing nonuniformities of the free stream velocity profiles in wind tunnels

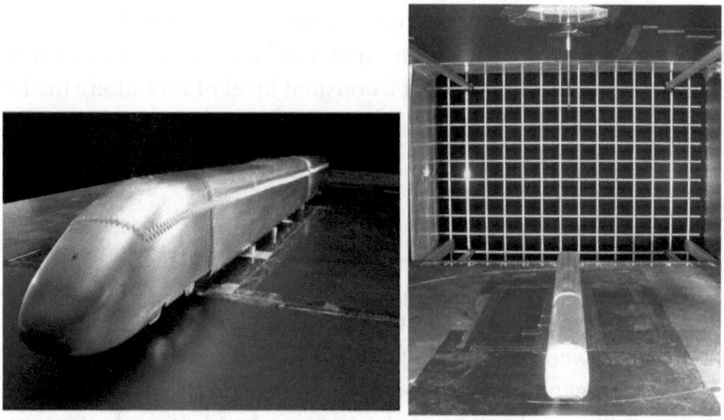

Fig. 3 *Left* Train model with tripping; *right* The contraction grid

(see [5] and references therein). The level of turbulence energy generated is directly proportional to the pressure drop across the grid. Thus, a compromise has to be found between the necessary turbulence generation and the admittable pressure loss. The opening ratio of grids should not be below $\gamma = 0.57$ for stable flow conditions [6]. In the present case it was intended to generate a turbulence level of at least 3 %. The grids should be easy to install and to remove, should not reduce the maximal achievable free stream velocity and should not generate long lasting wakes in the area were the train model is positioned. The grid properties tested within this campaign are shown in Table 1. The rods were of squared cross section with sharp edges.

The results from the fine grid are not reported here, as the generated pressure loss was too high. The axial distance of the contraction grid to the train was chosen based on the assumption that a sufficiently uniform flow is achieved at about 10 mesh widths M [5]. This configuration has the advantageous characteristic of combining a contraction with turbulence generation which leads according to Comte-Bellot and Corrsin [7] to a longer lasting isotropic turbulence.

When turbulence is generated through a local device the energy is dissipated when convected down stream. The decay of isotropic turbulence was modeled by Frenkiel [8] and can be described by:

$$FST = c_{FST} \left(\frac{x}{b}\right)^{-5/7} \tag{2}$$

Table 1 Grids properties; b grid bar width, M Mesh width, γ opening ratio

Grid name	b (mm)	M (mm)	γ
Fine grid	12	67	0.67
Close grid	12	134	0.83
Contraction grid	14	153	0.83

where c_{FST} is an empirical constant depending on the properties of the grid. The theoretical decay curve from Frenkiel is shown in Fig. 9. Thus, it can not be expected that the generated turbulence will be at a constant level of *FST* along the train model.

3 Results

In this section the results of the tripping, of the close grid and of the contraction grid are presented. The application of the tripping, which directly tripped the boundary layer on the train showed a very satisfactory effect. In Fig. 4 a maximum deviation $\Delta c_{Mx,lee}$ of 2 % is shown for the region of interest (ROI) between yaw angles of 30° and 70°. However, the application of the tripping band depends very much on the person who puts it on the train, thus this method is not recommended as a mean for differently shaped trains but rather as reference case in the context of this campaign.

The effect of the close grid is shown in Fig. 5. Compared to the baseline case this configuration improves the behavior quite well. The maximum value of $\Delta c_{Mx,lee}$ has decreased from more that 20 % to around 5 % in the ROI. The curves of the absolute values collaps quite well and the maximum values are around $c_{Mx,lee} \approx 7$. In the baseline case the maximum values of $c_{Mx,lee} \approx 8.7$ for $Re = 170{,}000$ and 250,000. This means, that through increased FST the values $c_{Mx,lee}$ are lowered to values of the higher Reynolds number $Re = 320{,}000$. The same trend is also observed for the tripping case in Fig. 4.

Profiles of *FST* for the close grid in Fig. 6 at different axial positions show that at $x = 0$ mm (head of the train), the wakes of the grid rods are clearly visible. Even further down stream at $x = 575$ mm the foot prints of the wakes are still visible. The FST values vary from ≈ 20 to 5 %. Obviously the grid is too close to the model. Therefore a new grid was made and installed in the contraction which increases the distance between grid and head of the train from 44 to 161 cm (Fig. 2).

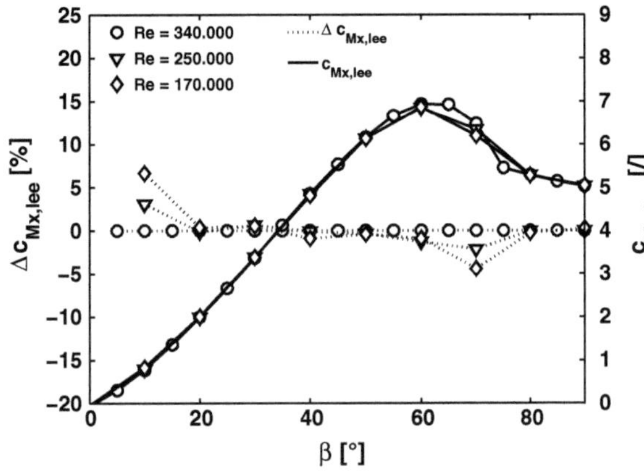

Fig. 4 Influence of Reynolds number on $c_{Mx,lee}$: tripping on the model

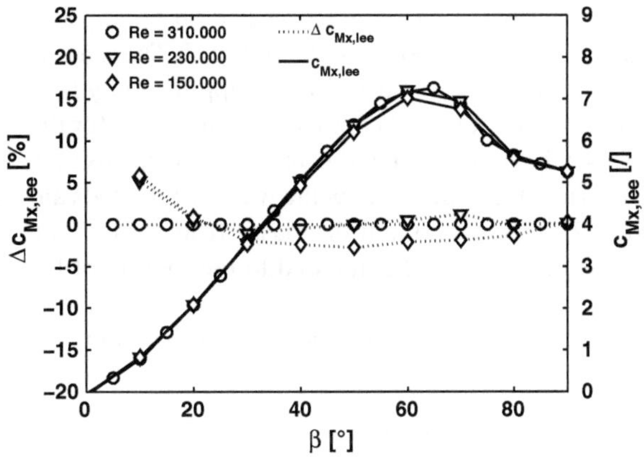

Fig. 5 Influence of Reynolds number on $c_{Mx,lee}$: close grid

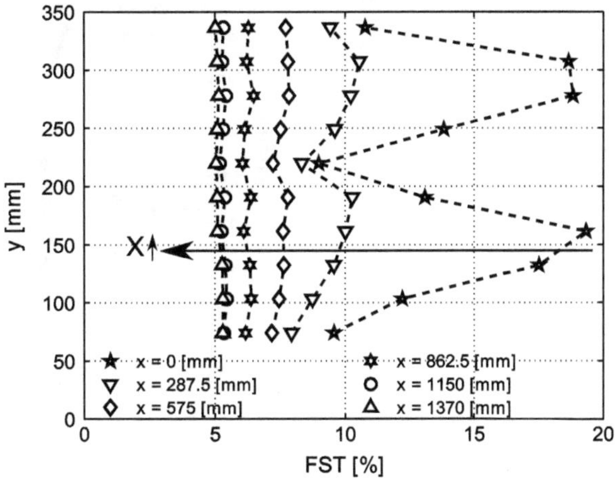

Fig. 6 Profiles of FST at various axial positions: close grid

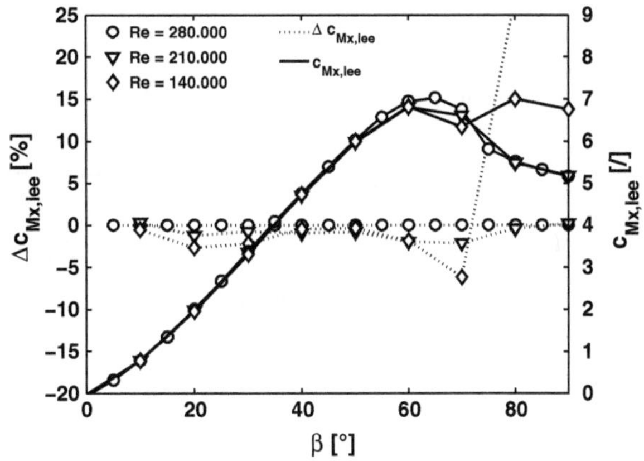

Fig. 7 Influence of Reynolds number on $c_{Mx,lee}$: contraction grids

The effect of the contraction grid on $c_{Mx,lee}$ and $\Delta c_{Mx,lee}$ for the three Reynolds numbers considered is shown in Fig. 7. The maximum deviation in the ROI is about 6 % which is the same as for the tripping. Only for yaw angles $\geq 70°$ the differences between the Reynolds numbers in particular for the lowest $Re = 140,000$ increase. This behaviour is not understood yet and needs to be further studied. The *FST* values shown in Fig. 8 don't show the wakes of the rods any more which is very satisfactory. The *FST* at $x = 0$ mm decreases from $\approx 4\,\%$ at to $\approx 2.8\,\%$. The decay of *FST* is shown in Fig. 9 and compared to the theorie of Frenkiel. The slope compares very well to the theory. The off set between the theory and the measurement is probably due to the remaining contraction ratio behind the grid.

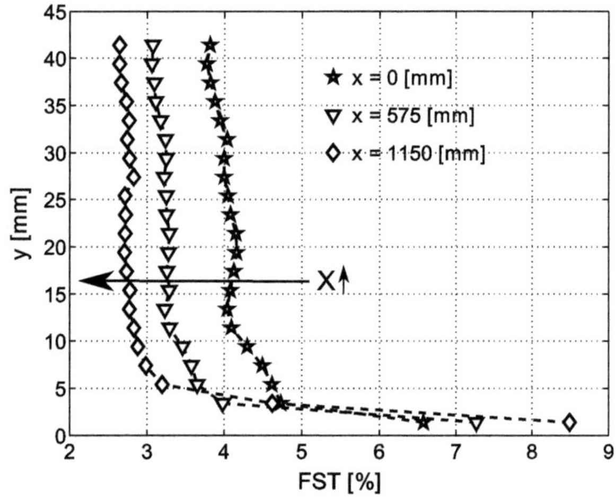

Fig. 8 Profiles of FST at varios axial position: grid in contraction; $Re = 270.000$

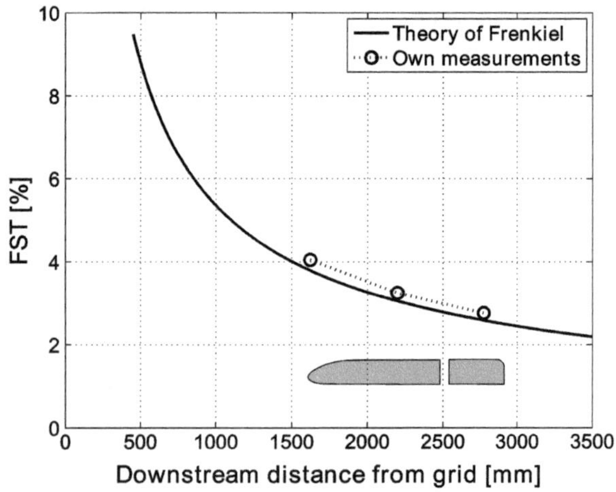

Fig. 9 Stream wise decay of FST for the contraction grid and comparison with theory

4 Conclusions

The Reynolds number dependency of the rolling moment coefficient in the large wind tunnel of the Hermann-Föttinger Institute was successfully reduced by the means of a rectangular grid in the contraction. A free stream turbulence level between 3 and 4 % is generated through a grid placed within the contraction with only low pressure losses and no traces of wakes. This simple method enables inexpensive studies and fast shape optimizations of high speed trains at the scale of 1:25 in the considered wind tunnel. Via a transfer function the obtained safety relevant rolling moment coefficients can be used as a prediction for coefficients measured in wind tunnels with much higher Reynolds numbers such as cryogenic wind tunnels.

It is believed that the unstable and transitional behavior of the leeward vortex is reduced through the increased turbulence level. However, this assumption remains to be investigated in detail.

References

1. Copley, J.M.: The three-dimensional flow around railway trains. J. Wind Eng. Ind. Aerodyn. **26**, (1987)
2. Rolén, C., Rung, T., Wu, D.: Computational modelling of cross wind stability of high-speed trains. In: ECCOMAS, European Congress of Computational Methods in Applied Sciences and Engineering (2004)
3. Orellano, A., Schober, M., Bussman, C.: On side wind stability of high-speed trains. In: WCCM V Fith World Congress on Computational Mechanics, Vienna, Austria (2002)
4. Orellano, A., Schober, M.: Aerodynamic performance of a typical high-speed train exposed to cross-wind. In: 4th WSEAS Transactions on Fluid Mechanics, pp. 379–386 (2006)
5. Roach, P.E.: The generation of nearly isotropic turbulence by means of grids. Int. J. Heat Fluid Flow **8**(2), 82–92 (1987)
6. Bradshaw, P.: The effect of wind-tunnel screens on nominally two-dimensional boundary layers. J. Fluid Mech. **22**(4), 679–687
7. Comte-Bellot, G., Corrsin, S.: The use of a contraction to improve the isotropy of gridgenerated turbulence. J. Fluid Mech. **25**(4) (1966)
8. Frenkiel, F.N.: The decay of isotropic turbulence. Trans. ASME. J. Appl. Mech. (1949)

Evaluation of the Cross Wind Velocity Through Pressure Measurements on Train Surface

F. Cheli, L. Mariano, D. Rocchi, P. Schito and G. Tomasini

Abstract The paper presents the results of a research activity studying the possibility to evaluate the cross wind velocity acting on a running train though surface pressure measurements. The research exploits the know-how developed in the aerospace field on FADS (Flush Air Data sensing Systems) and performs a complete design of the system. In particular, the choice of the number and the position of pressure taps on the surface of the train leading car is optimized, in terms of measurement system sensitivity and robustness, using a neural network approach and wind tunnel test results. A calibration of the system is performed and validated using wind tunnel tests on a moving train model.

1 Introduction

The operating conditions that are critical for a train running under cross wind action are defined by CWCs (Characteristic Wind Curves). CWCs define the maximum wind speed that a vehicle, characterized by its dynamic and aerodynamic properties, can withstand in different conditions (straight line, curve, train speed, etc.). The information provided by CWCs are used to develop methodologies to reduce the overturning risk of railway vehicles subjected to cross-wind.

A possible approach is to define an alarm system that, comparing the measurements of the cross wind velocity along the line with the threshold values provided by the CWCs, manages the train speed, deciding to slow down the traffic, if potential dangerous environmental conditions occur. This traffic control system requires to know both the wind direction and speed, by using discrete measurements performed by anemometer stations installed along the railway lines and to have numerical models, able to extrapolate the wind field both in time and space. In fact, it is necessary to predict the dangerous events in advance in order to have time to slow down a train

F. Cheli · L. Mariano · D. Rocchi · P. Schito (✉) · G. Tomasini
Politecnico di Milano, via La Masa 1, 20156 Milano, Italy
e-mail: paolo.schito@polimi.it

A. Dillmann and A. Orellano (eds.), *The Aerodynamics of Heavy Vehicles III*,
Lecture Notes in Applied and Computational Mechanics 79,
DOI 10.1007/978-3-319-20122-1_9

and to be sure that a measurement performed in a fixed position, is representative of what the vehicle will encounter in another neighbouring positions.

Different approaches could be developed if the wind speed measurements were performed directly on board of the running train experiencing the cross wind conditions.

The aim of the research is to develop a measuring device to estimate the cross-wind direction and speed on an operating train using pressure measurements performed by taps located on the train surface close to the vehicle head.

A similar technology is adopted in aerospace field and it is known as flush air data sensing system (FADSs).

The first prototypes of FADS system were developed and tested on the NASA X-15 hypersonic aircraft [1, 2]. However, many aeronautical applications of the FADS system started in the early 1980s [3–5]. Recently the FADS system was implemented on the NASA Dryden F-18 Systems Research Aircraft [6]. This system uses 11 pressure taps and was tested at speeds up to Mach 1.6, angles of attack up to 80° and angles of sideslip $\pm$ 20°. Other applications of the FADS system are reported in [7–9].

From the aerospace applications three ways to calibrate FADSs are proposed using:

1. suitable aerodynamic numerical models with known structure and unknown coefficients, based on complex flow physics theory (nonlinear regression algorithms);
2. basic lookup tables, based on calibration data;
3. neural network approaches (NNs).

Model based methods are faster but they require a detailed knowledge of the model that is easily available only in simple cases. The second method is the least efficient as it can result in high memory usage and slow execution times.

Neural networks (NNs) allow to model linear or nonlinear systems without the need of a detailed knowledge of the system so they are suitable for the present application where the pressure taps are located on a surface with a geometry far from simple conical or hemispherical cases.

In this paper, for the first time, the application of a FADS system on the nose of a train at relative wind speed up to Mach 0.15 is investigated. In particular, starting from steady tests performed in the wind tunnel of Politecnico di Milano on a 1:15 scale model, a multi-layer perceptron (MLP) Neural Network approach [7, 8, 10–13] is used to develop and calibrate the FADS system.

Wind tunnel tests on a 1:15 scaled model train have been performed in different conditions of wind velocity and angle of attack. Calibration functions have been defined by the correlation between the incoming wind characteristics and the variation of the pressure measured on different locations on the upwind and downwind surfaces. Wind tunnel steady and quasi-steady tests have been performed to obtained data useful both to identify the most suitable pressure taps for the implemented algorithm and to calibrate and validate the system.

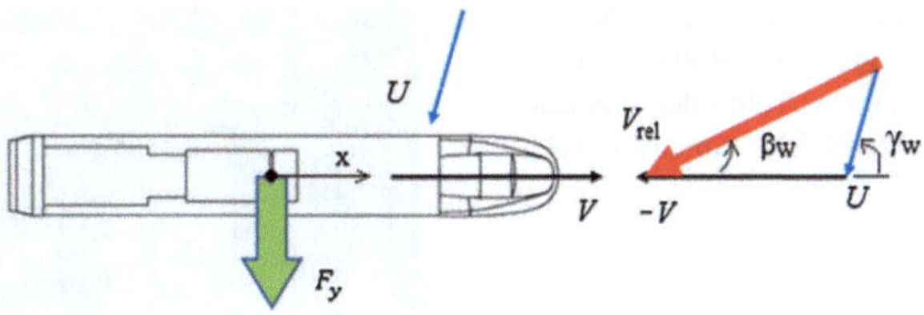

Fig. 1 Triangle of velocity for a moving train under cross-wind conditions

2 Air Data System Model

A train that is moving with a speed V in a cross wind condition with the air blowing at a speed U from the γ_w direction, experiences a relative wind speed V_{rel} with an angle of attack β_w as reported in Fig. 1.

The parameters that have to be estimated by the system are the angle of attack β_w and the relative velocity V_{rel}. These parameters have to be related to the pressure distribution on the vehicle surface that can be generally described as a vector of functions (denoted by $\boldsymbol{f}$) depending from the relative wind speed (V_{rel}) and the corresponding orientation (β_w):

$$\mathbf{p} = \boldsymbol{f}(V_{rel}, \beta_w) \tag{2.1}$$

The function $\boldsymbol{f}$ is known in discrete points from wind tunnel tests for several values of the parameters (V_{rel}, β_w). In fact, for each wind tunnel test, corresponding to the points $V_{rel}{}^i$, $\beta_w j$, a pressure distribution is measured.

The analytical problem can be therefore expressed as in Eq. 2.2:

$$(V_{rel}\beta_w) = \boldsymbol{f}^{-1}(\mathbf{p}) \Rightarrow \begin{bmatrix} V_{rel} \\ \beta_w \end{bmatrix} = \begin{bmatrix} f_V^{-1}(\mathrm{p}) \\ f_\beta^{-1}(\mathrm{p}) \end{bmatrix} \tag{2.2}$$

The air data parameters V_{rel}, β_w are estimated by solving Eq. 2.2 using the available surface pressure measurements $\mathbf{p}$ to extrapolate the continuous form of $\boldsymbol{f}^1$ with a Neural Network technique.

3 Experimental Set-Up

Steady and quasi-steady tests have been performed in the wind tunnel of Politecnico di Milano on a 1:15 train scale model (Fig. 2). The results of the first kind of tests are used to train the neural networks while the results of the second kind of test are used to validate the results.

All the tests were performed on a Single Track Ballast and Rail scenario (STBR, [14]) positioned on a turning table allowing for changing the angle of attack.

Fig. 2 Train scale model Single Track Ballast & Rail scenario (reference scenario of the CEN standard [14])

During the steady state tests, the pressure distribution on the model surface was measured on the still train model while during the quasi-steady tests, the pressure distribution was measured or on a still model, varying the wind speed, or on a rotating model (varying the angle of attack), by keeping the wind speed constant. For validation purposes, a combined variation of wind speed and angle of attack was investigated by rotating the model and, contemporary, varying the wind speed.

64 pressure taps are arranged in a symmetrical manner, on seven vertical sections, as reported in Fig. 3, for the first three planes that will be considered in this work.

The pressure taps adopted for this FADS system belong to the sections A, B and D (Fig. 3), positioned on the train nose since, in this zone, there are the highest pressure gradients due to its particular surface morphology.

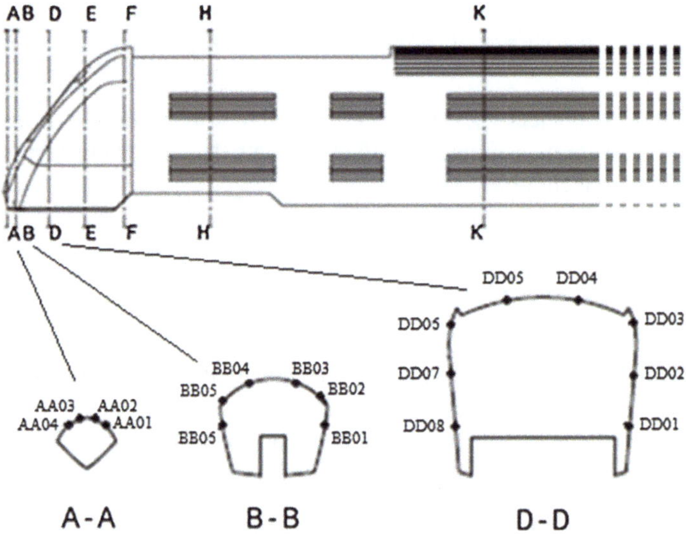

Fig. 3 Leading car: position of pressure taps in three nose sections

All the pressure measurements on the model have been recorded through a pressure acquisition system (PSI-System 8400), using two 2.5 PSI scanners (ESP-32). This type of scanners is thermally compensated and, consequently, no calibration procedure is needed before the acquisition.

The sampling frequency was equal to 200 Hz and a low pass filter was implemented to compensate any noise corruption, with a cut-off frequency of 1.5 Hz for the average values estimation.

Wind speed measurement was carried out through a pitot tube connected to a differential manometer.

Tests have been performed for angles of attack ranging between ±30° and wind speed ranging between 10 and 50 m/s. Figure 9 reports a map of the β_w and V_{rel} configurations considered during steady tests.

4 Pressure Taps Analysis: Required Characteristic and Selection

The choice of the pressure taps is a crucial point in the design of FADSs.

As a general rule the pressure taps, used by the algorithm, must have the following characteristics:

- high sensitivity to the variation of both the angle of attack β_w and the relative wind velocity V_{rel};
- substantial independence, in the range of interest, from Reynolds number;
- surjective relation between the pressure values measured $\mathbf{p}_\mathrm{n}$ and the estimated variables V_{rel} and β_w of the functions.

While the sensitivity of pressure measurements to the variations V_{rel} and β_w is mandatory since it affects the accuracy of the estimations, the verification of the pressure coefficients independence from the Reynolds number is necessary to make predictions on the pressure data in full scale. This verification has been carried out by observing the corresponding pressure coefficients measured at different wind velocities: the pressure coefficients of taps AA01, AA02, AA03, AA04, BB01, BB02, BB05, BB06, DD04 and DD05 do not significantly depend on the Reynolds number (the corresponding graphics are not shown for limited length reasons).

The sensitivity of all pressure taps to the variation of V_{rel} is automatically verified because, for each tap, under the assumption of incompressible flow, it is valid the following relationship:

$$p - p_\infty = \frac{1}{2}\rho C_p V_{rel}^2 \tag{4.1}$$

where $p - p_\infty$ is the differential pressure (referred to the static pressure p_∞) measured by a general tap.

4.1 Selection of the Pressure Taps

The most simple choice for the set up of the algorithm for the estimation of V_{rel} and β_w is to adopt only two pressure taps. The advantage of this choice is that the graphic showing the pressure variation domain (PVD) can be easily plotted through a bi-dimensional representation. To increase the robustness of the estimation, the system will use more than one couple, to obtain more estimations of the same variables.

By the analysis of the PVD, it is possible to evaluate the sensitivity of the couple under consideration, both to V_{rel} and β_w variation. Furthermore, it is possible to verify that, for a given vector of pressure values $\mathbf{p}_\mathrm{i}$ (bi-dimensional for a couple of taps), there is just one couple of V_{rel} and β_w values that is the existence of the functions f_V^{-1} and $f_\beta{}^{-1}$ (surjective property).

For a couple of symmetric taps with respect to the longitudinal plane, the corresponding PVD should be in theory symmetric with the same sensitiveness to the angles of attack with opposite sign.

In Figs. 4, 5 and 6 it is possible to observe the PVD for the taps AA01-AA04, BB02-BB05 and DD04-DD0, for different values of V_{rel} and β_w. Pressures measured during tests with the same angle β_w are aligned on the same colored line while pressures measured during tests with the same wind speed V_{rel} have similar distance from the virtual center on which all the colored lines converge, that corresponds to zero wind speed.

It is possible to check the sensitivity of the considered taps to the variations of the angle of attack and of the wind speed, by observing:

- for β_w, the angular distance among the 17 lines, corresponding to the 17 tested angles of attack;
- for V_{rel}, the distance among the symbols on the same line, corresponding to the 5 wind speeds.

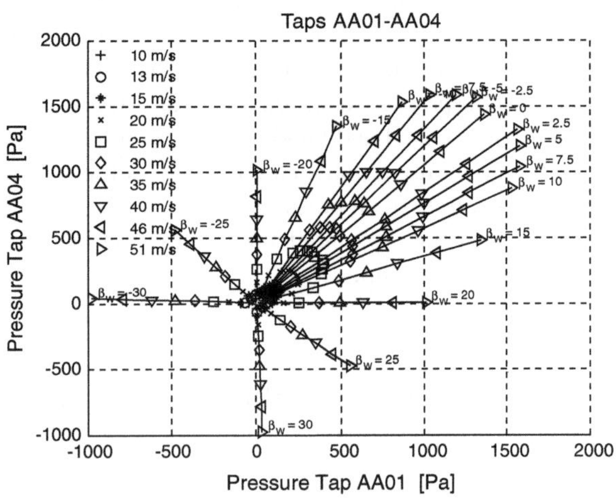

Fig. 4 Pressure variation domain for the tests performed by the taps AA01-AA04

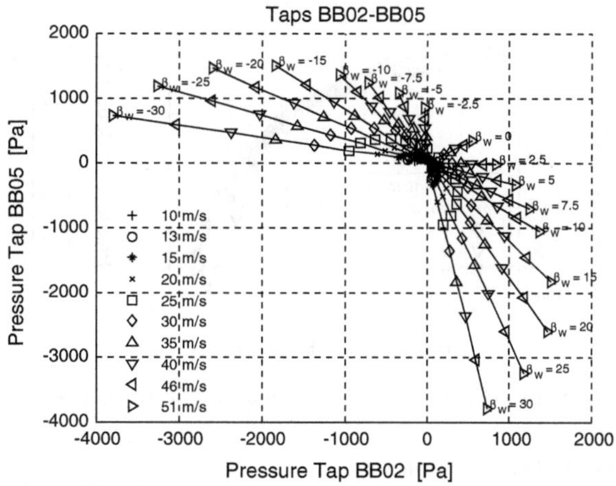

Fig. 5 Pressure variation domain for the tests performed by the taps BB02-BB05

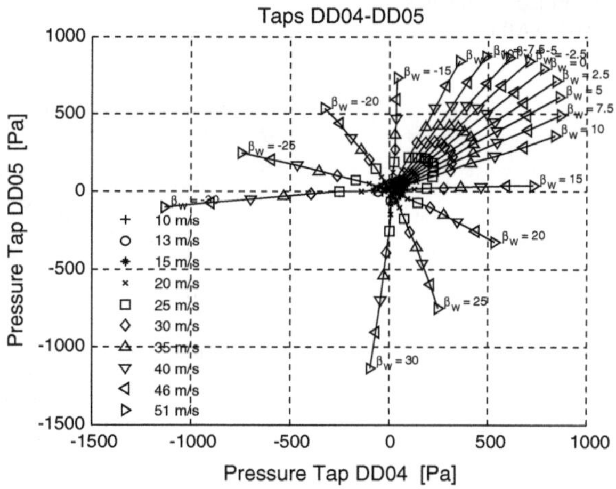

Fig. 6 Pressure variation domain for the tests performed by the taps DD04-DD05

While for the previous cases, the selected couples show a good sensitivity for each examined combinations of V_{rel} and β_w, Figs. 7 and 8 report two typical examples where the considered couples are not suitable for the algorithm purposes.

The first example, reported in Fig. 7, shows a case where the functions f_V^{-1} and f_β^{-1} do not exist. In fact, it is possible to distinguish only 10 of the 17 lines. This means that, for the same combination of pressures, measured by the two taps, there are different combinations of V_{rel} and β_w (black line in the zoom).

In the second example, reported in Fig. 8, the functions $f_V{}^{-1}$ and $f_\beta{}^{-1}$ exist (because it is possible to discern all the 17 lines) but the considered couple present poor sensitivity to the variation of β_w for low angles of attack: in fact, the lines referred to the tests with angles of attack smaller than 10° are very close one another and small errors in the pressure measurements can lead to large errors in the estimation of the angle of attack.

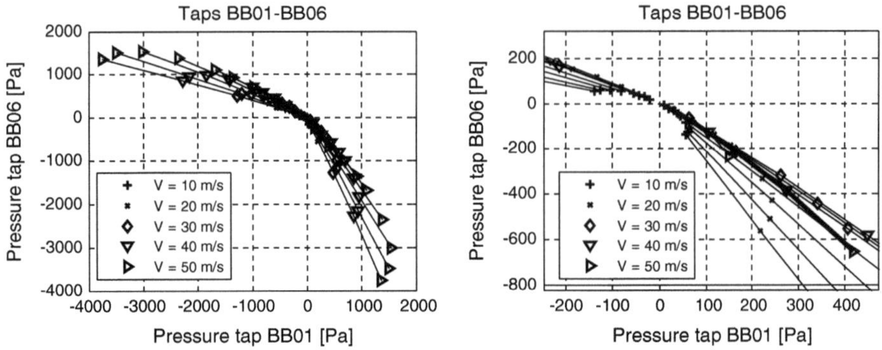

Fig. 7 PVD for the tests performed by the taps BB01-BB06 (*left*) and corresponding zoom (*right*)

Fig. 8 Pressure Variation Domain for the tests performed by the taps AA02-AA03

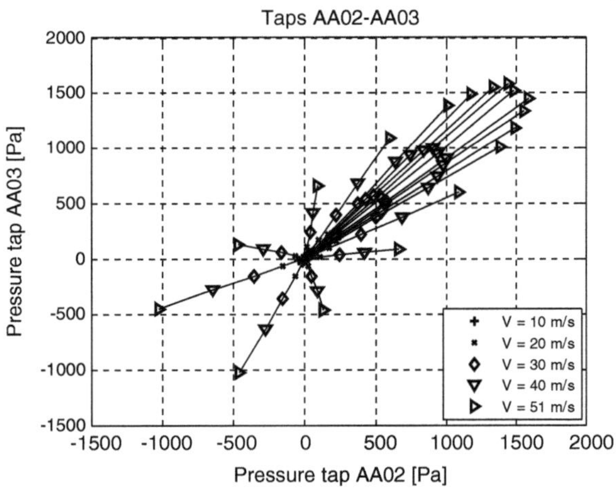

5 Neural Networks Training

5.1 Training Algorithm and Overfitting Problem

The neural network architecture adopted for the set up of the algorithm is the multy-layer perceptron [15]. There are no mathematical procedures to select the right number of layers and neurons for each layer but several attempts have to be performed in order to obtain a trade-off between both the network response accuracy and its generalization capability. The transfer functions of the neurons of the hidden layer are hyperbolic tangents: the non-linearity of these functions is useful to reproduce the non-linearity that is present in f^{-1}. On the contrary, each neural network output layer has just one neuron with linear transfer function.

The supervised learning training has been adopted to re-build the continuous functions which formed $\boldsymbol{f}^{-1}$. In this kind of learning, a set of examples (training set) of desired network behavior is provided to the network:

$$(\mathbf{p}_1, \mathbf{t}_1), (\mathbf{p}_2, \mathbf{t}_2), \ldots (\mathbf{p}_N, \mathbf{t}_N) \tag{5.1}$$

where $\mathbf{p}_n$ is an input vector to the network, and $\mathbf{t}_n$ is the corresponding target output vector.[1]

By comparing the network outputs to the targets, the weights and biases of the network are adjusted (training phase), in order to move the network outputs closer to the targets. The goal is reached minimizing a performance function depending on the errors between the network outputs and the targets.

One of the problems that may occur, during neural network training, is called "overfitting". In this case, the error on the training set is driven to a very small value but when new data are presented to the network the error is large. The network has recorded the training examples, but it is not able to generalize.

For a set of only two pressure taps, as that under consideration, the two functions of vector $\boldsymbol{f}^{-1}$ are 2-variables functions: as a consequence, they can be easily plotted through a tri-dimensional graphic as surfaces, allowing to directly verify the overfitting absence.

In order to limit the overfitting phenomena, it is possible to use two methods: Early Stopping and Regularization methods [15]. For this application it has been found that the latter works better than the former. For this reason, the Bayesians regularization algorithm [15] has been used to train neural networks: in the Regularization method, the performance function, that has to be minimised, depends both on the errors between the networks outputs and the targets and on the parameters of the networks (weights and biases).

5.2 Data Managing

The available data from steady wind tunnel tests have been divided into two sets: 70 % of data are used for network training ("training set") while the remaining 30 % for validation ("test set"). In fact, a good training is carried out when the errors made on the "training set" have the same order of magnitude than that made on the "test set". In this case, the trained neural network has good generalization properties.

The "training set" must be selected in order to uniformly cover the whole domain of interest. To reach this goal, the first 50 % of the "training set" data has been uniformly selected and the remaining 20 % in a random way. The result is showed in Fig. 9.

---

[1] In this application the vector $\mathbf{p}_n$ contains the pressure values measured by the selected pressure taps while the vector $\mathbf{t}_n$ contains the two variables that have to be estimated: V_{rel} *and* β_w.

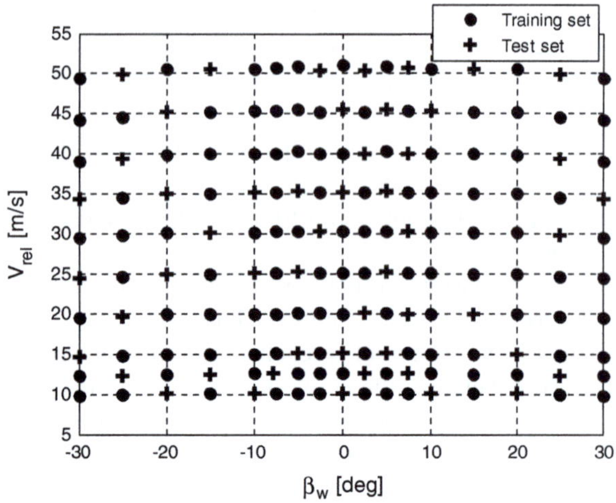

Fig. 9 Combination of angles of attack and wind speeds for which pressure values are available: "training set" (*circle*) and "test set" (*cross*)

5.3 Neural Networks: Response and Architecture

The pressure taps AA01-AA04, BB02-BB05 and DD04-DD05 have been selected to train two neural networks in order to define the functions $f_V{}^{-1}$ and $f_\beta{}^{-1}$ re-building.

The optimal number of layers and neurons, giving both good response accuracy and no overfitting phenomena, have been found for both the functions.

Overfitting absence can be verify by observing $f_V{}^{-1}$ and $f_\beta{}^{-1}$ trends. Outside the training range, the pvd functions have been forced to a constant value to make easier the representation. Examples of overfitting phenomena for both $f_V{}^{-1}$ and $f_\beta{}^{-1}$ are reported in Figs. 10 and 11.

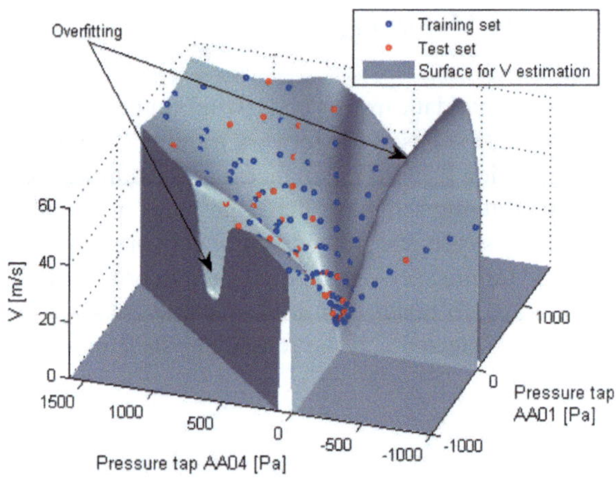

Fig. 10 Estimation surface for V_{rel}. Function $f_V{}^{-1}$ which takes as input the pressure values measured by the couple AA01-AA04. Response with overfitting

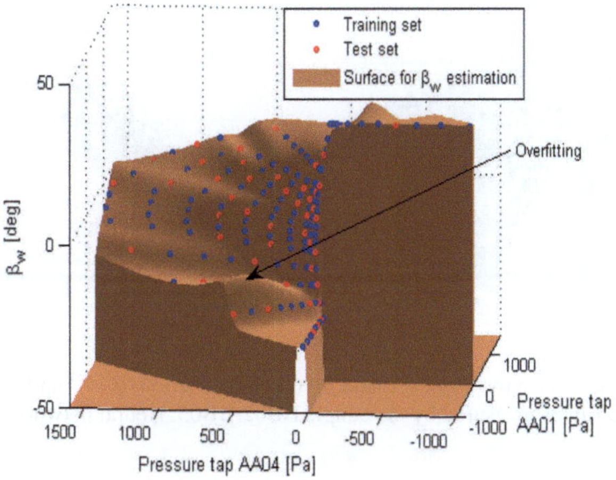

Fig. 11 Estimation surface for β_w. Function $f_\beta{}^{-1}$ which takes as input the pressure values measured by couple AA01-AA04. Response with overfitting

As a result of the neural network training, the functions $f_V{}^{-1}$ and $f_\beta{}^{-1}$ taking as input pressure values measured by the couples AA01-AA04, BB02-BB05 and DD04-DD05 are reported respectively in Figs. 12, 13 and 14.

Neural network for $f_V{}^{-1}$ function uses 2 hidden layers with 4 neurons each for couple AA01-AA04, 3 hidden layers with 4 neurons each for couple BB02-BB05 and 3 hidden layers with 3 neurons each for couple DD04-DD05.

Neural networks for $f_\beta{}^{-1}$ use 3 hidden layers with 4 neurons each for all the 3 couples of taps.

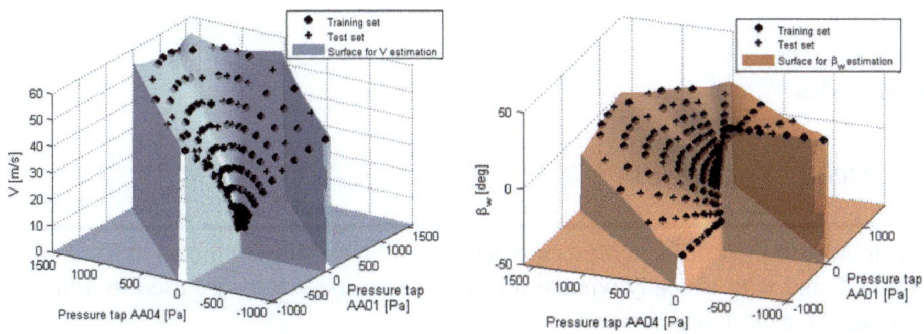

Fig. 12 Estimation surface for V_{rel} (*left*) and β_w (*right*). Functions $f_V{}^{-1}$ and $f_\beta{}^{-1}$ which take as input the pressure values measured by the couple AA01-AA04. Neural network responses without overfitting

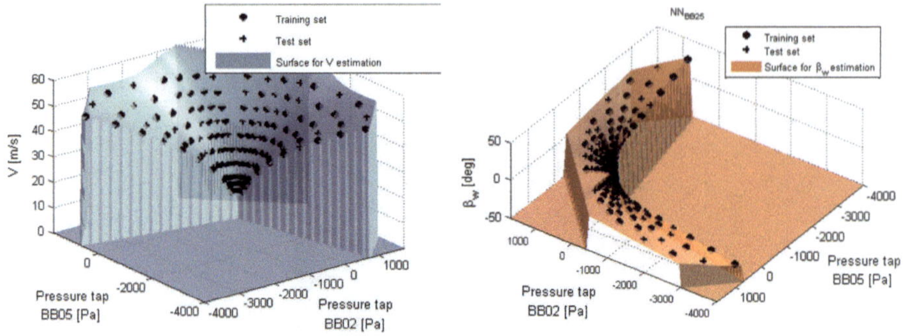

Fig. 13 Estimation surface for V_{rel} (*left*) and β_w (*right*). Functions $f_V{}^{-1}$ and $f_\beta{}^{-1}$ which take as input the pressure values measured by the couple BB02-BB05. Neural network responses without overfitting

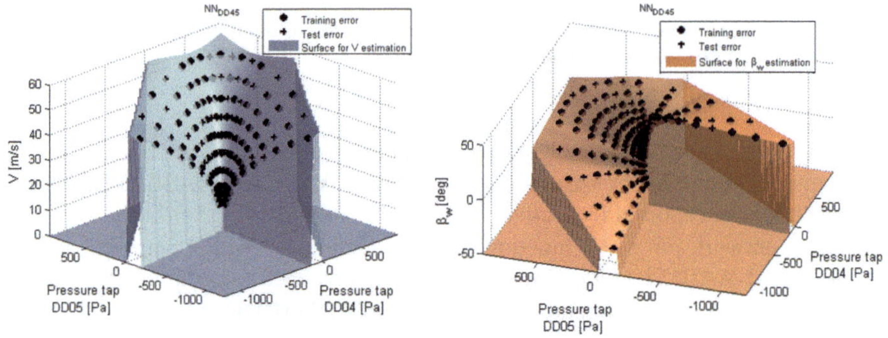

Fig. 14 Estimation surface for V_{rel} (*left*) and β_w (*right*). Functions $f_V{}^{-1}$ and $f_\beta{}^{-1}$ which take as input the pressure values measured by the couple DD04-DD05. Neural network responses without overfitting

6 Results

6.1 Steady Test Results

As already described, steady tests are performed on a still train model at fixed angle of attack and with a steady wind speed. Results of these tests are used to train and validate the neural networks.

In Figs. 15, 16 and 17 V_{rel} (*left*) and β_w (*right*) estimation errors are reported.

In each case, the training errors have the same order of magnitude than the test errors. This means that the trained neural networks have good generalization capability. The maximum estimation error has an absolute value of 0.84 m/s for V_{rel} and 0.75° for β_w.

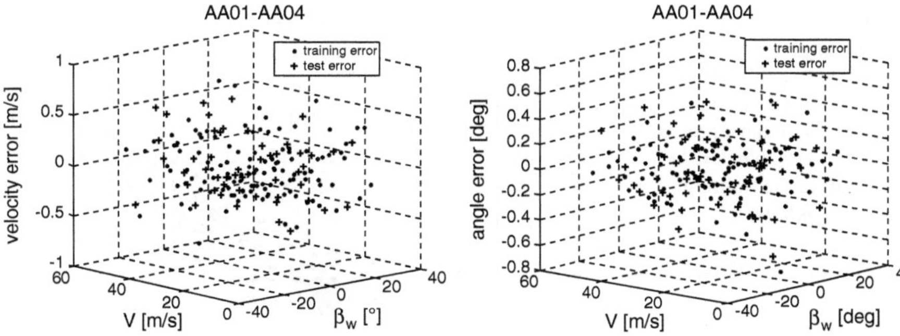

Fig. 15 V_{rel} (*left*) and β_w (*right*) estimation errors for neural networks which take as input the pressure values measured by the couple AA01-AA04

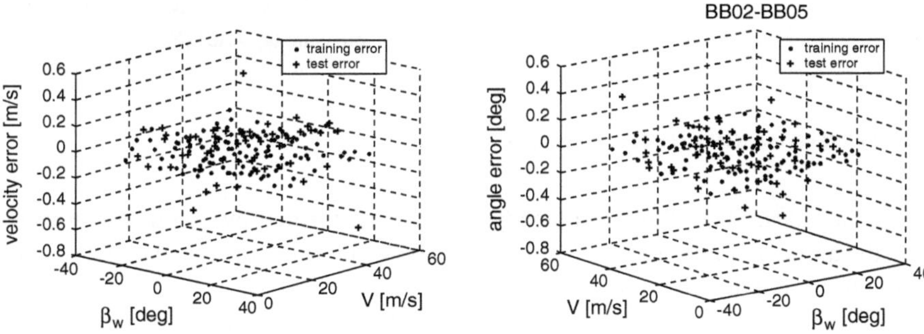

Fig. 16 V_{rel} (*left*) and β_w (*right*) estimation error for neural networks which take as input the pressure values measured by the couple BB02-BB05

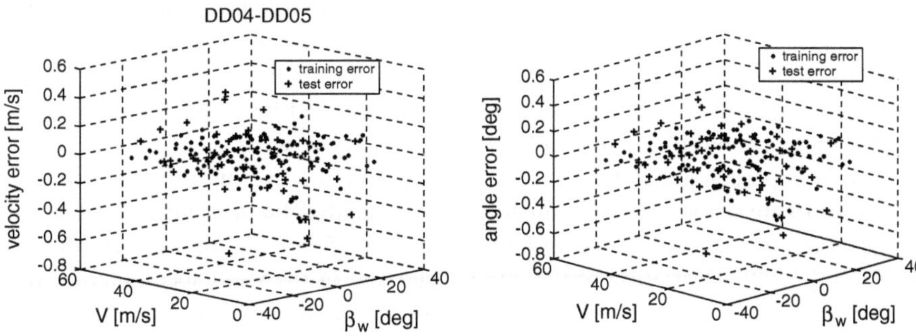

Fig. 17 V_{rel} (*left*) and β_w (*right*) estimation error for neural networks which take as input the pressure values measured by the couple DD04-DD05

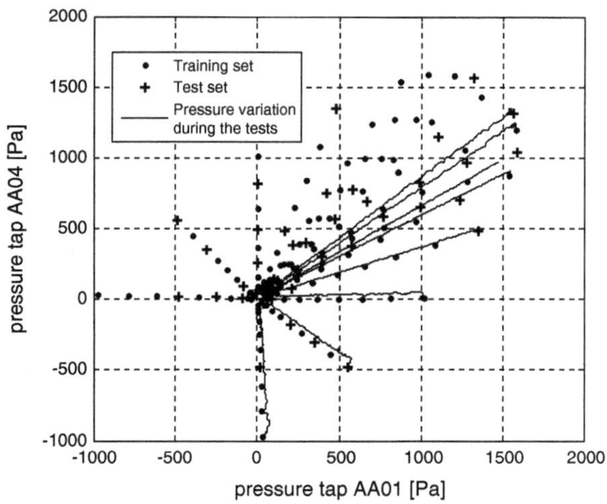

Fig. 18 Quasi-steady test, constant angle of attack with continous variation of the wind speed, AA01 and AA04 pressure taps: pressure measurements (*black solid lines*) within the training domain (*symbols*)

6.2 Quasi-Steady Tests Results

As already described, the quasi-steady tests are performed by continually varying both the angle of attack and the wind speed. The variation of the two parameters may be imposed individually or contemporary.

In Fig. 18, solid lines represent the trend followed from the pressures measured during quasi-steady tests by taps AA01 and AA04 in test with constant angle of attack and variable wind speed. Dots are, on the contrary, pressures measured during steady tests.

It is important to note that at equal V_{rel} and β_w there is a deviation between pressures measured during steady and quasi-steady tests. It has been shown in fact that, in unsteady flow fields, flow inertial effects can cause errors between the instantaneous static and total pressure in the flow and those predicted by FADSs. However, this error is proportional to the inverse of the square of the wind speed so it is usually negligible in percentage in the range of speed where the overturning risk is high.

In Fig. 19 are reported the time histories of the results obtained during the test with β_w equal to 30° and wind speed variable from 50 to 15 m/s. In Fig. 19a a comparison between the actual variation of the wind speed and of the wind speed predicted by the FADS is presented. The same comparison is reported in Fig. 19b as far as the angle of attack is concerned. It is possible to observe the overfitting absence since the neural network response is quite smooth.

Furthermore, while the absolute wind speed error is almost the same during the test (Fig. 19c), angle of attack error increases decreasing the angle of attack. For low wind speed, in fact, as previously asserted, the error in pressure between steady and quasi-steady tests became in percentage higher as can be seen in Fig. 19d even if it is always lower than 1°.

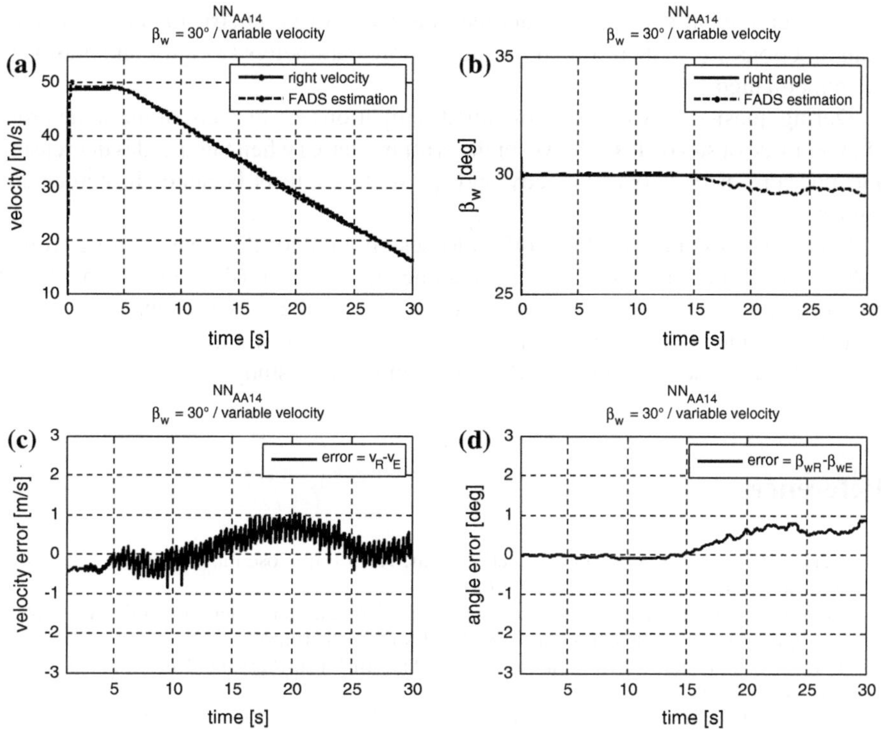

Fig. 19 Quasi-steady tests, taps AA01 and AA04, angle of attack $\beta_w = 30°$, time histories of: estimated and actual wind speed (**a**), estimated and actual angle of attack (**b**), wind speed error (**c**) and angle of attack error (**d**)

7 Conclusions and Future Work

Experimental data have been used to train multilayer perceptron neural networks. It has been demonstrated that the estimation problem of the variables V_{rel} and β_w can be solved by using only two suitably chosen pressure taps. Using only 2 inputs in neural networks allows to visualize the estimation surface and to obtain neural network responses with good generalization capabilities and no overfitting problems as also demonstrated by quasi-steady test results.

Three couples of pressure taps have been identified. For each of these couples, two neural networks, one for the velocity and the other for the angle of attack estimation, have been trained. The geometric simplicity of these functions allow to reach accurate results having neural networks with few layers and neurons.

Therefore, they require a limited computational power to be implemented. The maximum estimation error is 0.84 m/s in absolute value for V_{rel} and 0.75° for β_w.

In general, it has been found that functions $f_\beta{}^{-1}$ are geometrically more complex than functions $f_V{}^{-1}$ therefore they require more parameters (weights and biases) to be implemented.

During quasi-steady tests, the obtained estimations errors were of the same order of magnitude of steady tests, however the error increases when the speed is decreasing and pressure error between steady and quasi-steady tests becames in percentage higher.

Future development of this work is to design an algorithm that, starting from a redundant estimation of V_{rel} and β_w parameters obtained using several couples of pressure taps, is able to find anomalies or failures in the pressure taps functioning, so as to exclude malfunctioning taps from V_{rel} and β_w estimation process.

Finally, a full scale validation of the system is under study.

References

1. Cary, J.P., Keener, E.R.: Flight evaluation of the X-15 Ball-Nose Flow-Direction sensor as a airdata system. NASA TN D-2923 (1965)
2. Wolowicz, C.H., Gosett, T.D.: Operational and performance characteristics of the X-15 spherical hypersonic flow direction sensor. NASA TN D-3076 (1965)
3. Larson, T.J., Siemers III, P.M.: Subsonic tests of an All-Flush-Pressure-Orifice air data system. NASA TP 1871 (1981)
4. Larson, T.J., Whitmore, S.A., Ehernberger, L.J., Johnson, J.B., Siemers III, P.M.: Qualitative evaluation of a flush air data system at transonic speeds and high angles of attack. NASA TP-2716 (1987)
5. Larson, T.J., Moes, T.R., Siemers III, P.M.: Wind tunnel investigation of a flush airdata system at Mach numbers from 0.7 to 1.4. NASA TM-101697 (1990)
6. Whitmore, S.A., Davis, R.J., Fife, J.M.: In flight demonstration of a real time flush airdata sensing system. NASA TM 104314 (1995)
7. Rohloff, T.: Development and evaluation of neural network flush air data sensing systems, Ph.D. thesis, Department of Mechanical Engineering, University of California (1998)
8. Crowther, W.J., Lamont, P.J.: A neural network approach to the calibration of a flush air data system. Aeronaut. J. **105**, 85–95 (2001)
9. Brown, E.N., Friehe, C.A., Lenschow, D.H.: The use of pressure fluctuations on the nose of an aircraft for measuring air motion. J. Clim. Appl. Meteorol. **22**, 171–180 (1983)
10. Rediniotis, O., Vijayagopal, R.: Miniature multihole pressure probes and their neural network based calibration. AIAA J. **37**, 666–674 (1999)
11. Rediniotis, O., Chrysanthakopoulos, G.: Application of neural networks and fuzzy logic to the calibration of the seven-hole probe. J. Fluid. Eng. Trans. ASME **120**, 95–101 (1998)
12. Calia, A., Denti, E., Schettini, E., Di Fusco, C.: Ricostruzione degli angoli di incidenza e di derapata per velivolo mediante elaborazione dei dati aria basate su reti neurali. Università degli studi di Pisa. Tesi di laurea specialistica in ingegneria Aeronautica (2006)
13. Maggiani, G.: Sviluppo di metodologie basate su reti neurali per l'elaborazione dei dati aria dei velivoli. Università degli studi di Pisa. Tesi di laurea specialistica in ingegneria Aeronautica (2008)
14. European Standard prEN 14067–1
15. Demuth, H., Beale, M., Hagan, M.: Neural Network Toolbox™ 6. Matla® User's Guide (2009)

Part IV
Train Acoustics

A Simplified Model of the Wave Generation Due to Train-Tunnel Entry

Sebastian Wagner, Klaus Ehrenfried and Andreas Dillmann

Abstract The compression wave generated when a high-speed train enters a tunnel at Mach numbers smaller than 0.4 can be described in good approximation by a linear theory of an inviscid fluid. The wave equation for the acoustic potential becomes the governing equation. It is solved by a three dimensional boundary element method in time domain which forces a vanishing normal component of the velocity at the tunnel wall. It is assumed that the elements are compact in time. This leads to a linear equation in which a special matrix-vector multiplication has to be evaluated for every time-step. The aim is to create a fast method which sets as few constraints on the geometry as possible but nevertheless gives an accurate description of the wave propagation. In a first step the elements are assumed to be rectangles and an infinitely thin cylindric tube of finite length is taken as the geometry of the tunnel. The train is modeled by a single moving mass source of monopole type. It defines a semi-infinite body whose shape slightly changes when entering the tunnel. The results of this simple model along with the comparison with analytical solutions and experimental data are shown and discussed.

1 Introduction and Motivation

When a high-speed train enters a tunnel it induces a nearly plain pressure wave running through the tunnel. Two effects occur: on the one hand the wave is reflected at the open endings, on the others hand the *tunnel boom* can occur at the far end (schematically shown in Fig. 1). The latter is an explosion-like pressure-wave leaving the

S. Wagner (✉) · K. Ehrenfried · A. Dillmann
German Aerospace Center (DLR), Institute of Aerodynamics and Flow Technology,
Bunsenstraße 10, Göttingen, Germany
e-mail: sebastian.wagner@dlr.de

K. Ehrenfried
e-mail: klaus.ehrenfried@dlr.de

A. Dillmann
e-mail: andreas.dillmann@dlr.de

A. Dillmann and A. Orellano (eds.), *The Aerodynamics of Heavy Vehicles III*,
Lecture Notes in Applied and Computational Mechanics 79,
DOI 10.1007/978-3-319-20122-1_10

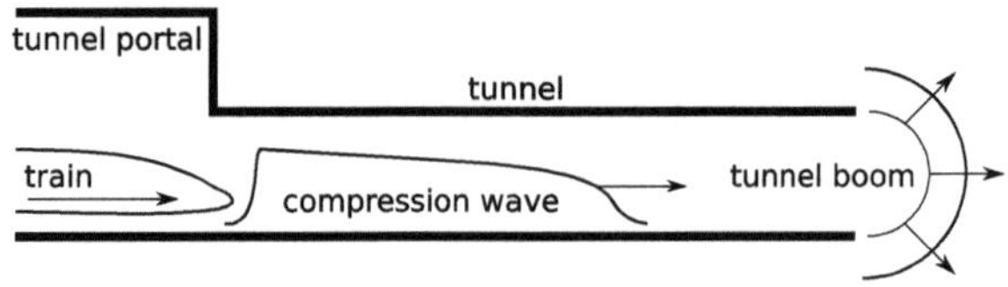

Fig. 1 Tunnel entry schematic

tunnel. It can damage the tunnel and surrounding structures. This effect is comparable to the sonic boom produced by an airplane and the noise is therefore disturbing for local residents. The amplitude of this pressure wave can be reduced by decreasing the speed of the train. But, since high-speed trains should keep their velocity also in tunnels structural modifications are necessary to decrease the pressure-gradient of the wave front which is proportional to the amplitude of the sonic boom ([8]).

These phenomena are the topic of various experimental and numerical investigations of which the TRANSAERO project ([9]) should be mentioned here. In addition Michael Howe has developed analytic methods to describe the wave propagation ([4, 5, 7] etc.). He has shown that solutions of the wave-equation can give a proper description of the main effects. This linear theory is based on several assumptions. On the one hand specific properties of the fluid are assumed, e.g. inviscidness. On the other hand strong restrictions on the occurring wavelengths are necessary to keep the situation accessible for analytical calculations. In addition the tunnel geometry has to be kept generic.

In the present paper the train-tunnel entry is also described in a linearized theory (Sect. 2), i.e. the same equation is solved. Instead of solving the equation analytically a boundary element method (BEM) is introduced to keep the possibility of a more general tunnel geometry (Sect. 3). Also the restrictions on the occurring wavelengths are relaxed. As a first step a simple model of train and tunnel is chosen to compare the BEM results with analytical and experimental results (Sect. 4). Aim is to check how well the main characteristics of the pressure wave can be reproduced with the simple model and in which way numerical errors occur or the discretization sets constraints for the model (Sect. 5).

2 Governing Equations

Consider an inviscid, compressible fluid. The conservation laws for mass, momentum and energy are in mathematical formulation the equation of continuity

$$\frac{\partial \rho}{\partial t} + \nabla \cdot (\rho \mathbf{u}) = m, \tag{1}$$

Euler's equation

$$\rho \frac{\partial \mathbf{u}}{\partial t} + \rho \mathbf{u} \cdot \nabla \mathbf{u} = -\nabla p \tag{2}$$

and a dependence between pressure p, density ρ and entropy S based on the equation of state for the fluid. The velocity field is denoted by $\mathbf{u}$ and m is an arbitrary mass-source. Velocity, pressure and density are decomposed into a constant part (ρ_0, p_0) for the fluid at rest and fluctuations $(\mathbf{u}', p', \rho')$, i.e.

$$\mathbf{u} = \mathbf{u}' \quad p = p_0 + p' \quad \rho = \rho_0 + \rho' \tag{3}$$

Because the fluid is inviscid it is possible to assume an isentrop, irrotational flow. If in addition only velocities small against the speed of sound in the undisturbed fluid c are considered,[1] the conservation laws in their linearized form lead to a wave-equation for the acoustic potential ϕ, namely[2]:

$$\frac{1}{c^2}\frac{\partial^2 \phi}{\partial t^2} - \underline{\Delta} \phi = \zeta \tag{4}$$

Occurring mass-sources are denoted by ζ. The pressure and velocity field[3] can by calculated by

$$\mathbf{u}' = \nabla \phi \quad \text{and} \quad p' = -\rho_0 \frac{\partial \phi}{\partial t} \tag{5}$$

where the second equations is result of a chosen gauge of the potential.

The solutions of the wave-equation have to fulfill boundary conditions. For the case of solid walls this is a vanishing mass-flux through the wall, i.e.

$$\mathbf{n} \cdot \nabla \phi \stackrel{!}{=} 0 \tag{6}$$

where $\mathbf{n}$ is the normal vector of the wall. Without walls in finite distance the solutions of the wave equation can be written via Green's function in free space as[4]:

$$\phi(t, \mathbf{x}) = \frac{1}{4\pi} \int \mathrm{d}^3 y \int \mathrm{d}\tau \, \frac{\zeta(\tau, \mathbf{y})}{|\mathbf{x} - \mathbf{y}|} \, \delta\left(t - \frac{|\mathbf{x} - \mathbf{y}|}{c} - \tau\right) \tag{7}$$

3 BEM in Time Domain to Solve the Wave-Equation

For a solid wall (boundary) Ω in finite distance the wave equation can be solved by a boundary element method (BEM) in the time domain.

---

[1] In [3] it is shown, that these requirements are sufficient.

[2] $\underline{\Delta}$ denotes the Laplacian.

[3] From now on p' is called *pressure* and $\mathbf{u}'$ is called *velocity*.

[4] The integral run over the whole space and all times. $\delta(\cdot)$ denotes the delta-distribution.

Additionally a mass-source strength ζ_w located at the wall is added to the right side of the wave equation, i.e.

$$\frac{1}{c^2}\frac{\partial^2\phi}{\partial t^2} - \Delta\phi = \zeta_s + \zeta_w \tag{8}$$

becomes the governing equation. ζ_s are the known mass-sources from above. ζ_w has to be chosen in a way that the boundary-condition at the wall is fulfilled. With this additional term on the right side the wave-equation can be solved by applying equation (7) as no additional boundary-condition has to be fulfilled. In a time dependend situation, ζ_w also has to be time dependend.

To calculate ζ_w numerically it is necessary to discretize Ω (and the time). In each of the boundary elements ζ_w is assumed to be constant in space, i.e. the boundary condition is fulfilled in an integral sense over each element (and not longer pointwise as in the continuous case). In addition it is assumed that the elements are compact in time, i.e. all occurring wavelengths are large in comparison with the size of the elements. This leads to an equation for the sourcestrength (per area) σ_w in element k for the timestep j:

$$\sigma_w^k(t_j) = \frac{\Phi_k(t_j) - \sum_{i\neq k}\left(A_{ki}\,\sigma_w^i(t_j - \chi_{ik}) + B_{ki}\,\dot{\sigma}_w^i(t_j - \chi_{ik}')\right) + \frac{B_{kk}}{\Delta_t}\,\sigma_w^k(t_{j-1})}{A_{kk} + \frac{B_{kk}}{\Delta_t}} \tag{9}$$

In this equation the integral mass-flux through element k induced by the pregiven ζ_s is denoted by $\Phi_k(t_j)$. The matrices A and B describe the interaction between the elements for near- and farfield. Δ_t is the timestep size and the matrices χ and χ' contain the information about the time in which the information propagates between the elements. The finite propagation speed (here speed of sound in the undisturbed fluid) leads to a time order, i.e. for the calculation of $\sigma_w^k(t_j)$ only σ_w^i from previous timesteps are necessary.[5] A derivation of this equation and an analysis of the numerical properties of this numerical method are in detail shown in [10]. There is also pointed out that for the calculation of the matrices A and B large numerical effort is necessary. For plain rectangular elements analytical integration can be used when calculating A and B, therefore the elements have a rectangular shape as a first step.[6]

If the matrices are known more or less a matrix-vector multiplication has to be evaluated every timestep to calculate σ_w^k for all k. Afterwards the potential ϕ can be calculated by using equation (7) with $\zeta = \zeta_s + \zeta_w$.

---

[5] $\sigma_w^i(t_j - \chi_{ik}) + B_{ki}$ and $\dot{\sigma}_w^i(t_j - \chi_{ik}')$ are just symbolic. They have to be replaced by numerical interpolatation respectively a discrete differentiation.

[6] In [2] is shown, that also for triangles a (very complex) analytical integration can be performed. This would allow more general geometries.

4 Tunnel with Circular Cross-Section

A simple model of the train and the tunnel is used to investigate the interaction from an acoustic point of view. The tunnel is modeled by a surface Ω which is an infinitely thin duct with a finite length L and a circular cross-section $A = R^2\pi$ (see Fig. 2). By construction the boundary-condition is fulfilled just on the inner-side of the wall. Therefore the surrounding surface Ω is not closed. The walls necessary for the closure of Ω are neglected as a first step. (The influence of this construction is partly discussed in Sect. 5.) As a consequence the surrounded space continues into the outer space at the ends of the duct.

The train is modeled by a single moving mass source of monopole type, i.e. $\zeta_s = q_0\delta(\mathbf{u}_0 t + \mathbf{x}_0)$ where $\mathbf{u}_0$ is the (constant) velocity of the source and $\mathbf{x}_0$ is its starting-point. It defines a stream-line body which has a contour comparable to a generic train (see Fig. 3).

4.1 Deformation of the Stream-Line Body

As no explicit boundary-condition has to be fulfilled on that stream-line body, deformations caused by the boundaries are possible. As an example the contour of a train entering the tunnel is shown in Fig. 4. It is obvious that the shape of the train is compressed by the tunnel walls. In Fig. 5 the deformation is illustrated quantitatively. A_0

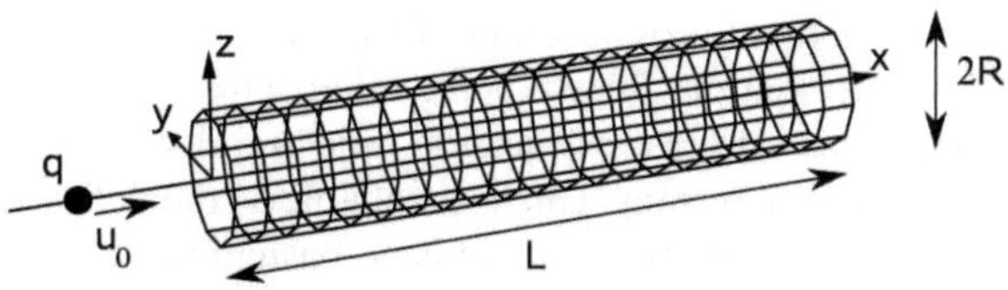

Fig. 2 Tunnel geometry (schematical) and nomenclature

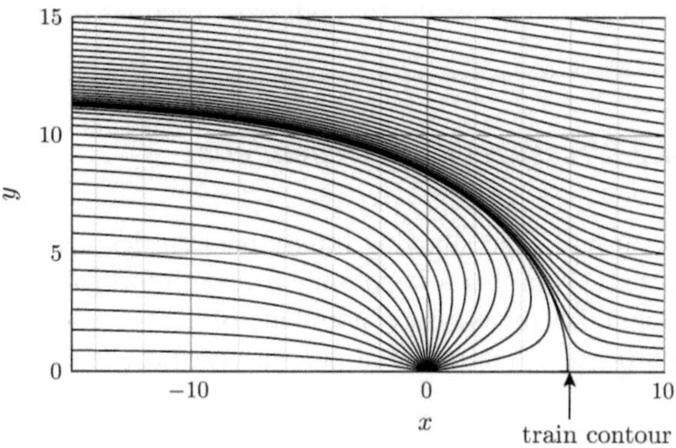

Fig. 3 Stream-lines of a source of monopole type in the origin with constant incoming flow in x-direction and contour of the train (rotation symmetry around x-axis)

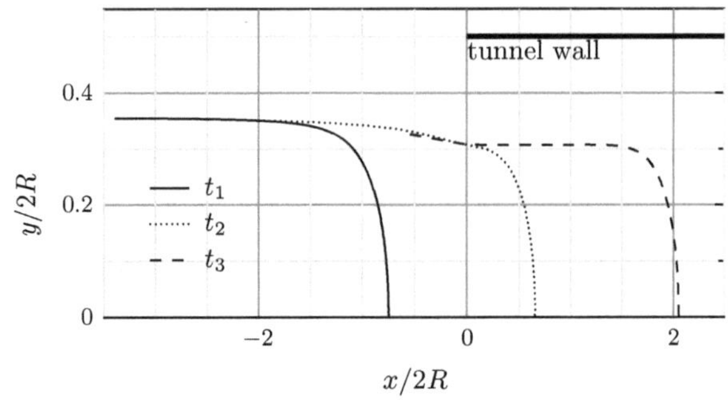

Fig. 4 Example of the stream-line deformation by the tunnel. Shape of the train at different times before and after entering the tunnel

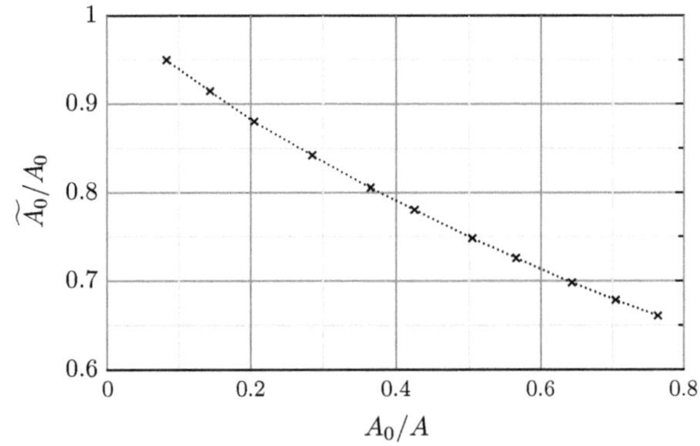

Fig. 5 Relative deformation in dependence of the cross-section ratio for a fixed train speed

is the maximal cross-section of the train in free space and $\widetilde{A_0}$ is the cross-section after deformation. Shown is the relative deformation of the cross-section $\widetilde{A_0}/A_0$ in dependence of the cross-section ratio A_0/A of train and tunnel. The velocity of the train is kept constant. This is resonable as the deformation process is not part of the investigation. Therefore there is no basic time-scale and consequently the complete relative deformation is velocity-independent. It is observed that even for small cross-section ratios[7] the deformation should not be neglected. In Sect. 4.3 the comparison with analytical results for both cross-sections is performed.

4.2 Pressure-Time Dependence

As usual when studying train-tunnel entries the pressure-time dependence is analyzed. From this it is possible to gain information about the strength of the tunnel boom, i.e. the pressure amplitude of the wave leaving the tunnel, because it is proportional to the maximal pressure gradient (see [8]). In Fig. 6 the main content of

[7] As they occur in tunnels with two tracks ([1]).

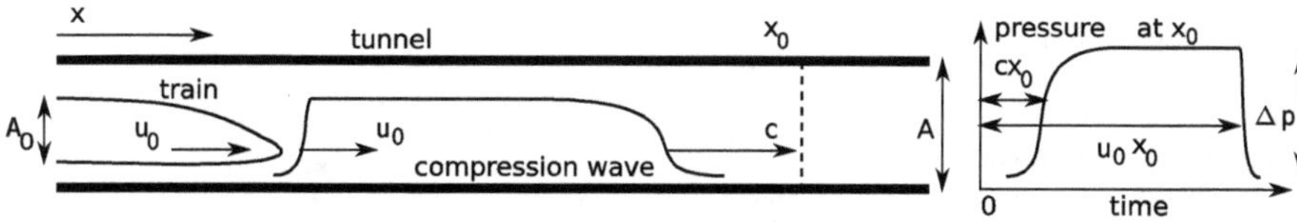

Fig. 6 Tunnel entry: processes and nomenclature (schematic)

Fig. 7 Pressure-time dependence at different postions x_0 inside the tunnel for $M = 0.2$ and $\widetilde{A_0}/A = 0.15$

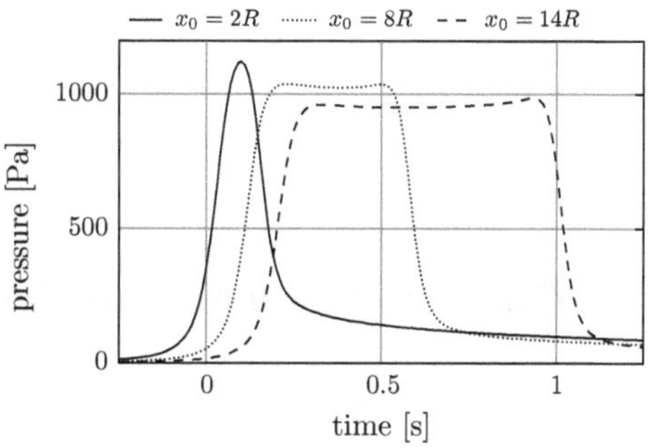

the pressure-time dependence (in linear theory) is shown schematically (motivated by [4]). Figure 7 shows the calculated pressure for different positions x_0 in the tunnel where the x-axis points along the tunnel and has its origin at the beginning of the tunnel. The pressure was calculated very close to the wall with $M = 0.2$ and $\widetilde{A_0}/A = 0.15$. Here $M := u_0/c$ denotes the Mach-number calculated with the train velocity u_0 and the speed of sound in the undisturbed fluid c. It is obvious that the expected characteristics can be found in the plot. But also an error is visible, namely the decreasing pressure amplitude Δp. It is expected that in a linear theory the wave amplitude stays constant when propagating along the tunnel (see [6]). The reason for the decrease is a numerical problem (and not a problem in the physical model). It occurs due to a small leakage along the tunnel. This is understandable if one considers a duct consisting of many rings. In each ring the boundary condition is violated because of the discretization, namely mass leaves the duct. Therefore, the wave running through the duct looses amplitude in every ring and far insight the tunnel so much mass got lost that the error is quite large. A justification for this argument is shown in [10]. It is obvious that the error can be reduced by using finer grids. But this would mean much more numerical effort. The grid is kept here because this error is not a large problem: in the end the quality of a tunnel portal with respect to the strength of the tunnel boom can be evaluated with the help of the pressure profile at a point near the tunnel entry. In addition in a non-linear theory (as the common fluid dynamic is) there are effects like wave steepening that are not included in the linearized theory. Therefore the wave propagation along the tunnel anyway cannot be sufficiently described by the linear theory.

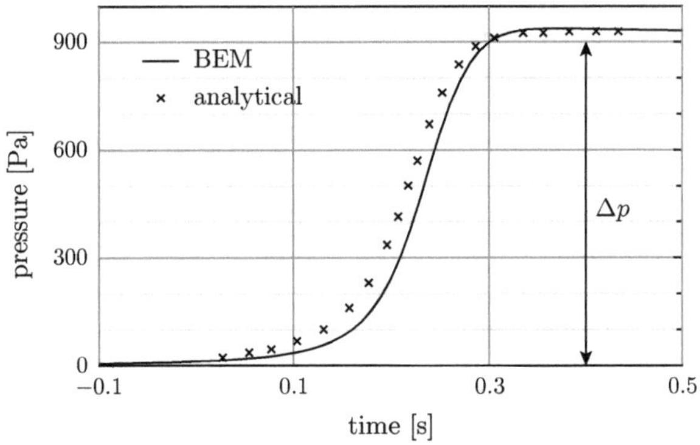

Fig. 8 Pressure-time dependence far inside the tunnel ($x_0 = 16R$) in comparison with analytical result (Howe) for $M = 0.2$ and $\widetilde{A_0}/A = 0.15$

4.3 Comparison with Analytical and Experimental Results

As a first quality check of the BEM results they are compared to analytical results which are based on the same governing equations. Michael Howe has treated the train-tunnel interaction by using compact Green's functions, i.e. wavelength are assumed to be large in comparison with the size of the tunnel portal.[8] With this method a pressure profile far inside the tunnel is calculated for the same model of train and tunnel as presented here. He has shown that up to $M = 0.4$ the linear theory fits roughly to experiments with a generic geometry of train and tunnel (see [7]). In Fig. 8 his results (taken from [5]) are compared with the BEM results (again at $M = 0.2$, $\widetilde{A_0}/A = 0.15$). The BEM results seem to fit quite well but there are some differences. The first visible discrepancy is the beginning of the pressure rise. In Howe's results the pressure seems to rise earlier. The reason for this could be that for Howe's tunnel the boundary condition is fulfilled at both sides of the wall. Therefore a different behavior of the wave generation is possible. In Sect. 5 the influence of a different condition at the entry is discussed.

Another discrepancy is the pressure amplitude Δp. The pressure is calculated at $x_0 = 16R$. Reviewing Fig. 7 it gets obvious that at this point the pressure amplitude in BEM already decreased about 200 Pa due to the discretization error. Therefore the pressure amplitude from BEM is higher than that of the analytical calculations. Howe derives

$$\Delta p = \frac{\rho_0 u_0^2}{1 - M^2} \frac{A_0}{A} \left(1 + \frac{A_0}{A}\right) \tag{10}$$

for the pressure amplitude if $M < 0.4$ and $\frac{A_0}{A} \ll 1$ (shown in [5]). In Fig. 9 the BEM results for $x_0 = 10R$ for different train speeds and train cross-sections are

---

[8]Remembering that for the BEM wavelength large in comparison with the size of the boundary elements have to be assumed, this is a much stronger restriction.

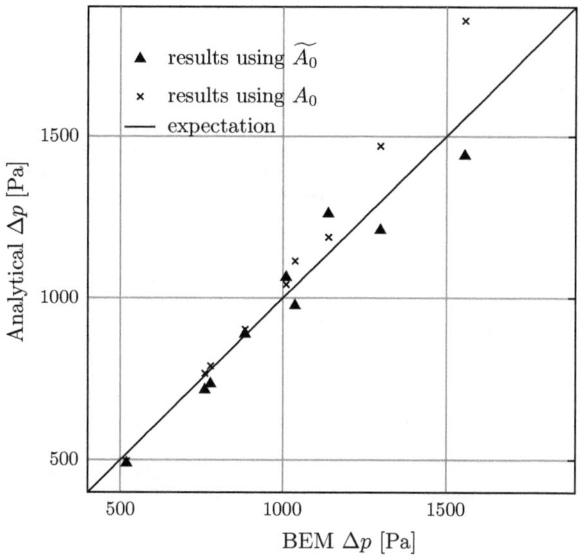

Fig. 9 BEM results for pressure rise Δp in comparison with analytical result (Howe) for A_0 and $\widetilde{A_0}$

compared to the according results when using the formula. There are two cases. In the first case the deformed cross-section $\widetilde{A_0}$ is used in the formula and in the second case the free field cross-section A_0 is used. It is observed that the BEM results fit quite well to the analytical results. For larger Δp the results using $\widetilde{A_0}$ seem to fit better. It is therefore concluded that just the deformed train is responsible for the wave generation. It should be mentioned that the deformation also occurs in Howe's theory but is neglected there.

As Howe's calculations are based on the same linear equations a comparison with experimental fullscale results taken from [1] is interesting. The parameters are $M = 0.2$, $\widetilde{A_0}/A = 0.15$ and $A = 10.6\,\mathrm{m}^2$. The tunnel in the experiment has two tracks, therefore the entry is not symmetric and the measurements are on the trainside tunnel wall. Nevertheless in the simulation symmetry is assumed to be able to use the deformed cross-section (which is then also symmetric). Again the pressure-time dependence is measured for different x_0 inside the tunnel. In Fig. 10 experimental results and BEM results are compared. It is observed that in spite of the (in parts) large differences the main characteristics of the experimental results can be found also in the BEM results. Recalling that here a very simplified numerical method based on a linear theory with a deformable train is compared to fullscale results for a special tunnel geometry and a solid train with special shape it is concluded that the results fit quiet well. This gives reason to continue the development of the simplified model.

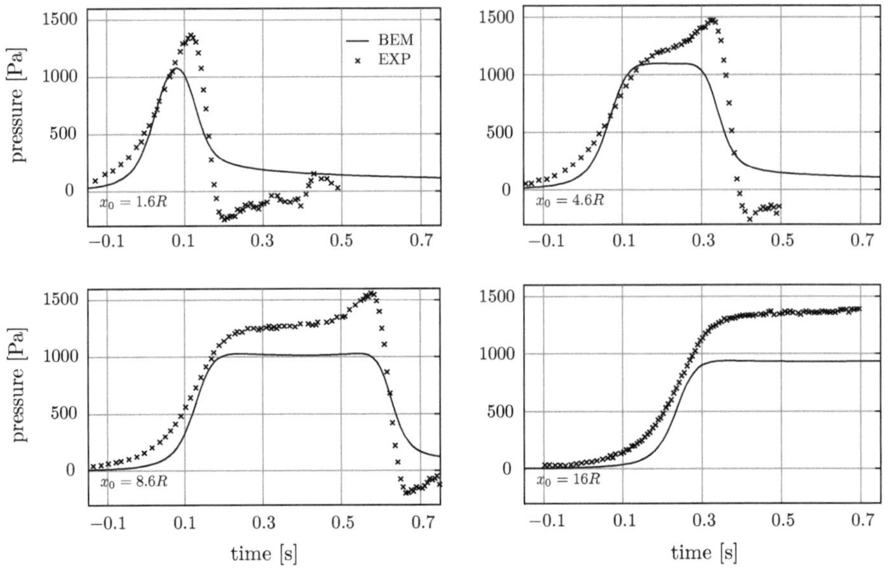

Fig. 10 Comparison with experimental results at different positions x_0 inside the tunnel for $M = 0.2$, $\widetilde{A_0}/A = 0.15$

5 Tunnel with Quadratic Cross-Section and Flanged Portal

A first step to modify the model of the tunnel is a change of the boundary condition in the front section of the portal. Therefore the influence of a front wall (attached to the duct at the side where the train enter the tunnel is investigated. As the analytical solutions for rectangular elements should be used, again it is necessary to have a grid made of rectangulars. Since this is not easy as long as the tunnel has a cicular cross-section, it is necessary to change the shape of the cross-section from circular to rectangular (or here quadratic as a special case). The change to a quadratic cross-section has nearly no influence of the pressure-time dependence inside the tunnel. This is shown as an example in Fig. 11. Compared are the two different shapes for the same size of the cross-section, namely $\pi R^2 = A_{circ} = A_{quad} = d^2$. Used is $\widetilde{A_0}/A = 0.15$ at $M = 0.2$ and the point for comparison is $x_0 = 5d$. The integral deviation for the shown time interval is approximately 2 %. It is therefore concluded that it is reasonable to investigate the influence of the front wall for the quadratic cross-section and afterwards conclude that the influence should be comparable for a circular cross-section.

In Fig. 12 the pressure variation in case of an attached front wall with different size is shown. The other parameters are taken from the test above. Obviously the front wall and its size strongly influences both the pressure gradient and the maximal pressure inside the tunnel. This is reasonable because of a completely different behavior of the fluid leaving the duct. When the train enters the tunnel the compression wave travels

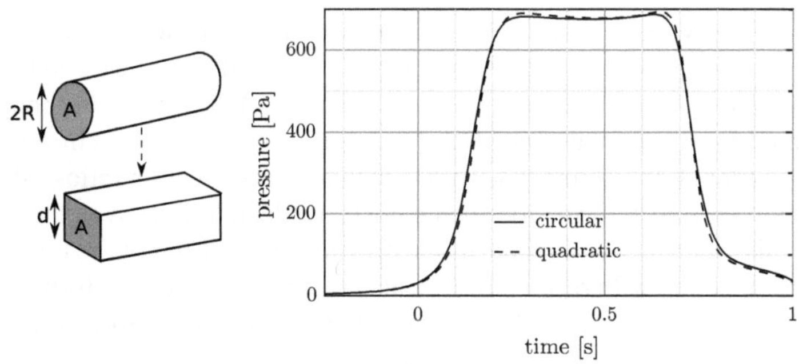

Fig. 11 Comparison between circular and quadratic cross-section (example)

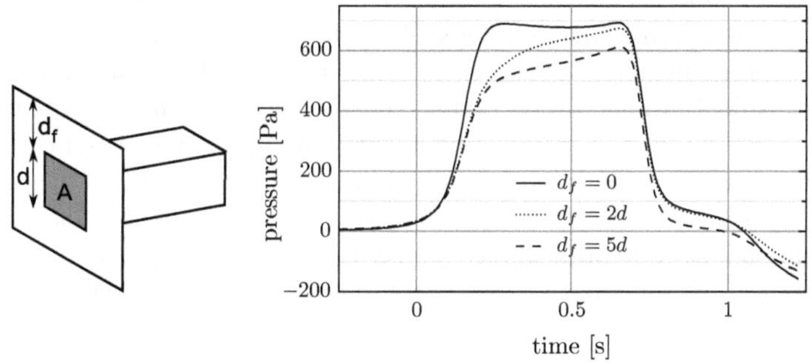

Fig. 12 Comparison between different sizes of the front wall ($x_0 = 5d$, $M = 0.2$, $\widetilde{A_0}/A = 0.15$)

in the same direction as the movement of the train. But some fluid is also pressed in the opposite direction outside the tunnel. By the added boundary condition the fluid is forced to change its direction to be parallel to the frontwall. This is not physical in a not-linear world since there vortices will be generated and the flow separates. From this point of view the somehow less correct boundary at the front fits better to the reality as the comparison with the experiment above shows.

6 Conclusions and Outlook

A boundary element method has been developed to investigate the wave propagation caused by a train entering a tunnel. The instationary situation is described in a linear theory with simple models of the train and the tunnel. The objective was to keep the tunnel geometry as general as possible. As the BEM is still under development just first tests with generic tunnel geometries have been performed to check the quality of the simple model by comparison with analytic results (based on the same linear theory) on the one hand, and fullscale experimental data on the other hand.

The train is modelled by a single mass source of monopole type which defines a deformable semi-infinite streamline body. No additional boundary condition has to be fulfilled on the train. In a first step the deformation of the streamline body induced by the tunnel has been investigated in dependence of the cross-section ratio of the train and the tunnel. It is found that even for a small cross-section ratios (like in usual two track tunnels) the deformation should not be neglected. To answer the question in which way the deformation influences the pressure rise in front of the train inside the tunnel, a comparison with results from an analytic calculation is performed. It shows that for large train velocities or large cross-section ratios the deformed train has the cross-section which fits to the analytic results. Still the influence of the deformation process is not completely understood, but it is obvious that there is one.

Also the comparison with analytic and experimental results is done with respect to the pressure-time dependence at different positions inside the tunnel. Looking at the BEM results reveals that mass is lost along the tunnel. It can be shown that this is a numerical problem and is not caused by the model itself. Comparison with analytic results far inside the tunnel show good agreement apart from the time of the pressure rise. The reason might be a different boundary condition in the analytic calculation. In BEM the boundary condition is fulfilled just on the inner side of the tunnel wall (and not inside and outside as in the analytic calculation). To check the influence of a different boundary condition at the entry, a front wall is attached to the tunnel. The investigation of its influence has just begun but it seems to be clear that it strongly influences the pressure wave inside the tunnel. Even the size of the front wall seems to be important. Interesting is that the comparison with experimental data reveals a better agreement without a frontwall, i.e. the curious construction with inner space passing into outer space. But all in all the comparison with experimental data shows some discrepancies. Nevertheless the main characteristics of the pressure behavior can be reproduced. It should be pointed out that effects related to the viscosity of the fluid or the non-linearity of the equations can not be reproduced with the simple model. Therefore special experiments with generic geometries should be performed to investigate the influence of the geometrical differences on the BEM results.

It should be emphasized that the investigation of the train-tunnel entry in this simple model are still the beginning. There are a still a lot of open questions and issues which justify to continue this work. These are numerical issues on the one hand, like further development of the solver to make it more effective and to make general geometries possible. On the other hand the investigation of the quality of the model has just begun. Special experiments for comparison would deliver more information about how well the simple model can describe reality. In addition it might be useful to integrate special modifications afterwards like Howe does ([7]).

As a conclusion the combination of linearized equations solved by a boundary element method seems useful to describe in a simplified way the wave propagation during train tunnel entry as the main characteristics from experiment and theory can be reproduced from the current point of view.

References

1. Bellenoue, M., Morinière, V., Kageyama, T.: Experimental 3-D simulation of the compression wave, due to train-tunnel entry. J. Fluid. Struct. **16**, 581–595 (2002)
2. Davey, K., Hinduja, S.: Analytical integration of linear three-dimensional triangular elements in BEM. Appl. Math. Model. **13**, 450–461 (1989)
3. Eckard, C.: Vortices and streams caused by sound waves. Phys. Rev. **73**, 68–76 (1948)
4. Howe, M.S.: The compression wave produced by a high-speed train entering a tunnel. Proc. R. Soc. Lond. **454**, 1523–1534 (1998)
5. Howe, M.S.: Review of the theory of the compression wave generated when a high-speed train enters a tunnel. Proc. Instn. Mech. Engrs. **213**(part F), 89–104 (1999)
6. Howe, M.S.: Hydrodynamics and Sound. Cambridge University Press, Cambridge (2007)
7. Howe, M.S., Iida, M., Fukuda, T., Maeda, T.: Theoretical and experimental investigation of the compression wave generated by a train entering a tunnel with a flared portal. J. Fluid Mech. **425**, 111–132 (2000)
8. Pierce, A.: Acoustics: An Introduction to its Physical Principles and Applications. Acoustical Society of America (1989)
9. Schulte-Werning, B., Grégoire, R., Malfatti, A., Matschke, G. (eds.): TRANSAERO: A European Initiative of Transient Aerodynamics for Railway System Optimisation. Springer, Berlin (2002)
10. Wagner, S.: Potentialtheoretische Modellierung instationärer kompressibler Strömungen mit komplexen Randgeometrien (diploma thesis under review). Georg-August Universität Göttingen, III. Phys. Inst. (2010)

Investigations of Aeroacoustics of High Speed Trains in Wind Tunnels by Means of Phased Microphone Array Technique

Andreas Lauterbach, Klaus Ehrenfried and Sigfried Loose

Abstract The present study focuses on the analysis of the main aeroacoustic sound sources of a high speed train, measured in a wind tunnel. The experiments using a 1:25 Inter City Express 3 model were carried out in two different wind tunnels: The Aeroacoustic Wind tunnel (AWB) of the German Aerospace Center (DLR) in Brunswick and in the Cryogenic wind tunnel (DNW-KKK) of the DNW (German Dutch wind tunnels) in Cologne. The AWB is a Goettingen type wind tunnel with open test section which is surrounded by an anechoic chamber. The advantage of this facility is its low background noise level and its nearly anechoic test section. The maximum Reynolds number, based on the wind speed and the width of the train, achieved is 0.46 million. In order to obtain higher Reynolds numbers a second measurement campaign has been conducted in the cryogenic wind tunnel, using another array for cryogenic in-flow applications. The DNW-KKK enables higher Reynolds numbers up to 3.7 million by cooling down the fluid to 100 K. The DNW-KKK has a closed test section and the microphone array is mounted on a side wall inside the wind tunnel, and therefore the measurements are affected by the turbulent boundary layer. Drawback of this facility is that it is not optimized for aeroacoustic experiments and reflexions as well as the high background noise level can disturb the results. Differences of the two different experimental setups on the results and primarily, influence of the Reynolds number on the aeroacoustic of a high speed train will be discussed.

1 Introduction

Airplanes as well as any kind of ground vehicles produce aerodynamic noise which normally depends on the velocity. Especially trains became faster in the last few decades. Modern High speed trains allow operational speeds up to 350 km/h and

A. Lauterbach (✉) · K. Ehrenfried · S. Loose
Department Fluid Systems,
Institute of Aerodynamics and Flow Technology, German Aerospace Center (DLR),
Bunsenstrasse 10, 37073 Goettingen, Germany
e-mail: andreas.lauterbach@audi.de

A. Dillmann and A. Orellano (eds.), *The Aerodynamics of Heavy Vehicles III*,
Lecture Notes in Applied and Computational Mechanics 79,
DOI 10.1007/978-3-319-20122-1_11

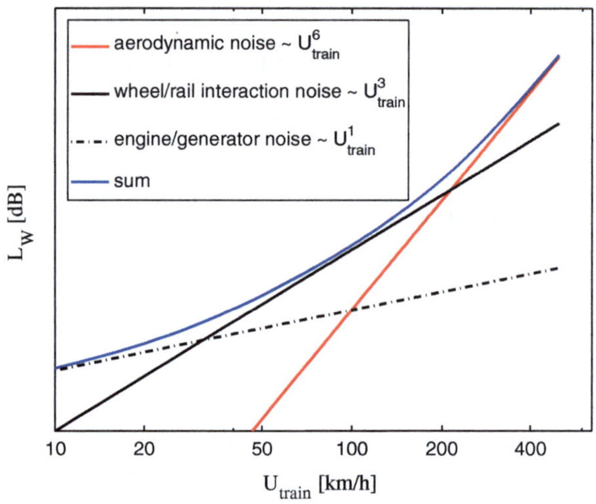

Fig. 1 The relation between sound pressure levels of different sources of sound and the velocity of the train, from [9]

for this range of velocities aerodynamic noise is dominating and exceeds all other sources of sound like engine/gearbox noise, noise from aggregates, general noise from the bogies and interaction between wheel and rail. Figure 1 (from [9]) depicts the relations between sound pressure levels of different sources of sound and the velocity of the train.

In general, traffic noise is a problem for the residents which live close to the traffic route and decrease the comfort for the passengers inside the vehicle. For the reduction of the noise the detailed knowledge of the distribution and the properties of the sound sources is necessary. In the field of acoustics of high-speed trains both, investigations on the full-scale vehicle as well as testings in wind tunnels on models has been done. Full-scale tests have the advantage, that the measurement can be conducted under real conditions and appropriate Reynolds numbers. The disadvantage is that these testings only can be performed when the train is already in operation. It is not possible to analyse acoustics during the design process. Nevertheless, most of the earlier publications report on full-scale testings and have their focus on the so-called wheel-rail noise (see Barsikow et al. [5, 6]). This kind of noise is generated by the mechanical interaction between the wheels and the rails. It depends on the conditions of rails and wheels like roughness or damping features of the bogies and the driving speed, of course. As already mentioned, at higher velocities aerodynamic noise prevails and in the last years aerodynamic noise became more and more the focus of interest. At the same time microphone array technique became an effective tool for the localisation and quantification of sound. This technique allows a detailed study of wheel-rail noise and aerodynamic noise, like noise from the pantograph, noise from the gaps between two cars and noise from components like antennas, and so on (see Martens et al. [19]).

To predict the aeroacoustic properties during the design process, wind tunnel testings on down-scaled models are fundamental. Yamazaki et al. [26] used beamforming techniques for the investigation of aerodynamic noise of a simplified train model in the wind tunnel. The scale of the model is 1:8 and a maximum Reynolds number of about 2 mio can be achieved. They focused on noise of the bogie cavities and the gaps between the coaches. Based on this measurements, modifications for noise reduction were applied. Afterwards the noise reduction arrangements were verified qualitatively in a wind tunnel as well as by full-scale testings. Other wind tunnel studies just point out investigations of parts of trains. One of the main topics is the acoustic of the pantograph (see [14, 21]).

Still an open issue is a quantitative aeroacoustic study of the whole train, considering a model in a wind tunnel. Desirable is an identification of all aeroacoustic sources, a validation of the aeroacoustic measurement in order to allow a comparability to full-scale testings. But the influence of the model scale and the Reynolds number on high speed train aeroacoustics is still an open issue.

The aim of the paper is to do a first step for the development of aeroacoustic scaling laws for model train measurements. First, the most important sources of sound need to be identified. In the next step the influence of the Mach and the Reynolds number are examined.

Therefore, measurements on a 1:25 scaled model of the Inter City Express 3 (ICE3) in two different wind tunnels have been conducted. The main aeroacoustic sources are identified in the Aeroacoustic Wind tunnel of the DLR in Brunswick (AWB). This wind tunnel provides nearly perfect conditions due to its low background noise level and its almost anechoic test section. The drawback is the limitation of the Reynolds number achievable in this facility. In order to obtain higher Reynolds numbers, which is important for the investigation concerning aeroacoustic scaling, a second measurement under cryogenic conditions is presented. By cooling down the fluid one can increase the Reynolds number by a factor of $\sim$5. Cooling does not only effect the aerodynamics. To allow a quantitative comparison of measurements acquired at different temperatures it is important to be aware of the effect of temperature on acoustics. This will be discussed in this paper as well.

2 Methods

2.1 Aerodynamic Scaling

The Reynolds Number is an important non dimensional parameter in the scope of aerodynamic scaling and can be seen as the ratio of inertia forces and viscous forces. This coefficient can describe the state of the flow (e.g. whether it is laminar or turbulent):

$$Re = \frac{U_\infty \cdot L \cdot \rho}{\mu}. \tag{1}$$

U_∞ denotes the free stream velocity, ρ the density, L a characteristical length and μ the dynamic viscosity. Independent of the scale of the model, for a constant Reynolds number one can expect the same flow topology as long as the shape of the model is the same. This holds as long as compressibility effects are not relevant and the Mach number is small enough:

$$Ma < 0.3. \tag{2}$$

The Mach Number is defined by the free stream flow velocity U_∞ divided by the speed of sound c:

$$Ma = \frac{U_\infty}{c} \tag{3}$$

The speed of sound c is a function of temperature T. For ideal gases the relation is:

$$c(T) = \sqrt{\kappa \cdot R \cdot T} \tag{4}$$

The adiabatic exponent $\kappa = 1.4$ is constant for diatomic gases. The top plot in Fig. 2 depicts the relation between speed of sound and temperature. Also the density of the fluid is a relevant parameter for the Reynolds number which is a function of temperature. The ideal gas law provides the relation between density, pressure and temperature:

$$\rho(T) = \frac{p}{R \cdot T} \tag{5}$$

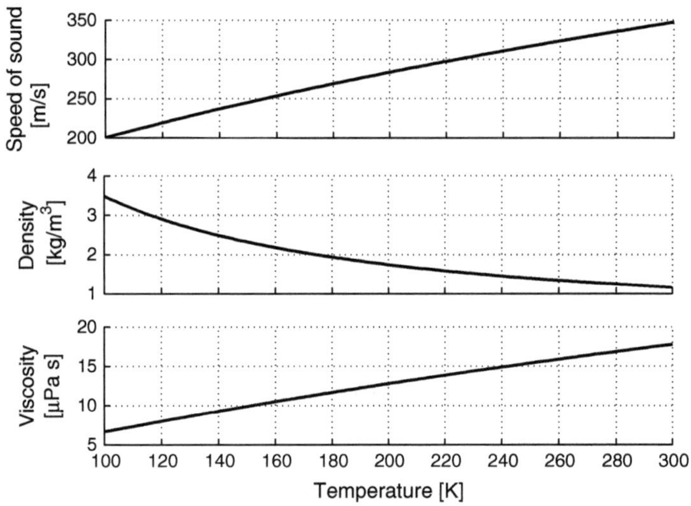

Fig. 2 Temperature dependence of the Reynolds number relevant parameters speed of sound, density and viscosity for diatomic gases

Table 1 Constants for Sutherland's formula for air and nitrogen

Gas	C (K)	T_0 (K)	μ_0 (1×10^{-6} Pa · s)
Air	120	291.15	18.27
Nitrogen	111	300.55	17.81

For a constant pressure the relation between temperature and density is shown in Fig. 2, the middle plot. Furthermore, the dynamic viscosity is an important parameter of the Reynolds number which is also a function of temperature. It can be computed using Sutherland's formula:

$$\mu(T) = \mu_0 \cdot \frac{T_0 + C}{T + C} \cdot \left(\frac{T}{T_0}\right)^{3/2} \tag{6}$$

C, μ_0 and T_0 are constants and depend on the medium. Table 1 provides all constants for nitrogen and air. The temperature dependence of the viscosity is sketched in Fig. 2, bottom plot.

Using Eqs. 4–6 the Reynolds Number (1) can be written as

$$Re = \frac{c(T) \cdot Ma \cdot L \cdot \rho(T)}{\mu(T)} = \frac{\sqrt{\frac{\kappa}{R}} \cdot Ma \cdot L \cdot p \cdot (T + C) \cdot T^{-2}}{\mu_0 \cdot (T_0 + C) \cdot \left(\frac{1}{T_0}\right)^{3/2}} \tag{7}$$

Due to its temperature dependency the density and the viscosity will increase the Reynolds number with decreasing temperature. In contrast, the speed of sound will lead to a decrease but this effect is over—compensated by density and viscosity effects.

Finally, the relation between Reynolds number, Mach number and temperature can be expressed as follows:

$$Re \propto Ma \cdot (T + C) \cdot T^{-2}. \tag{8}$$

In cryogenic wind tunnels it is appropriate to specify the temperature and the Mach number. The Reynolds number versus Mach number for different temperatures is depicted in Fig. 3a. The computation is valid for a 1:25 scaled model, with a characteristic length of $L = 0.12$ m, based on the width of the train, in a nitrogen atmosphere. By cooling down from room temperature to $T = 100$ K one can increase the Reynolds number by a factor of about 5.

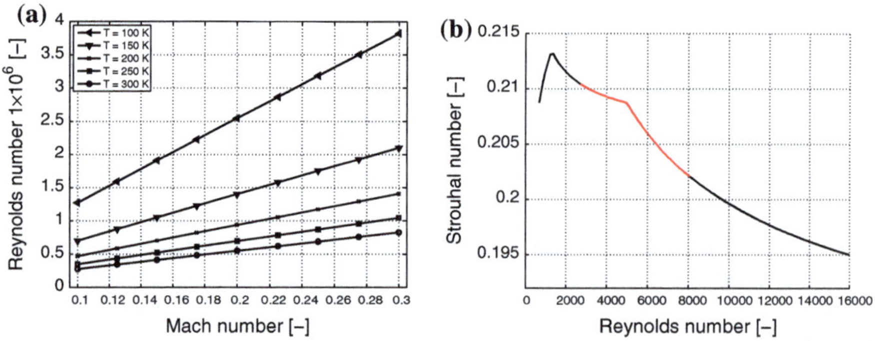

Fig. 3 **a** Renumber versus mach number for different temperatures of the fluid. **b** Strouhal number as a function of Reynolds number. The *red* span corresponds to the Reynolds number range of a cylinder with $L = 2\,\text{mm}$

2.2 Aeroacoustic "cryo"-Scaling

The effect of cooling down on the aerodynamics is discussed in the previous section. Here, the influence of temperature on the acoustics is focused. Ahlefeldt et al. [1] carried out measurements under cryogenic conditions on a cylinder in cross flow. This well known aeroacoustic sound source was chosen in order to gain experiences using the microphone array technique under such drastic conditions with temperature as lower as 100 K. Later, this technique has been used for a more complex source, a half model of the airplane Dornier 728 (see [2]).

Mainly, acoustics can be influenced in two ways: in amplitude and in observed frequencies. First, some aspects concerning observed frequencies are pointed out. As demonstrated later, there are aeroacoustic sources where the frequency is not influenced by the flow velocity within the regarded velocity range at a constant temperature. It seems that the wavelength λ is directly connected to a typical length scale or a volume. This behavior can be described by cavity mode excitation. For such types of sound sources the frequency will be a function of the temperature, because the speed of sound c changes (Eq. 4).

$$f = \frac{c(T)}{\lambda}. \tag{9}$$

For the comparison of data acquired at different temperatures this effect must be considered. In this particular case it is reasonable to use a dimensionless frequency which is nondimensionalized by the quotient of the speed of sound and a characteristically length:

$$f_{norm} = \frac{f \cdot L}{c} \tag{10}$$

This formulation is very similar to the well known Strouhal number:

$$Sr = \frac{f \cdot L}{U_\infty}. \tag{11}$$

The Strouhal number is a common coefficient to describe another type of sound sources which show a linear dependence between frequency and flow velocity. A famous example for this situation is the Kármán's vortex street which develops behind a cylinder in cross flow. The frequencies in the wake only show a weak Reynolds number dependence. Figure 3b depicts this dependence in terms of the Strouhal number. The data are based on the work from Fey et al. [11]. For a wide range of the Reynolds number the Strouhal number is within an interval of $Sr = 2 \pm 10\,\%$. Later the Strouhal number is used as dimensionless frequency for analysing frequency spectra of the pantograph, because the situation is comparable to a cylinder in cross flow: the pantograph consists of several cylindrical elements of different orientations and diameters.

Also the amplitude can be influenced by the temperature, which is exemplarily shown for an well-known aeroacoustic source, the cylinder in cross flow. Lighthill's aeroacoustical analogy [18] laid the fundamentals of the prediction of aeroacoustic sound generation of a free flow. Curle [8] extended Lighthill's equations to account for solid boundaries. Based on Curle's work, Phillips [22] deduced an equation, which enables the prediction of the intensities of the Aeolian tones emitted by a cylinder in cross flow. The assumption is made, that the fluctating aerodynamic force acting upon the cylinder, which can be seen as the origin of the sound generation, is proportional to the dynamic pressure. Phillips found:

$$\bar{p}^2 \propto A Sr^2 \rho^2 c^4 Ma^6. \tag{12}$$

The square of the acoustic pressure depends on the 6th power of the Mach number, on the 4th power of the speed of sound and on the square of the Strouhal number. The variable A is introduced as a factor which combines Reynolds number dependent effects, namely the influence of different correlation length along the cylinder and the dependence of the flow conditions on the fluctuating lift force. The correlation length will decrease with increasing Reynolds number, because three-dimensional instabilities will rise and phase deviations along the cylinder will increase. The variable A is constant, as long as there is no Reynolds number effect on the sound generation. The question is, in how far the observed sound pressures of the dipole source are influenced, only by changing the temperature. For both paramters one can find the following dependencies in literature:

$$\rho(T) = \frac{p}{R \cdot T} \Rightarrow \rho^2 \propto \frac{1}{T^2} \tag{13}$$

$$c(T) = \sqrt{\kappa \cdot R \cdot T} \Rightarrow c^4 \propto T^2. \tag{14}$$

The pressure p, as well as the specific heats κ are set constant here. A dimensional analysis reveals, that the effects of varying the temperature on density and speed of sound compensate each other, as long as results are discussed in terms of the Mach number for a source mechanism, which has no Reynolds number dependence. This scaling concept can be seen to be valid also for other sources, which can be be described as "force-sources" with dipole characteristic, where the sound generation is attributed to a fluctuating aerodynamic force. Furthermore, all results discussed in this paper are presented in terms of sound power levels in order to allow a comparabillity of data acquired at different fluid temperatures.

2.3 Beamforming

For the sound localistion and quantification in aeroacoustic experiments the Delay and Sum Beamforming method with a microphone array is applied. The principle is based on the summation of retarded microphone signals. For more details see [13]. Initially, standard beamforming in the frequency domain was performed.

$$A = \frac{\mathbf{e}^{\dagger}\mathbf{W}\mathbf{R}\mathbf{W}^{\dagger}\mathbf{e}}{M^2} \tag{15}$$

Thereby, **R** denotes the cross-spectral matrix, **W** a weighting matrix, M the number of microphones and **e** the steering vector. $(.)^{\dagger}$ indicates the complex conjugate and transposed vector or matrix. Especially in noisy surroundings one can achieve a higher signal-to-noise ratio by subtracting the diagonal elements [20], the auto spectra, of the cross-spectral matrix.

In order to improve the spatial resolution of the beamforming algorithm especially at lower frequencies and to enable more quantitative results the CLEAN SC algorithm introduced by Sijtsma [24, 25] is used. Reflections as well as side lobes of the point spread function, which can appear as ghost sources are subtracted.

2.4 Phase Calibration

Besides other data the Delay and sum Beamforming algorithm requires the accurate positions of the array microphones. The microphone array which is used for out of flow applications (see Fig. 5a) is a filigree construction in order to avoid strong interaction with the sound field. The accurate positions of the microphones are not given by the setup itself. To achieve the required accuracy it is necessary to measure the exact positions. Therefore a calibration procedure has been developed which works similar to the well known global positioning system (GPS): Several acoustic "satellites" and a reference microphone enable the localisation of the array microphones with an accuracy better than 1 mm. For more details see Lauterbach et al. [17].

2.5 Bi Clean

Measurements in closed test sections may be disturbed by reflections, strong background noise and mode excitation. The Bi Clean algorithm, introduced by Ehrenfried et al. [10, 16] detects plane waves in the array data, which may belong to wind tunnel modes or which have their origin far away from the test section. This contribution is filtered out, what improves the signal-to-noise ratio thereby.

2.6 Flow Correction for Closed Test Sections

The cryogenic wind tunnel is a Goettingen-type wind tunnel with closed test section. The microphone array is mounted on a side panel inside the wind tunnel boundary layer. From the origin of the sound in the centre of the test section to the microphones the sound gets convected by the flow. A correction to the steering vector is applied which assumes a uniform flow to take the modification of travel time due to the convection into account. For more details see Koop [15].

2.7 Shear Layer Correction for Open Test Sections

In the aero-acoustic wind tunnel the array is positioned outside the flow and the sound which is emitted by the model has to propagate through the wind tunnel shear layer. Phase shifts and variations in amplitude, induced by refraction on the shear layer and convection of the sound wave by the flow, are corrected according Amiet [3, 4]. This correction is applied by a modification of the steering vector.

2.8 Focused Spectra

Besides noise maps, which map the distribution of sound sources, an integration technique described by Brooks et al. [7] is applied which enables the computation of sound pressure level spectra for specified scan areas.

3 Acoustic Measurements

For all experiments described in this paper a 1:25 scaled Inter City Express 3 (ICE3) model is used. It is designed for aerodynamic testings and therefore it has a low level of itemisation, which is common for such type of models. For aerodynamic testings

Fig. 4 The 1:25 scaled ICE3 model. Relevant details are the bogies, the gap between the head car and the first coach and the pantograph

the shape of the model is more important than every single detail. For aeroacoustics this may differ, and single components like antennas, cooling intakes or equipment on the roof are important sources of sound. Nevertheless, the main important details are reproduced, namely the bogies, the pantograph and the gap between head car and first coach. For all considerations described later concerning the Reynolds number, as characteristically length scale the width $L = 0.12\,\text{m}$ of the train is chosen (Fig. 4).

3.1 Measurement in the Aeroacoustic Wind Tunnel (AWB)

At first the main aeroacoustic sources are identified by measurements carried out in the Aeroacoustic Wind Tunnel facility (AWB) of the German Aerospace Center (DLR) in Brunswick [23]. It is an closed circuit Goettingen type wind tunnel with open test section. The whole air duct is lined with noise absorbing foam and the test section is sourrounded by an anechoic chamber. The nozzle diameter is $1.2 \times 0.8\,\text{m}$ and the maximum wind speed is $U_\infty = 65\,\text{m/s}$. Figure 5b depicts the facility and Fig. 5a shows the setup in the test section. The train model is installed on a splitter plate, which has an elliptical leading edge and a sharpened trailing edge. This plate is positioned 10 cm above the lower edge of the nozzle in order to peel off the wind tunnel shear layer. At the leading edge of the splitter plate a new boundary layer is

(a) **(b)**

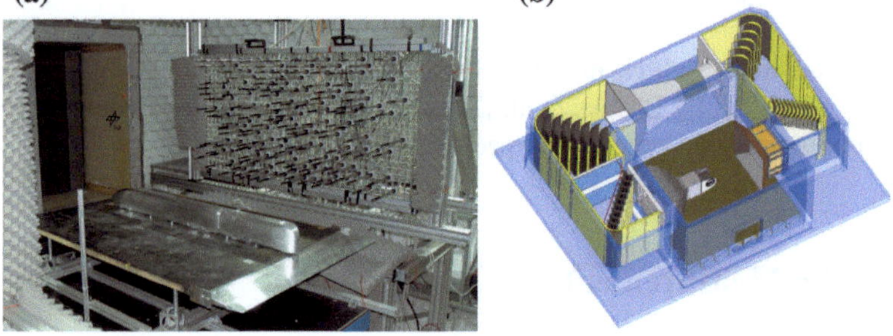

Fig. 5 Overview of the setup and the wind tunnel. **a** Setup in the AWB: inside the test section on a splitter plate the model of the ICE3 is installed. In the background outside the flow the microphone array is mounted. **b** The AWB. The test section is highlighted by the *dark green* region

formed which is thinner than the wind tunnel boundary layer. The aim is to keep the thickness of the boundary layer thin in order to ensure a reasonable simulation of the flow underneath the train. In the wind tunnel experiment one expects a Poisseuille flow between splitter plate an train underframe. In contrast to that in the full-scale world the train penetrates the fluid, which is at rest and and a turbulent Couette flow develops. Only a moving belt can provide a better simulation of the full scale world in the wind tunnel, or maybe a mirror model (see Grunwald [12]). However, experiences have shown that the setup on the splitter plate is a good compromise in the sense that it leads to reasonable results, at least for aerodynamic testings.

Outside the flow beside the splitter plate the microphone array is installed. The sound propagates through the flow and through the wind tunnel shear layer. For the evaluation of the array data a shear layer correction is necessary, as described in the previous Sect. 2.7. The layout of the 143 microphones is numerically optimized in order to achieve a broadband design up to 30 kHz.

Figure 6 shows source maps for three different one third-octave bands with mid—frequencies of 2.0, 3.15 and 5 kHz, measured at a flow velocity of $U_\infty = 60$ m/s. The sound pressure level in dB is colour-coded. Obviously, for the different frequency bands different source mechanisms are active. For the lower frequencies the third bogie is the main aeroacoustic source, but for frequencies higher than 5 kHz the pantograph becomes dominant. Also, the gap between head car and first coach and the other bogies are visible in the source maps. Noticeable is that the first bogie is a much stronger source compared with the other bogies. This can be explained by the characteristics of the boundary layer: at the head of the train close to the stagnation point the boundary layer is thin what may lead to a strong excitation of the source mechanism. Furthermore, at the front the ICE3 has a spoiler which guides the accelerated flow into the bogie cavity.

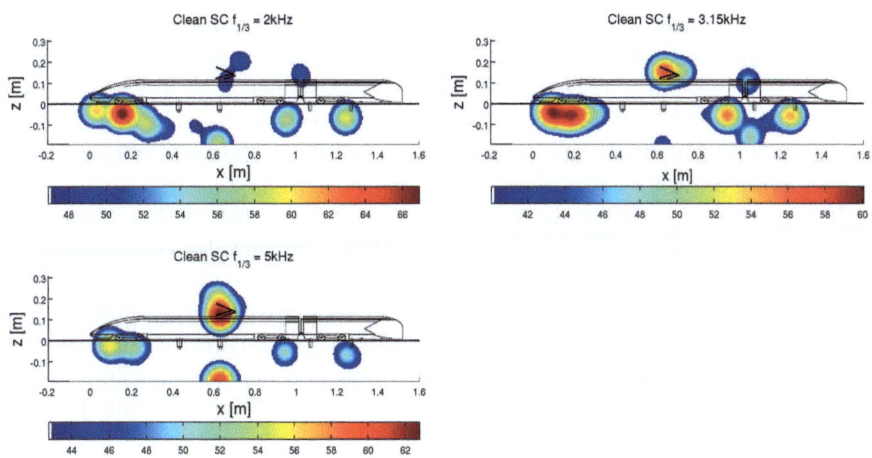

Fig. 6 Third octave source maps of the ICE3, measured in the AWB at $U_\infty = 60$ m/s

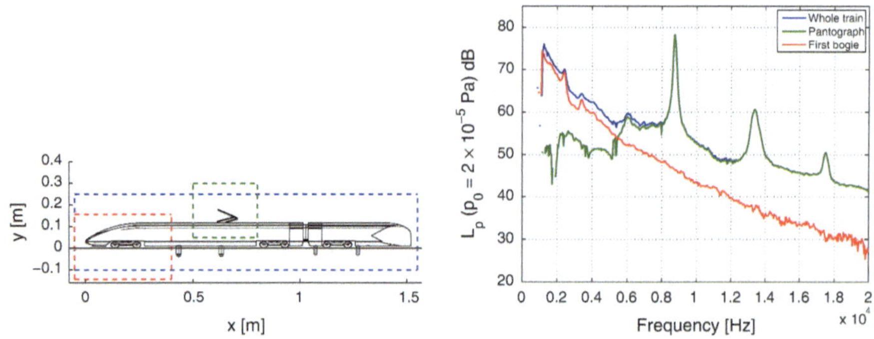

Fig. 7 Focused spectra for three different scan areas: the whole ICE3 train, the pantograph and the first bogie. The Flow velocity is $U_\infty = 60\,\mathrm{m/s}$

A more quantitive representation is given by the focused spectra, depicted in Fig. 7. The three plots show spectra, which belong to the areas of the whole train (blue), the pantograph (green) and the first bogie (red). The spectra show, that for higher frequencies than 5 kHz the pantograph is the main source. The pantograph spectrum has strong tonal components. The spectrum of the first bogie also contains tonal components in the low frequency region <4 kHz, but its overall shape is smoother and the sound pressure level declines more for higher frequencies, comparable to typical turbulence spectra.

In order to investigate the characteristics of the sound genereration of the first bogie and the pantograph, in the next step spectra are shown for different flow velocities between $20 < U_\infty < 60$ m/s, what corresponds to Reynolds numbers between $0.153 \times 10^6 < Re < 0.456 \times 10^6$. Figure 8a depicts spectra of the first bogie. The overall shape of the spectra does not change significantly with increasing flow velocity and the two peaks in $f = 2455\,\mathrm{Hz}$ and $f = 3442\,\mathrm{Hz}$ are not flow velocity depended. Obviously, certain cavity modes are activated and within the treated velocity range no other modes can be excited. Measurements at higher Reynolds numbers are essential to find out if higher acoustic modes can be excited, or whether they persist. At this point is questionable if it is possible to extrapolate these results to the full scale world.

In Fig. 8b the spectra for the same velocity range are plotted. The pantograph noise differs completely from the bogie noise. Here, a strong velocity dependence is observed. Not only the overall level increases, also the shape is velocity dependent. In order to find out if there is a linear dependence between frequency and flow velocity the spectra are plotted over the Strouhal number (Eq. 11), as depicted in Fig. 8c. The pantograph consists of several cylindrical elements of different diameters between 0.1 and 4 mm and different orientations to the flow. As characteristically length a mean diameter $L = 2\,\mathrm{mm}$ is chosen.

It turns out that there is a linear dependence between flow velocity and frequency. In the spectra in Fig. 8c several peaks at different Strouhal numbers appear and the Strouhal numbers are nearly constant for different flow velocities. In some cases the

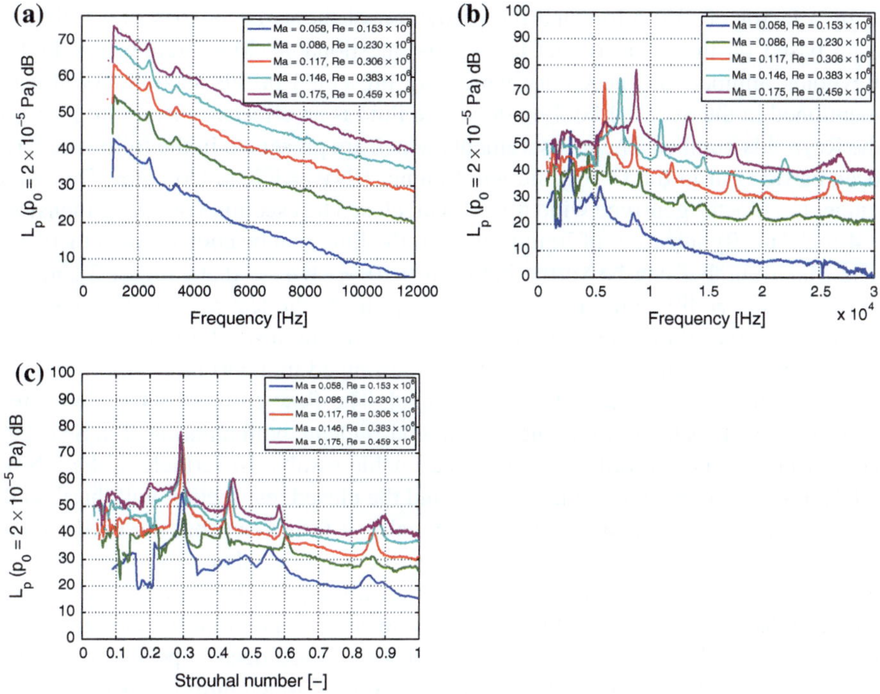

Fig. 8 Spectra for different flow velocities, measured in the AWB. **a** Spectra of the first bogie. **b** Spectra of the pantograph. **c** Pantograph spectra, plotted over strouhal number

Strouhal numbers increase slightly with increasing flow velocity and sometimes it is the other way around. This behavior is in accordance with the Strouhal number—Reynolds number relation which is shown in Fig. 3b. For thin elements of the pantograph the corresponding Reynolds number might be smaller than $Re = 1750$ and therefore within the ascending range of the Strouhal number. Above $Re = 1750$ a increase in flow velocity will lead to a decrease of the Strouhal number. The Reynolds number span for $L = 2\,\text{mm}$ for the treated flow velocities is plotted red.

For both, the aerodynamic induced noise from the first bogie and the pantograph one observes a similar power law between overall sound pressure level and flow velocity. For the first bogie one obtains $L_P \propto U_\infty^{6.22}$ and for the pantograph $L_P \propto U_\infty^{6.12}$.

3.2 *Measurement in the Cryogenic Wind Tunnel (DNW-KKK)*

The measurements in the aeroacoustic wind tunnel AWB described in the previous Sect. 3.1 allow detailed aeroacoustic investigations for Reynolds numbers up to 0.46×10^6. Compared with a full scale train which operates at $Re > 16 \times 10^6$ this is by a factor of 30 to low and the prediction of the aeroacoustics of a full scale train based on this wind tunnel testings is questionable. Measurements at higher Reynolds numbers are necessary which are presented in this section. As already mentioned in the Sect. 2.1 cooling down the fluid is an effective method to increase the Reynolds number. The measurements which are presented in the following have been carried out in the Cryogenic wind tunnel DNW-KKK in Cologne. This wind tunnel is a Goettingen-type wind tunnel with closed test section with the dimensions 2.4×2.4 m. By injection of liquid nitrogen the fluid can be cooled down to 100 K. The Mach number can be varied between $0.10 < Ma < 0.30$. Figure 2 shows a sketch of the facility and a photo of the setup inside the test section. For the same reason as in the AWB, again the model is mounted on a splitter plate. The microphone array consisting of 144 microphones, arranged in logarithmic spiral arms, is mounted on the side wall of the wind tunnel. The flow correction for closed test sections as well as the noise reduction method Biclean, both mentioned in Sect. 2.6, needs to be applied in order to obtain reasonable results. Nevertheless, the DNW-KKK has a high background noise level and the closed test section with hard walls causes a reverberant environment. Additionally, it turnes out that the used splitter plate produces more self noise compared with the splitter plate of the aeroacoustic wind tunnel. Here, all the mountings underneath the plate are exposed to the flow. All these facts lead to a loss of the signal-to-noise ratio and limit the measurements under these conditions. Therefore, in the following third octave spectra are shown. The Fig. 10a, b show spectra of the first bogie for the Mach numbers between 0.10 and 0.30 for two different temperatures $T = 200$ K and $T = 100$ K. Please note that the data are plotted over the nondimensionalized frequency f_{norm} (Eq. 10). For both temperatures we obtain a similar behavior by changing the Mach number. Within the measuring accuracy the sound pressure levels are the same. The dependency between sound pressure level and flow velocity (or equivalent: in terms of the Mach number, as far as the temperature is hold constant for one measurement series) is $L_p \propto U_\infty^6$, in each case for the both temperatures. This result is also in a good agreement with the experiments in the aeroacoustic wind tunnel at lower Reynolds numbers and with a different experimental setup. Further, noise from the bogie does not show a velocity dependence—this effect is also in accordance with the measurements described before (Fig. 9).

For a detailed investigation of the influence of the Reynolds number on the aeroacoustic source mechanism of the first bogie, measurements at a constant Mach number for different temperatures are plotted. That means, a variation of the Reynolds number is done by variation of the temperature of the fluid. The plots at $Ma = 0.2$ and different Reynolds numbers are shown in Fig. 10c. It turns out that there is no strong influence of the Reynolds number. Within the measuring accuracy the spectra lie on top of each other. The differences in the order of magnitude ± 3 dB can be explained by uncertainties due to the restricted signal-to-noise ratio of the measurements under these difficult conditions. There are no clear tendencies of the relation between sound pressure level and Reynolds number.

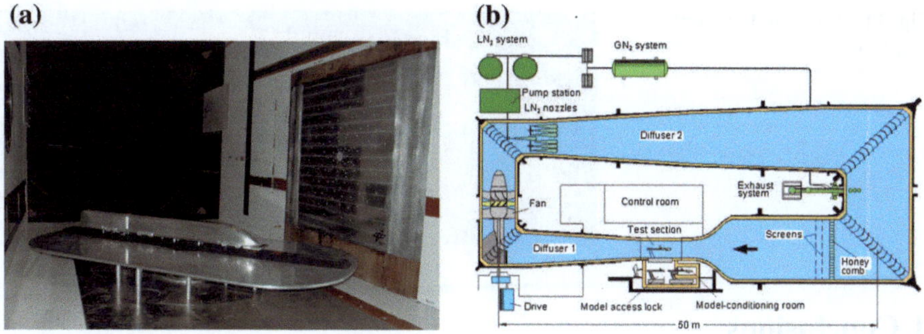

Fig. 9 **a** Setup in the test section of the DNW-KKK. The train model is mounted on a splitter plate. The microphone array is arranged behind, mounted on the wind tunnel wall. **b** Sketch of the DNW-KKK, drawing from www.dnw.aero

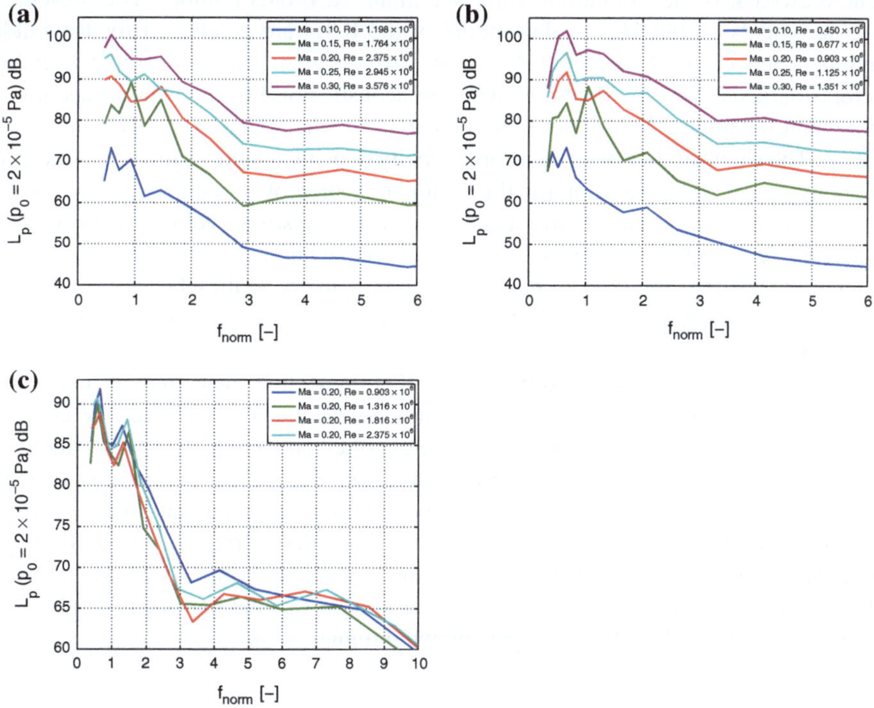

Fig. 10 Spectra of the first bogie, measured in the DNW-KKK facility. **a** T = 200 K constant. **b** T = 100 K constant. **c** Ma = 0.20 constant

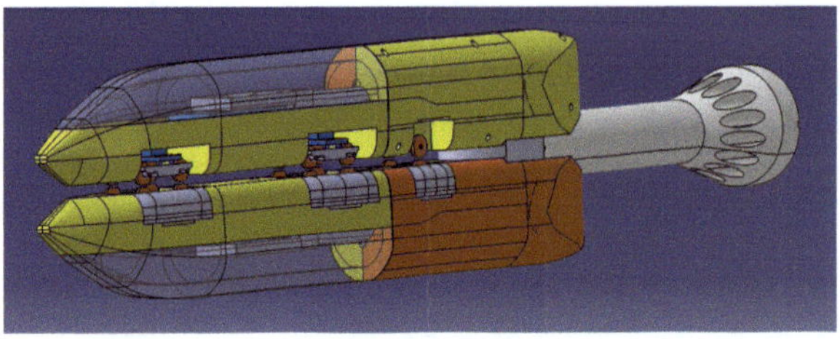

Fig. 11 The double model of a train

4 Conclusions

In this paper two measurement campaigns are presented: The first has been conducted in an aeroacoustic wind tunnel and the second one in a cryogenic wind tunnel. The aeroacoustic wind tunnel provides very good conditions for aeroacoustic purposes. The drawback is the limitation in the maximum Reynolds number. The presented measurements on the 1/25 scaled model of the ICE3 are conducted for Reynolds numbers between $0.15 < Re \times 10^6 < 0.46$. Two different sources of sound with different characteristics are identified: The aeroacoustic noise from the bogie section is dominant for frequencies $f < 6$ kHz and can be characterized by cavity mode excitation. The shape of the spectrum is not velocity dependendend within the investiged velocity range. The pantograph is the dominant source of sound above $f = 6$ kHz. It is more tonal noise and the frequency peaks show a strong velocity dependence. There are several peaks with nearly constant Strouhal numbers. These two sources of sound demonstrate that it is not possible to formulate one scaling law for all sources of sound of a train. It is essentiell to treat every single source separately and to take their characteristics into account.

In order to obtain higher Reynolds numbers a second measurement using the same train model in the cryogenic wind tunnel has been conducted. By cooling down the fluid one is able to increase the Reynolds number by a factor of about 5. Beside of this feature this wind tunnel enables the independent variation of Mach and Reynolds numbers. To assure a comparability of aeroacoustic measurements conducted at different temperatures a nondimensionalized frequency is introduced. The experiments show, that the noise of the first bogie is not Reynolds number dependent. The Mach number dependence is similar to what has been found in the previous experiments in the aeroacoustic wind tunnel.

5 Outlook

For further investigations a double model is constructed, which is depicted in Fig. 11. This model yields the following benefits:

- Better ground simulation: The flow underneath a full scale train can be characterized by a Couette flow, but in the wind tunnel simulations one will obtain a

Poisseuille flow. The flow between the plane of symmetrie and the bogie cavity should be more comparable to the situation of the full scale world.
- More signal-to-noise ratio: A splitter plate is not needed anymore. The measurements in the cryogenic wind tunnel show that the splitter plate is a strong source of sound, which disturb the measurements.

References

1. Ahlefeldt T., Koop L.: Microphone array measurements in a cryogenic wind tunnel. In: 15th AIAA/CEAS Aeroacoustics Conference, Miami, FL, USA. AIAA 2009–3185 (2009)
2. Ahlefeldt, T., Lauterbach, A., Koop, L.: Aeroacoustic measurements of a scaled half model at high reynolds numbers. In: 16th AIAA/CEAS Aeroacoustics Conference, Stockholm, Sweden. AIAA 2010–3748 (2010)
3. Amiet, R.K.: Refraction of sound by a shear layer. J. Sound Vib. **58**(4), 467–482 (1978)
4. Amiet, R.K.: Correction of open jet wind tunnel measurement for shear layer refraction. In: 2nd AIAA Aeroacoustics Conference (AIAA-Paper 75–532), Hampton, USA (1975)
5. Barsikow, B.: Das Rad-Schiene Geräusch eines Hochgeschwindigkeitszuges der deutschen Bundesbahn. DFVLR-FB 84–38: DFVLR. Abteilung Turbulenzforschung, Berlin (1984)
6. Barsikow, B., King III, W.F., Pfizenmaier, E.: Wheel/rail noise generated by a high-speed train investigated with a line array of microphones. J. Sound Vib. **118**(1), 99–122 (1987)
7. Brooks, T.F., Humphreys, M.W.: Effect of directional array size on the measurement of airframe noise nomponents. In: 5th AIAA/CEAS Aeroacoustics Conference, Bellevue, USA. AIAA-Paper 99–1958 (1999)
8. Curle, N.: The influence of solid boundaries upon aerodynamic sound. Proc. R. Soc., Lond. A **231**(1187), 505–514 (1955)
9. Dittrich, A.G., de Beer, F.G.: The applicability of prEn ISO 3095 for European legislation on railway noise, final draft v3.2. TNO-report HAG-RPT-010014 (2001)
10. Ehrenfried, K., Koop, L., Henning, A., Käpernick, K.: Effects of wind tunnel noise on array measurements in closed test section. BeBeC-2006-07. In: 1st Berlin Beamforming Conference, 22–23 Nov 2006
11. Fey, U., König, M., Eckelmann, H.: A new Strouhal—Reynolds—number relationship for the circular cylinder in the range 47 < Re < 200000. Am. Inst. Phys. **10**(7) (1998)
12. Grunwald, K.J.: Aerodynamic characteristics of vehicle bodies at crosswind conditions in ground proximity. NASA TN D-5935 (1970)
13. Johnson, D.H., Dudgeon, D.E.: Array Signal Processing: Concepts and Techniques. Prentice-Hall, Englewood Cliffs (1993)
14. King III, W.F., Pfizenmaier, E., Herrmann, I.: Schallquellen an Hochgeschwindigkeitsstromabnehmern und Möglichkeiten zur Reduktion. Eine Literaturübersicht. DFVLR-IB 92517–97, B1: DFVLR. Institut für Antriebstechnik, Abteilung Turbulenzforschung, Berlin (1997)
15. Koop, L.: Experimental Aeroacoustics. Chapter beam forming methods in microphone array measurements—theory, practice and limitations, pp. 1–78. Von Karman Institute for Fluid Dynamics, VKI LS 2007–01. ISBN 978-2-930389-70-2 (2007)
16. Koop, L., Ehrenfried, K.: Microphone-array processing for wind-tunnel measurements with strong background noise AIAA-2008-2907 (2008). In: 14th AIAA/CEAS Aeroacoustics Conference, Vancouver, BC, Canada, 5–7 May 2008
17. Lauterbach, A., Ehrenfried, K., Koop, L., Loose, S.: Procedure for the Accurate Phase Calibration of a Microphone Array (AIAA 2009–3122). In: 15th AIAA/CEAS Aeroacoustics Conference, FL, USA, Miami (2009)
18. Lighthill, M.J.: Waves in Fluids. Cambridge Mathematical Library, Cambridge (1978)

19. Martens, A., Wedemann, J., Meunier, N., Leclere, A.: High Speed Train Noise—Sound Source Localization at Fast Passing Trains. S.E.A, Deutsche Bahn AG, Sociedad Espanola de Acoustica (2009)
20. Mueller, T.J. (ed.): Aeroacoustic Measurements. Springer, New York (2002)
21. Pfizenmaier, E., King III, W.F., Schewe, G., Herrmann, I.: Windkanaluntersuchungen an einem Stromabnehmer für den Intercity-Experimental (ICE) der Deutschen Bundesbahn. DFVLR-IB 22214–85, B5: DFVLR. Abteilung, Turbulenzforschung, Berlin (1985)
22. Phillips, O.M.: The intensity of aeolian tones. J. Fluid Mech. **1**, 607–624 (1956)
23. Pott-Pollenske, M., Delfs, J.: Enhanced Capabilities of the Aeroacoustic Wind Tunnel Braunschweig (AIAA 2008–2910). In: 14th AIAA/CEAS Aeroacoustics Conference, Vancouver, British Columbia, Canada (2008)
24. Sijtsma, P.: CLEAN Based on Spatial Source Coherence. AIAA Paper 2007–3436 (2007). In: 13th AIAA/CEAS Aeroacoustics Conference, Rome, Italy, 21–23 May 2007
25. Sijtsma, P.: Tutorial: Improving resolution with CLEAN-SC. In: Bebec, 2nd Berlin Beamforming Conference, 19–20 Feb 2008. http://www.bebec.eu. GFaI, Gesellschaft zu Förderung angewandter Informatik e.V., Berlin (2008). http://www.bebec.eu
26. Yamazaki, N., Takaishi, T., Toyooka, M., Nagakura, K., Sagawa, A., Yano, H.: Wind tunnel tests on the control of aeroacoustic noise from high speed train. In: 9th International Workshop on Railway Noise, Munich (2007)

Part V
Truck Aerodynamics

Aerodynamics of Commercial Vehicles

Thorsten Frank and James Turney

Abstract In recent years, the strong pressure from both customers and government bodies toward improvement of fuel consumption and reduction of pollution has made it necessary for the manufacturers of road vehicles to pursue increasingly innovative solutions and novel technologies toward this end. The average heavy truck customer's sophisticated understanding of a vehicle's total cost of ownership cause the Heavy Truck Market to be driven by objective measures of vehicle performance to a larger degree than is present in most other road vehicle markets. The fuel consumption of a heavy truck is among the largest contributors to the vehicle's total cost of ownership. As such, the improvement of fuel consumption through reduction of aerodynamic drag is of particular interest to Heavy Truck Manufacturers and their customers. Today, the optimization of aerodynamics for commercial vehicles must not be restricted merely to the installation of devices on the roof of the cabin. Rather, the vehicle must be regarded as a complete system and optimized accordingly. That is to say that aerodynamic optimization today focuses on the entire vehicle rather than just the tractor unit. Aerodynamic optimization must be balanced against such considerations as styling, ergonomics and soiling. The interactions between these various considerations will be explained through the use of some different cases. For example, the subject of windshield rake angle as it relates to headroom and aerodynamics. In closing, an outlook regarding the future development of Aerodynamic commercial vehicles will be given.

T. Frank (✉) · J. Turney
Daimler AG, Mercedesstrasse 132/1, 70327 Stuttgart, Germany
e-mail: thorsten.frank@daimler.com

A. Dillmann and A. Orellano (eds.), *The Aerodynamics of Heavy Vehicles III*,
Lecture Notes in Applied and Computational Mechanics 79,
DOI 10.1007/978-3-319-20122-1_12

1 Introduction

This paper outlines key considerations regarding the aerodynamics of heavy trucks. Rather than offer a purely technical discussion of this topic, this paper also endeavors to explain why aerodynamic improvements are fundamentally important to truck manufacturers, their customers, and the environment.

2 Why the Aerodynamics of Heavy Trucks is Important

Today, heavy trucks greater than 16 Tonnes account for approximately 5 % of all European CO_2 emissions. A review of the development of CO_2 emissions by source as shown in Fig. 1 further shows that, while emissions by households, services, and industry as a whole have fallen or remained steady between 1990 and 2006, the emissions of the transport service sector (which includes rail, sea, and over the road shipping) have risen by 28 %. Based on present industry conditions, this trend is projected to continue through the year 2030 as shown in Fig. 2. When considered as a whole, these historical trends, future projections, and present figures, serve to explain why the reduction of these emissions is an important subject for the manufacturers of heavy trucks.

At present, Germany's Transportation Service Sector moves some 650 billion Tonne-Kilometers of goods each year, with over the road trucking accounting for roughly 75 % of this activity. By 2030, the annual Tonne-Kilometers are projected to nearly double, with over the road trucking accounting for the vast majority of the increase, as shown in Fig. 3. Based on these projections, it is clear to see that the majority of Germany's present and future CO_2 emissions for the Transport Service Industry are likely to originate from trucking.[1]

In addition to its significant influence on CO_2 emissions, aerodynamics also plays a major role in the profitable operation of heavy trucks. As shown in Fig. 4, fuel costs make up a full 30 % of the total life cycle costs for a European heavy truck. There is a high level of consensus among industry experts, heavy truck manufactures, and heavy truck customers that while diesel prices will fluctuate in the short term, the long term trend as almost certainly toward rising prices and by extension an increasing role for fuel costs in the total cost of vehicle ownership. Therefore, it is reasonable to anticipate that the need to optimize aerodynamics with the aim of reducing heavy truck fuel consumption will become even more prominent in the coming years.

[1] This conclusion based on trucking's position as the dominant transportation form, rather than any comparison between rail, trucking and sea shipping on a per Tonne-Kilometer basis.

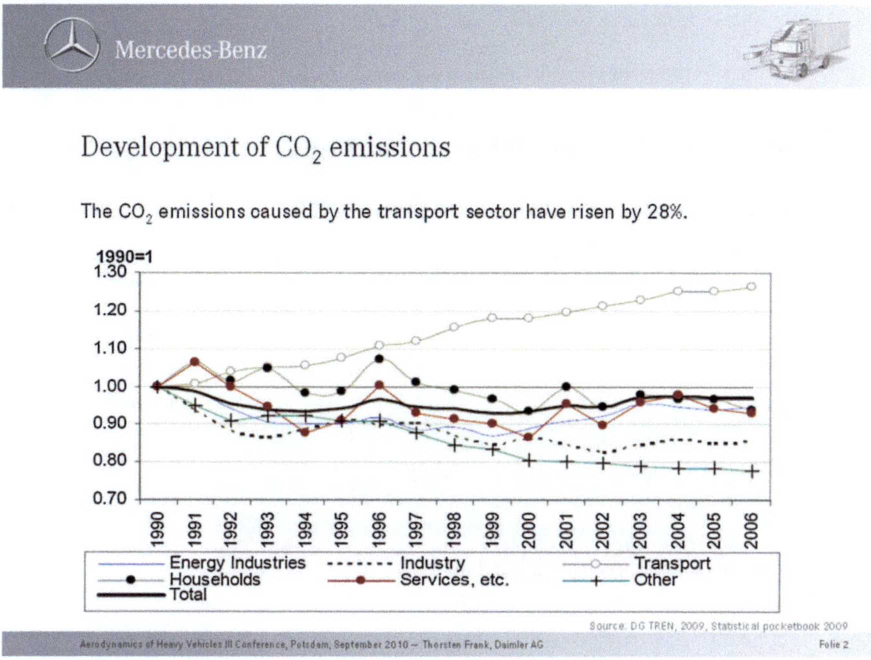

Fig. 1 Historical development of CO_2 emissions (Europe)

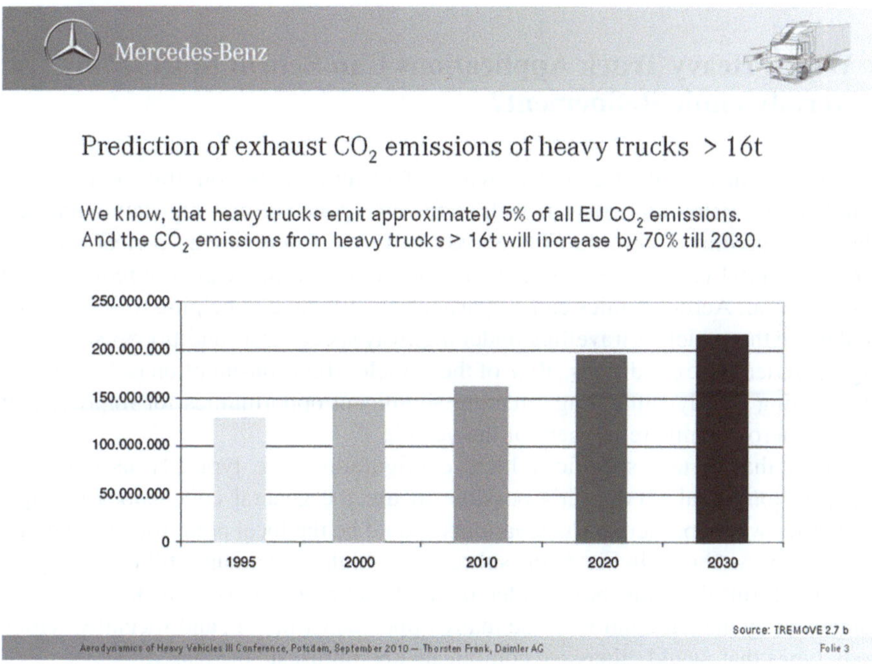

Fig. 2 Future projection of CO_2 emissions (Europe, heavy trucks)

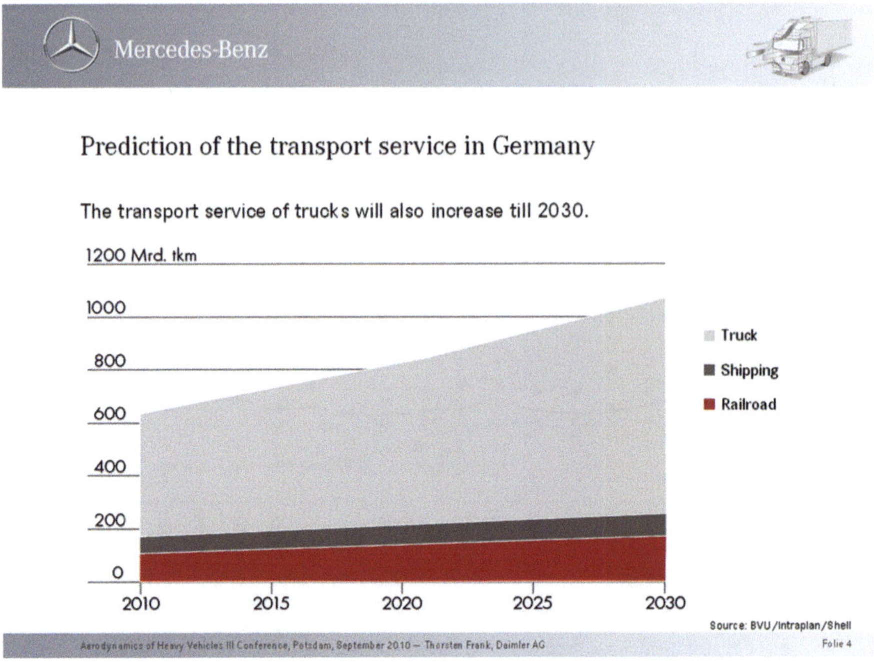

Fig. 3 Transport service sector emissions in Germany (Europe, heavy trucks)

3 Which Heavy Truck Applications Can Benefit Most from Aerodynamic Refinement?

When taken in sum, the trend of increasing CO_2 emissions from transport services coupled with rising fuel costs as well as the emergence of new legislation and regulations regarding allowable CO_2 emissions in the future, create a clear mandate for commercial vehicle manufacturers to take steps toward the goal of reducing fuel consumption. Aerodynamics can be particularly effective in helping to achieve this goal when the vehicle is travelling under highway speeds and conditions, as shown in Fig. 5. Under these conditions, 40 % of the vehicle's fuel consumption is dedicated to overcoming aerodynamic drag, creating significant opportunities for improvement through aerodynamic refinement of the vehicle.

Given that certain specific vehicle configurations are typically used for on-highway long haul service, it's possible to use the general conclusions of Fig. 5 to identify which particular truck variants should be the focal point for aerodynamic refinement. As shown in Fig. 6, the subset of on-highway variants still has a range of sizes and forms that must be considered, but the aerodynamicist's work is simplified somewhat by the elimination of the many tipper, construction, and specialty vehicle body types that would otherwise complicate and obfuscate the task.

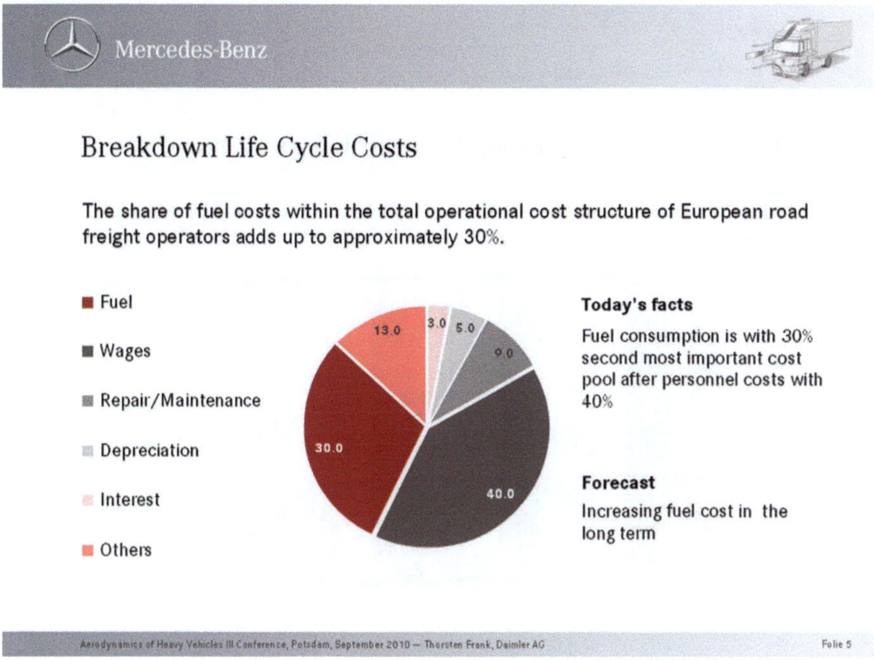

Fig. 4 Breakdown of life cycle costs (Europe, heavy trucks)

4 Key Areas and Opportunities for Refinement of Heavy Truck Aerodynamic Drag

4.1 The Importance of Cab Exterior Panels and Aerodynamic Equipment

The early days of aerodynamic development were marked by comparatively simple innovations such as the roof spoiler, which served to direct air up and over the leading edge of the trailer as shown in Fig. 7. While treatments like the roof spoiler were important initial steps, modern aerodynamic development must delve deeper into the many details of the cab itself.

A heavy truck cab, taken as a complete product, must meet not only aerodynamic requirements, but of course also requirements related to aero-acoustics, soiling, styling, packaging, and many others (these subjects are discussed in some detail later). To achieve such a balance, the individual components of the cab must be considered in the context of a complete system. Figure 8 outlines some key exterior components which influence aerodynamics.

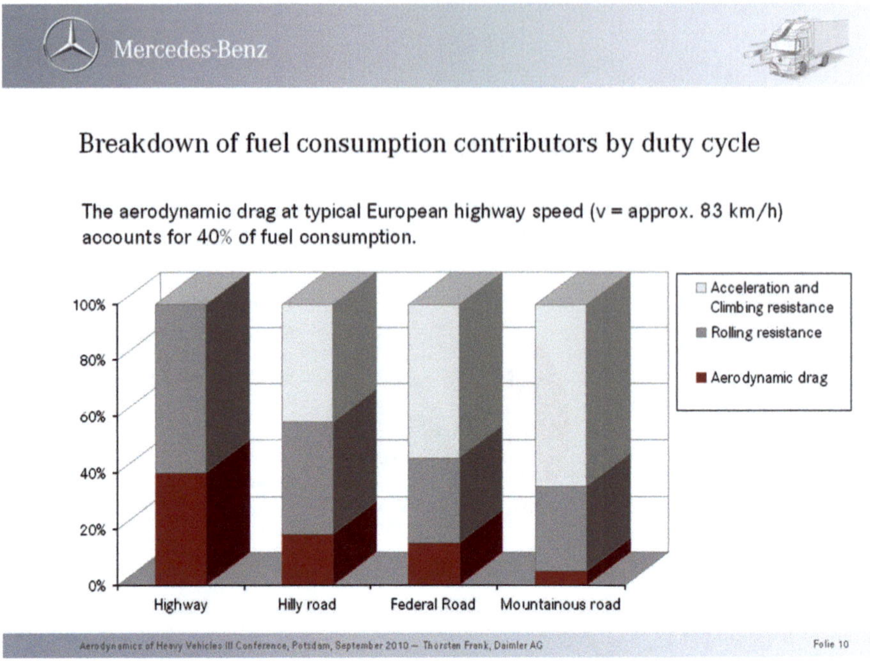

Fig. 5 Fuel consumption breakdown by duty cycle (83 km/h)

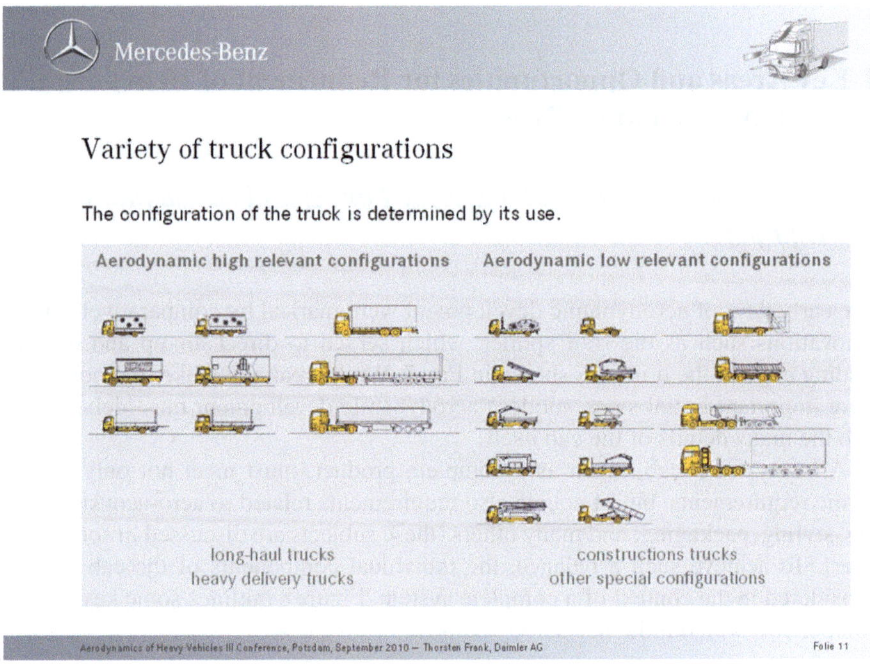

Fig. 6 On and off highway truck configurations

4.2 The Importance of Basic Cab Geometry

Before going into further detail regarding the aerodynamics of cab exterior components, it's helpful to speak about the importance of the basic cab dimensions. The specific combination of cab width, cab height, cab length, corner radius, sweep, side tapering, roof radius and windshield angle has an important influence on the aerodynamic behavior of the resultant cab. Width, height and length are generally defined by legal regulations. For example, the maximum height for most European trucks is 4 m. The other parameters are affected by the requirements of styling, ergonomics, packaging, and of course aerodynamics. In order to understand the interaction between these dimensions, it is necessary to consider them by means of a parametric study, as outlined in Figs. 9 and 10.

The three-dimensional chart in Fig. 10 shows the drag dependency between roof radius and windshield angle (green indicates a lower drag). For example, on the chart in Fig. 10 it can be seen that, with a 10° windshield angle a drag improvement for roof radii greater than 200 mm cannot be achieved. However, with a rising windshield angle an improved aerodynamic performance due to increasing the roof radii becomes possible. The message from these figures is clear, the combination of basic (and difficult to alter later) cab dimensions that are established at the early stages of product development must be considered very carefully so that the aerodynamicist does not find him or herself "designed into a corner" later in the project.

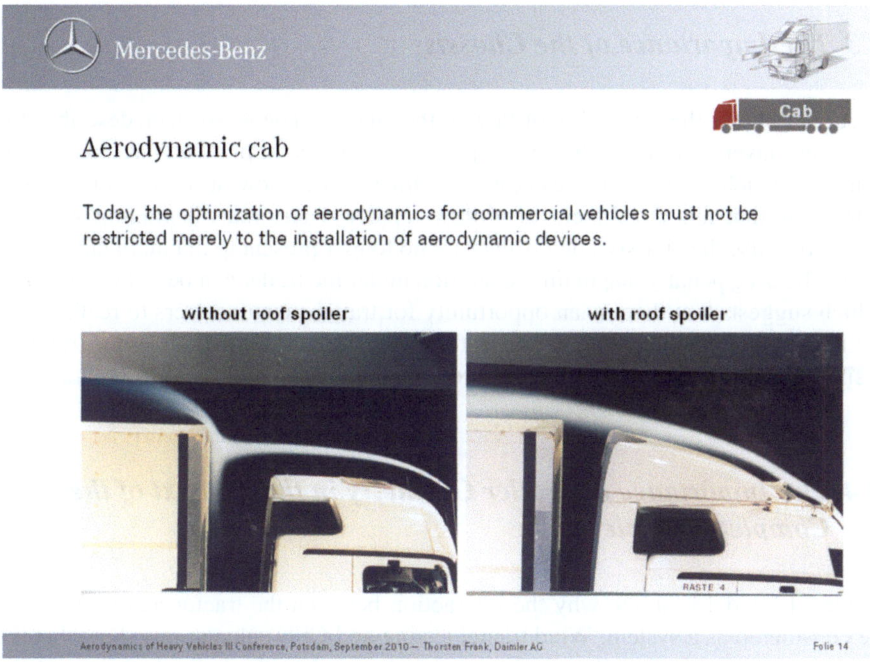

Fig. 7 Roof spoilers

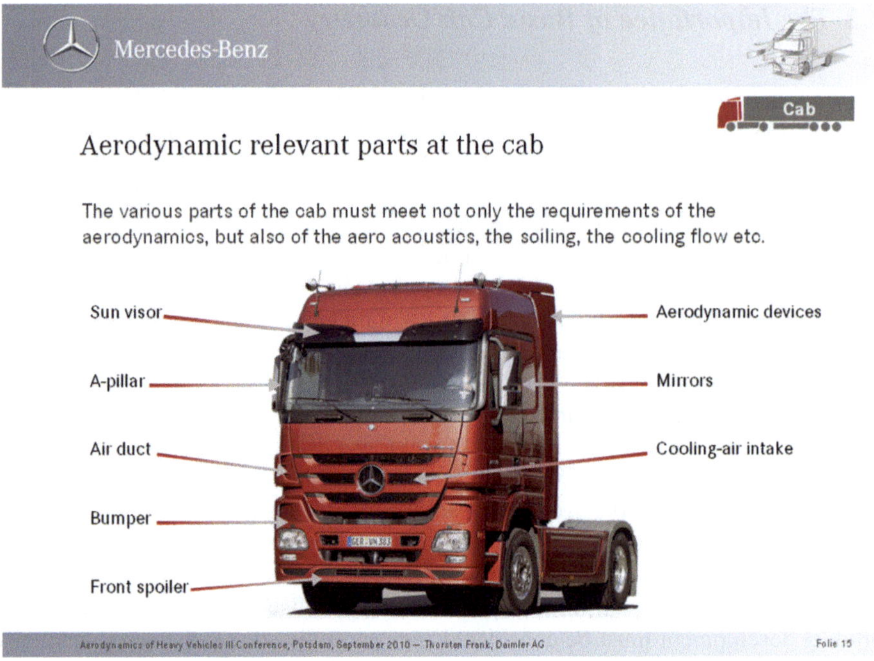

Fig. 8 Some key cab features for aerodynamic refinement

4.3 The Importance of the Chassis

Figure 11 shows the separation of flow at the chassis. The iso-surface describes the boundary layer between high and low pressure. The tractor pictured has chassis side fairings, which explains the resulting attachment of airflow at the tractor. Also of interest is that this attachment of airflow rapidly deteriorates beginning with the trailer because the chassis side fairings are no longer present to maintain laminar air flow. The drag penalty due to this separation under the trailer can be as high as 10 %, which suggests that there is an opportunity for trailer manufacturers to realize drag savings through the implementation of fairings that close the gap under the trailer in a similar fashion to those often present on the tractor.

4.4 The Importance of Trailer Geometry in the Context of the Complete Vehicle

Figures 12 and 13 outline why the interaction between the tractor and trailer must be considered as a system. Wind tunnel testing and CFD calculations show that the total drag of the semi-trailer combination is lower than the individual drag of the

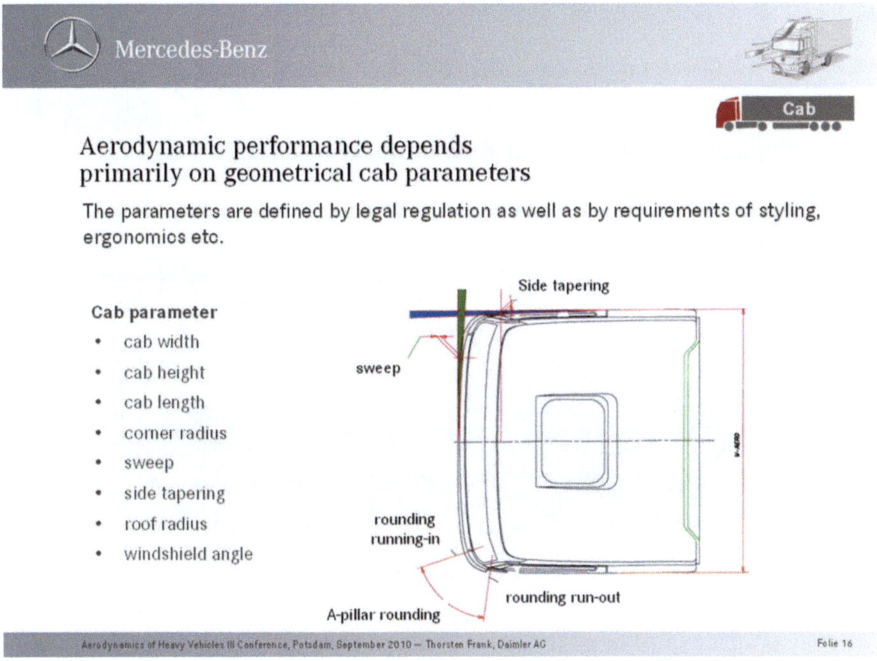

Fig. 9 Key cab dimensional parameters

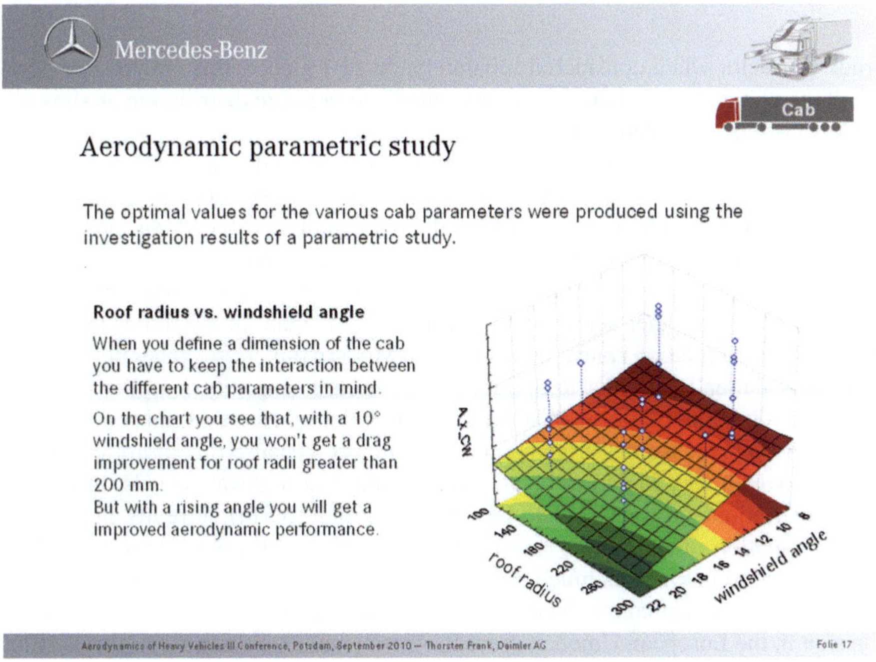

Fig. 10 Example of a cab dimensions parametric study (roof radius versus windshield angle)

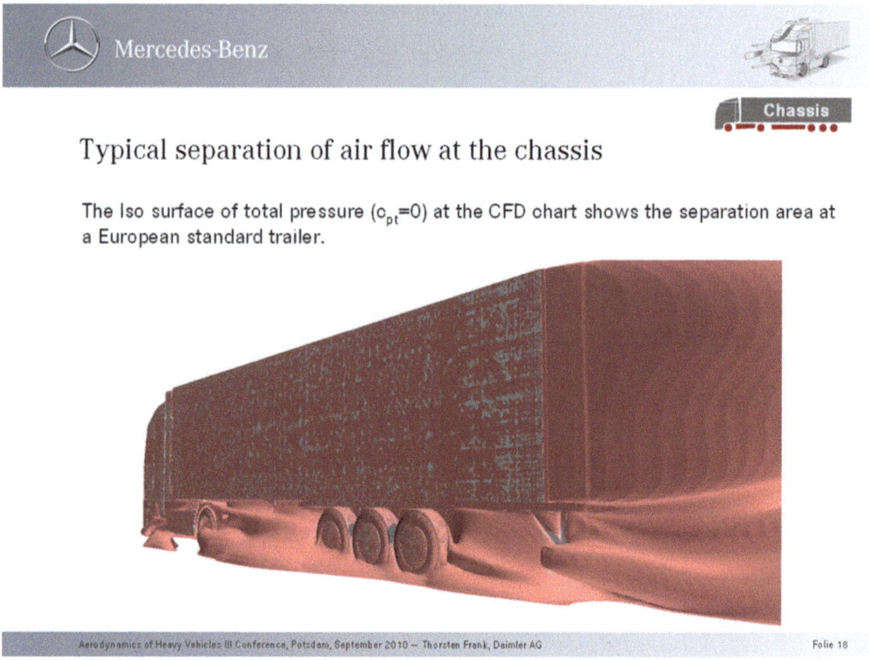

Fig. 11 Separation of airflow at the chassis, influence of chassis side fairings

truck or tractor when considered separately. At first glance, this would seem to be an illogical conclusion, but an examination of the pressure distribution as shown in slide 13 offers an explanation.

One item to note, is that the above information is derived from wind tunnel testing and CFD analysis on a particular vehicle configuration. Results will certainly vary depending on the configuration considered. For example, a cab without a roof spoiler would yield a considerably different distribution of drag between tractor and trailer.

One area of opportunity for improving trailer geometry is through the introduction of a "boat tail" or other aerodynamic treatment at the end of the trailer in order to reduce the recirculation zone it leaves behind. Modern trailers are generally designed with very flat ends in order to optimize interior cargo capacity, simplify construction, and for compatibility with standard loading docks. However, this optimization comes at the expense of aerodynamic performance because the abrupt end of the trailer leaves a very large turbulent wake behind it at highway speeds. In Europe, legal restrictions on total vehicle length (maximum 16.5 m) currently prevent the implementation of treatments that could reduce this wake. Such treatments have been shown to produce as much as an 8 % reduction on total vehicle drag, making the possibility of regulatory changes an important point of political and technical discussion in the European Union at present. Aerodynamic optimization of trailers has also been generally slow the world over because most tractor manufacturers do not

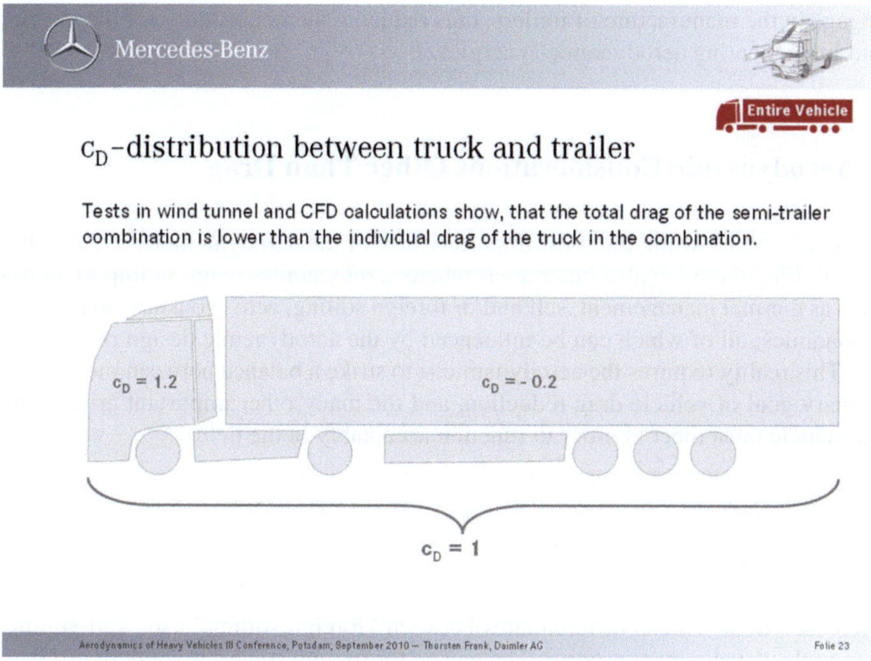

Fig. 12 Coefficient of drag, distribution between truck and trailer

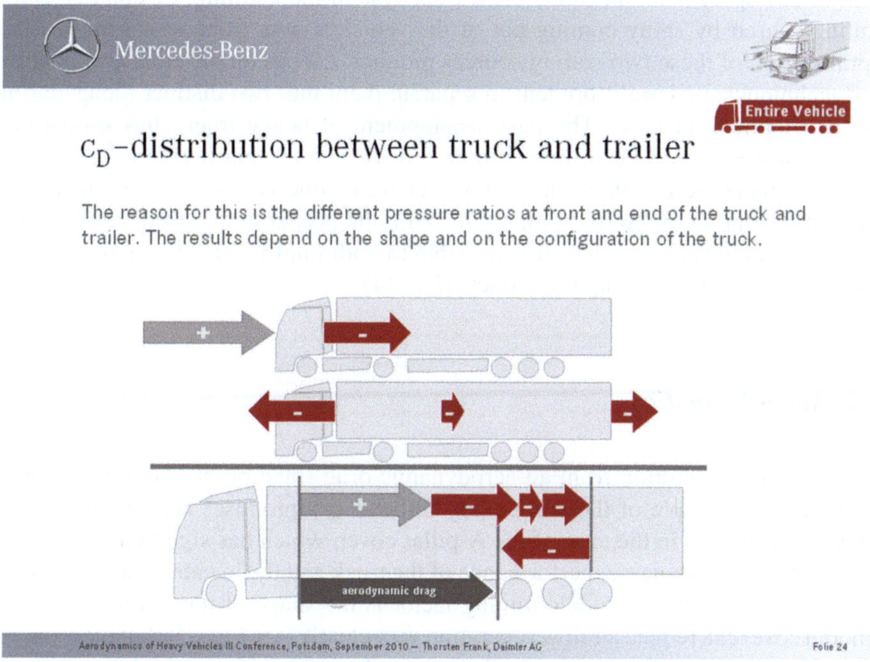

Fig. 13 Coefficient of drag, distribution between truck and trailer (continued)

engage in the manufacture of trailers, thus reducing the opportunity for discovering and implementing aerodynamic synergies.

5 Aerodynamic Considerations Other Than Drag

As noted earlier, while the most important task of the aerodynamicist is the reduction of drag, there are also other areas where aerodynamics plays an important role such as thermal management, self and/or foreign soiling, aero acoustics, styling, and ergonomics; all of which can be influenced by the aerodynamic design of the vehicle. This reality requires the aerodynamicist to strike a balance between meeting the primary goal of vehicle drag reduction, and the many other important criteria that the vehicle must meet in order to function acceptably in the field.

5.1 Soiling

Mercedes-Benz Trucks differentiates between "foreign soiling" and "self soiling" during the development process. The reason for foreign soiling is spray or dirt from a vehicle traveling in front of, next to, or coming from the opposite direction, as well as any spray from the vehicle's washer system. Self soiling, by contrast is the soiling caused by spray coming out of the vehicle's own front wheel house. The optimization of these two soiling sources requires a focus on different components and techniques, making it prudent to separate them into two distinct topics during the development process. The past development of heavy trucks has shown that most measures to reduce soiling of both kinds (such as splash guards or A-Pillar covers) penalize aerodynamic drag. However, the air ducts at the vehicle front corner panel can offer an exception to this rule. With the proper tuning of dimensions for the intake, outtake, and vanes, it is possible to both improve the aerodynamics and reduce vehicle soiling using these ducts (Fig. 14).

5.2 Aero-Acoustics

Often, the ideal solution from an aerodynamic drag stand point is not the quietest solution. An example of the particularly difficult compromises aero-acoustics can cause is evidenced in the case of the A-pillar cover, which has significant influence over both the aerodynamic performance of the truck and the in cab noise level experienced by the driver. The determining factor is the shape of the A-pillar cover; a smooth cover can reduce air flow separation very clearly (see Fig. 15), but the reduced separation serves to direct more air flow along the side window, which in turn generates vibrations that increase the noise in the cabin and can decrease the driver's comfort.

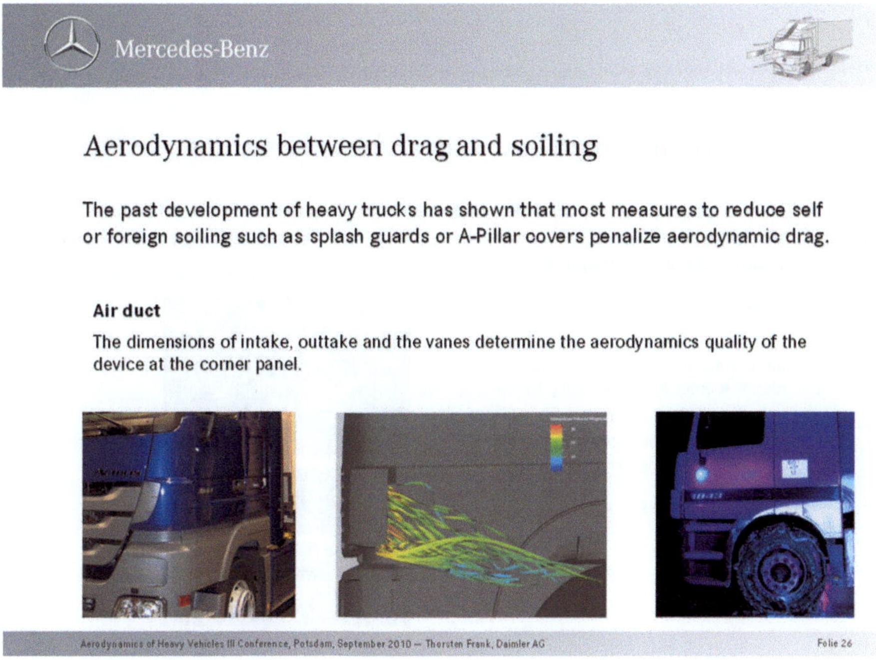

Fig. 14 Front corner panel air ducts

5.3 Styling

During the development of a new heavy truck Styling is one of the departments with a very significant indirect influence on the vehicle's aerodynamics. This is because the appearance of each curve and edge must blend into the overall styling theme of the vehicle. A typical example of a conflict between the priorities of aerodynamics and styling can be found at the corner radius. Styling generally likes to have curves with a decreasing radii because it looks much more dynamic, as shown Fig. 16. However, a decreasing radius leads to airflow separation and with it an increased aerodynamic drag. The need to compromise between vehicle styling and vehicle aerodynamics can be a particularly difficult point for development programs, because both considerations are so critical to the ultimate market success of the product.

5.4 Ergonomics

Many of the most aerodynamically efficient shapes for a heavy truck are entirely unsuitable for implementation due to the practical considerations of human ergonomics. Figure 17 shows an example of a very aerodynamic heavy truck shape, which

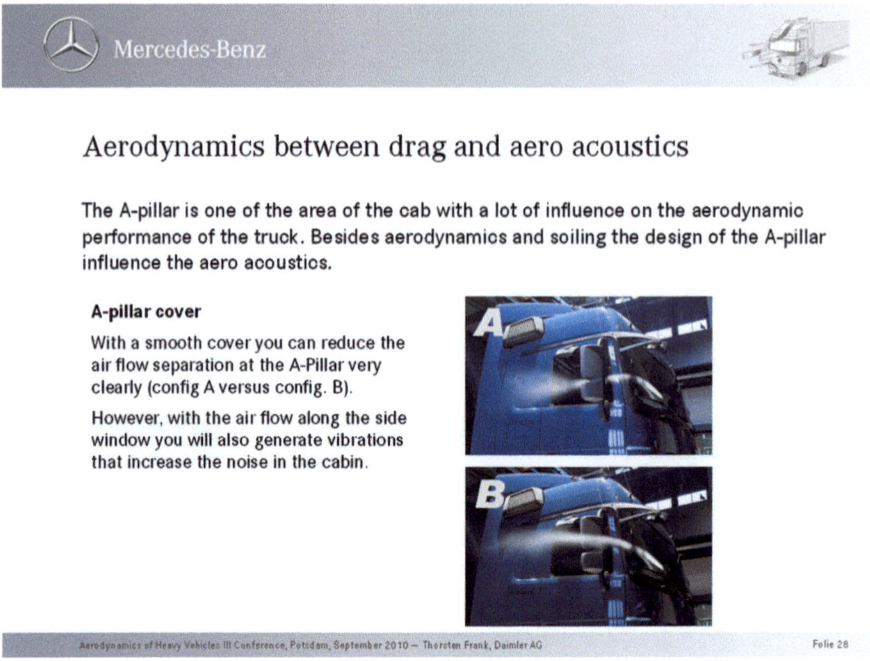

Fig. 15 A-Pillar cover geometry as an example of an aero acoustic compromise

would be unusable in the field for long haul applications due to the insufficient living area provided for the driver. Simple every day needs such as the ability to stand inside the cab to change clothing must be considered in the development of heavy trucks. Aerodynamic solutions must take these ergonomic considerations into account in order to ensure that the final vehicle design finds success in the market.

6 A Note About the Future

The days when a heavy truck could be aerodynamically refined by considering it as a series of individual parts (cab, trailer, spoiler, and etcetera) have passed. For most modern vehicles, the "low hanging fruit" has already been harvested, and the achievement of additional aerodynamic refinement hinges on the aerodynamicist's ability to consider the vehicle as a complete system and optimize it accordingly.

In particular, truck manufactures must begin to look not only at the whole tractor-trailer combination, but also at the vehicle's interaction with the road surface it is travelling over. Methodologies for simulation of the moving road conditions under both the truck and the trailer (meaning rotating wheels and a moving ground surface) will soon be necessary in order to further expand understanding and drive innovation

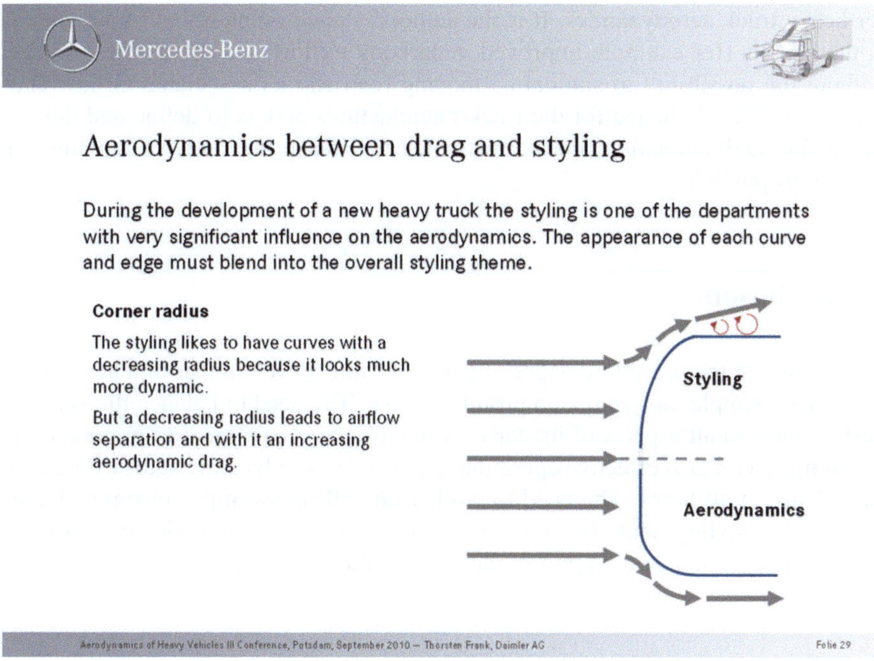

Fig. 16 Balancing visual dynamics and aerodynamics

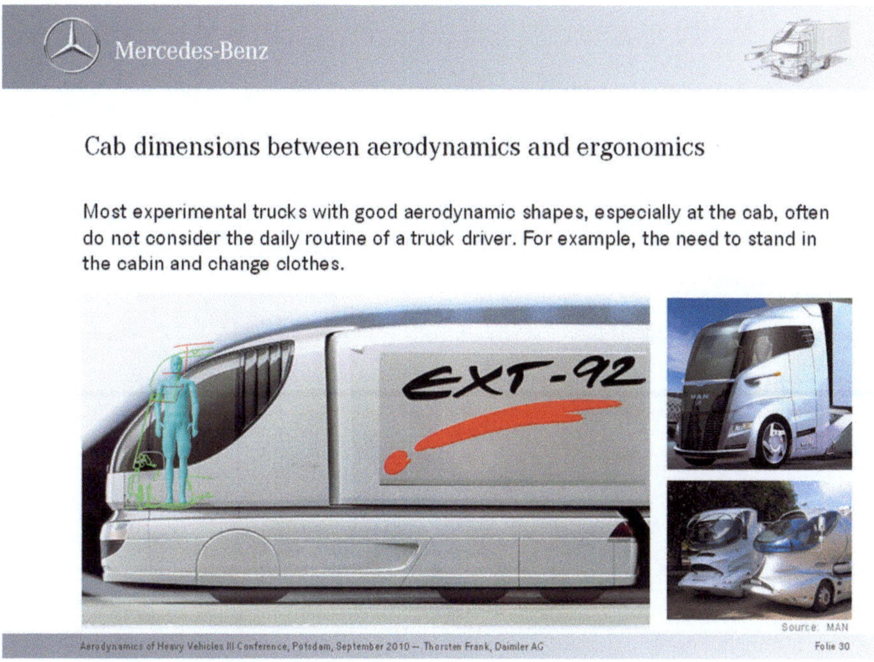

Fig. 17 Ergonomic considerations

for heavy truck aerodynamics. It is the authors' shared opinion that improvements at the chassis (for example improved underbody airflow) cannot be fully realized without the possibility to simulate a moving road and its associated airflow characteristics. The challenge for the truck manufactures now is to define and develop the testing and simulation procedures which will make the study of moving road conditions possible.

7 Conclusion

As outlined in the preceding pages, the aerodynamic development of a heavy truck is neither a simple, nor an uncompromising task. The need to balance the vehicle's performance in all aspects of its daily work with the requirement to minimize fuel consumption for cost effective operation requires the aerodynamicist to strike a careful balance at all times. The need to work with colleagues in power-train, chassis, ergonomics, styling, and elsewhere in order to achieve technologically innovative compromises is an essential part of the aerodynamic design process.

Wheel Housing Aerodynamics on Heavy Trucks

David Söderblom, Per Elofsson, Linus Hjelm and Lennart Löfdahl

Abstract Modern trucks have a reasonably optimised cab shape, and there exist several OEM and aftermarket devices for drag reduction for heavy trucks as well. To further reduce the aerodynamic drag major changes to the current layout of the vehicle are required, or the focus must be shifted from the cab and tractor trailer gap to other regions of the vehicle. The drag of the underbody, including wheel housings, wheels and engine compartment, represents a significant proportion of the aerodynamic drag and there has not been much investigation in this specific area on heavy trucks. To be able to reduce the fuel consumption and to fulfil the legislated emission standards for heavy trucks it is important to take all areas of the vehicle under consideration, and even though the individual improvements may be small, the total drag reduction will be substantial. In order to study the flow close to the vehicle underbody it is important to utilise the correct boundary conditions, that is, moving ground and rotating wheels. This work has focused on the flow in the front wheel housings. The flow field around the front wheels under the influence of ground simulation on a heavy truck of standard European configuration was investigated using numerical simulations. The in- and outflow to the wheel housing was located and the vortices originating from the front wheels were identified. This information was then used to identify which areas of the wheel housing having the greatest potential for aerodynamic improvements by changing the front wheel housing design. Furthermore, several wheel housing design parameters were defined, and their influence on the flow field and aerodynamic drag were investigated. Examples of these parameters are the shape of the wheel housing opening and implementation of wheel housing ventilation. It was found that there is potential for reducing the aerodynamic drag by applying these geometric changes to the wheel housing, and several of the configurations could be implemented on current production vehicles.

D. Söderblom (✉) · L. Löfdahl
Chalmers University of Technology, Göteborg, Sweden
e-mail: david.soderblom@chalmers.se

P. Elofsson
Scania, Södertälje, Sweden

L. Hjelm
Volvo 3P, Göteborg, Sweden

A. Dillmann and A. Orellano (eds.), *The Aerodynamics of Heavy Vehicles III*,
Lecture Notes in Applied and Computational Mechanics 79,
DOI 10.1007/978-3-319-20122-1_13

1 Introduction

Increasing fuel costs and demands for reduced emission levels have led to an increased interest in reducing the total resistance of heavy trucks. At 80 km/h, which is the typical speed for heavy trucks in long-haul operation, Lögdberg [1] showed that the aerodynamic drag becomes the dominant resistance force for a 40 ton tractor semi-trailer combination. Over the last 35 years the drag coefficient of European-type trucks has been reduced by more than 30 % [2]. So-far the aerodynamic optimisation of the tractor has focused mainly on the shape of the cab. With sufficient front corner radius and well designed roof- and side air deflectors the aerodynamic properties of the tractor cab can be considered as fairly good. It is difficult to further optimise the cab shape without influencing the drivers' comfort or having problems with the packaging levels of the tractor unit. Thus, it is important to consider other areas of the truck for further aerodynamic optimisation of the vehicle. It was noted in numerical simulations performed by the author that the underbody, which is considered to include the cooling air flow, engine compartment, chassis, suspension, wheels and wheel housings, can stand for as much as half of the total aerodynamic drag for some tractor trailer combinations.

When investigating the flow field around the wheels and underneath the vehicle, it is of most importance to utilise the correct boundary conditions, that is, moving ground and rotating wheels. This applies to both numerical and experimental investigations. It has been shown in experiments on passenger cars that the impact of ground simulation on the derived drag coefficient can be significant [3]. Major interference effects on the entire flow field due to the implementation of moving ground and rotating wheels was presented by Elofsson and Bannister [4] and Wäschle [5], which implied that a correct simulation of the ground motion not only affects the flow locally. The observed trend for heavy trucks was that the measured drag coefficient increased when ground simulation was utilised [6].

It has been observed in studies on passenger cars that there is a potential for reducing the aerodynamic drag by performing modifications to the geometry in the vicinity of the wheel and wheel housing [7, 8]. In these studies different methods for influencing the flow in the wheel housing were evaluated, either by guiding the inflow with spoilers or by alterations to the rim design. Cogotti [9] showed in experiments on a passenger car-like model that the drag decreased when the inner wheel housing radius and depth was reduced. It was concluded that the wheel housing volume had to be kept as small as possible in order to reduce the drag of the vehicle. Similar results were observed in a study by Fabijanic [10] where wheels and wheel housings were added to a basic vehicle model. The effect of varying wheel housing radius and depth was investigated with ground simulation, and it was shown that the drag of the model increased with larger wheel housing radius. However, the variation of the depth of the wheel housing did not seem to have a great influence on the measured drag coefficient. Both Cogotti [9] and Fabijanic [10] performed their measurements on simplified vehicle models with only a single wheel axle, and it was stated by

Cogotti [9] that it is possible that the results for a four wheel car body would be different and might depend on the specific vehicle shape.

The amount of published investigations regarding the flow field in the wheel housings of heavy trucks is very limited, and most work in this area has been carried out on passenger cars. It was therefore chosen to begin this project by studying the local flow field around the front wheels and in the wheel housings under the influence of moving ground and rotating wheels on a heavy truck before doing any modifications to the wheel housing geometry.

2 Wheel Housing Flow Field Characteristics

The flow field around the front wheels and in the vicinity of the wheel housings of a heavy truck was surveyed by numerical simulations. In these simulations a long-haulage truck of so-called standard European configuration was used, that is, a two axle tractor of COE-type (Cab Over Engine) with a three axle semi-trailer. The geometry is presented in Fig. 1. Two different commercial solvers were used for the CFD simulations. These were: EXA PowerFLOW and ANSYS Fluent. Moving ground and rotating wheels was utilised for all cases, and the simulations were carried out at 25 m/s (90 km/h). In this study the influence of crosswinds was also taken into account, and the importance of considering yawed flow conditions for heavy trucks is described later in this chapter. To achieve the yawed flow conditions, the domain was rotated five degrees around the vertical axis.

In the following chapter the flow field around the wheels and in the wheel housings, of the previously described geometry, is presented. It is important to note that the flow field can differ significantly from one vehicle to another and that the described flow field in this study may only apply to this specific vehicle.

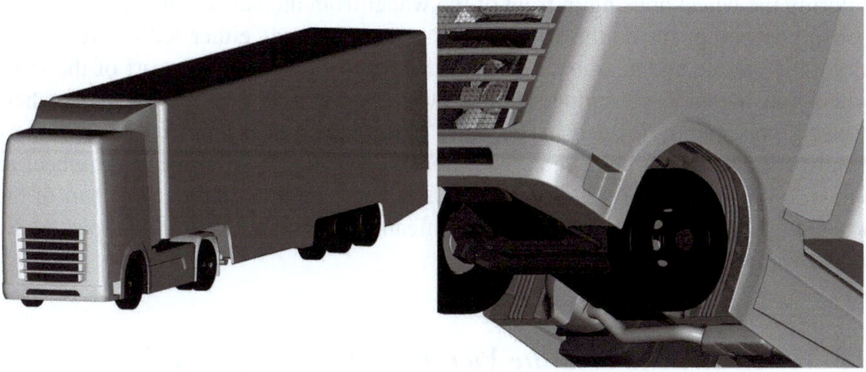

Fig. 1 Tractor trailer geometry used in the numerical simulations

Fig. 2 Path lines coloured by velocity magnitude visualising the inflow and outflow from the wheel housing

2.1 Characteristics of the Flow Entering and Exiting the Wheel Housing

The first step in this study was to identify the origin of the inflow to the wheel housing. It was seen that the flow entering the wheel housing was highly yawed compared to the free stream flow. As the flow passed underneath the bumper it was displaced by the vehicle underbody and was forced outwards to the sides of the vehicle. For this specific geometry, a large amount of the airflow entering the bumper grille had its exit underneath the vehicle, and this resulted in an even larger inclination of the flow approaching the front wheels. The flow that was entering the wheel housings came mainly from underneath the bumper skin, as seen to the left in Fig. 2, and from the inside of the wheel housing, through the engine compartment. There was no flow entering the wheel housing in front of the wheel from the outside through the wheel housing opening. Inside the wheel housing the flow was either led out rearwards along the engine and gearbox, or directed upwards along the rear part of the inner wheel housing under the influence of the wheel rotation. The outflow from the wheel housing was mainly located in the inner part of the wheel housing, as previously described, and out through the wheel housing opening behind the front wheel as shown to the right in Fig. 2. There was also a large outflow in the upper part of the wheel housing opening which was a significant contribution to the upper part of the wheel wake structure.

2.2 Vortex Structures in the Vicinity of the Front Wheels

It has been shown in previous studies [5, 11] on passenger cars that there exist several vortex structures around a wheel in a wheel housing originating from both the flow

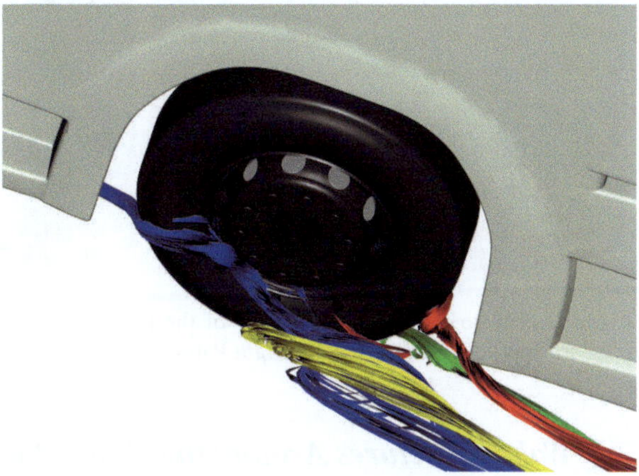

Fig. 3 Vortex structures, illustrated by coloured particle paths, originating from the flow past the front wheel under the influence of the wheel rotation

past the wheel and the wheel rotation. In the current study it was chosen to identify these vortex structures for a heavy truck in order to investigate whether there were any major differences in the local flow field compared to passenger cars.

When the surface of the wheel met the ground the fluid was forced around the corners of the tyre, and when the flow separated at the tyre shoulders in front of the tyre contact patch a pair of vortices were formed. Due to the yawed oncoming flow the inner vortex was suppressed by the increased momentum from the wheel rotation and ground motion. The outer vortex trail can be seen in Fig. 3 coloured in yellow. There was also a vortex formation in the rear part of the tyre contact patch by the backflow into the wake of the wheel in close ground proximity. The vortex was rolled up parallel to the wheel axle and was then pulled along by the free stream flow. In Fig. 3, in red, it can be seen that the angled incoming flow, and the flow between the wheel and the wheel housing, shift the vortex outwards. In the study by Wäschle [5] it was observed that the outer vortex trail dissipated due to the asymmetric flow field around the enclosed wheel, and the results in the current study were in good agreement with this behaviour.

Since the vortex coloured in yellow in Fig. 3 was relatively weak, the flow was able to reattach and reach the rear part of the tyre shoulder before it separated. This led to a formation of a pair of counter rotating vortices. In the work by Wäschle [5] it was described that for a stationary wheel the separated region was so large that the angled incoming flow was not able to reattach. In the current study it was seen that the innermost vortex was stronger and that it was energised by the yawed flow that separated on the rear tyre shoulder. This structure is visualised in Fig. 3 coloured in green. The outer vortex trail was quickly dissipated and not possible to visualise with particle paths. Coloured in blue in Fig. 3, is a vortex that was formed when the flow separated at the rim flange.

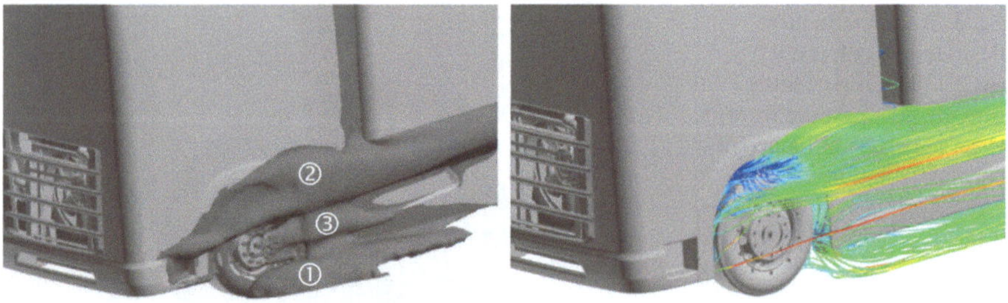

Fig. 4 Local flow field in the vicinity of the front wheel housing. *Left* Iso-surface of the total pressure coefficient equal to zero. *Right* Particle paths coloured by velocity magnitude

2.3 Wake Structures Around the Front Wheels

In Fig. 4 the local flow structures around the front wheel is presented by the iso-surface of the total pressure coefficient equal to zero and by particle paths coloured by velocity magnitude. The front wheel wake structure can be divided into three separate regions. The lower part ① originated from the flow separation at the tyre shoulder near the tyre contact patch and the vortex formations in this particular region. The lower part of the wheel wake was quite wide and this was due to the highly angled incoming flow. The wake increased in size by the outflow from the wheel housing downstream of the tyre contact patch. The upper part of the wheel wake ② originated from the outflow from the wheel housing that was located above and behind the front wheel. Due to the wheel rotation the flow did exit further upstream compared to the stationary case which resulted in a slightly larger upper part of the wheel wake structure. On the geometry used in this study there was also a contribution to the size of the upper part of the wheel wake originating from the flow separation at the boarding step. There was also a wake structure located between these two regions ③ arising from the flow past the rim and the outflow behind the front wheel. This structure will be referred to as the intermediate wake structure in this paper.

2.4 Influence of Crosswinds on the Wheel Housing Flow

When investigating the aerodynamics of heavy trucks it is of most importance to consider yawed flow conditions. This is due to the high crosswind sensitivity of the drag coefficient. When the vehicle is exposed to crosswinds, an increased amount of flow will pass through the tractor trailer gap and past the vehicle underbody. Due to the increased cross flow and the flow separation on the leeward side of the vehicle this will result in a significant increase of the drag coefficient of the vehicle. It was noted in five degrees yaw that the wheel wake structure on the windward side was suppressed, and that a large portion of the outflow was forced to exit underneath the vehicle instead of going out through the wheel housing opening. The shape of the wheel wake structure itself looked quite similar to the case at zero yaw, but the size of the wake was much smaller. On the leeward side of the vehicle, the wheel wake structure was much larger when the vehicle was yawed. This was both due to the flow separating at the vehicle underbody and exiting on the side of the vehicle and due to the increased outflow through the wheel housing opening. The lower part of the wake decreased in size on the windward side since the flow approaching the wheel was more head-on compared to the case at zero yaw. On the leeward side of the vehicle the lower part of the wheel wake structure close to the tyre patch increased in size.

3 Influence of Wheel Housing Design

When the flow field investigation was completed, the study of the influence of different wheel housing design parameters was started. Several design parameters were investigated in order to study the influence of the wheel housing design on the local flow field, as well as investigating the potential for reducing the drag by modifying the geometry. A large variety of design parameters were identified and the selection had to be limited in order to be able to perform detailed investigations of the parameters individually. It was chosen to investigate parameters that would be possible to implement on a heavy truck without interfering with the steering capability or the suspension travel, as well as being changes that were feasible on production vehicles. Therefore no extreme designs were investigated. It was chosen to focus on the front wheel housings instead of the entire underbody area of the vehicle since the flow approaching the front wheels is relatively undisturbed. There is also more room for modifications to the wheel housing design in the foremost part of the tractor compared to the rear wheel housings.

3.1 Wheel Housing Volume

To investigate whether the conclusions presented by Cogotti [9] were applicable for heavy trucks it was chosen to vary the wheel housing volume on the tractor trailer geometry used in this project. In order to change the wheel housing volume quantitatively the inner surface of the wheel housing was considered to be circular, as seen in Fig. 5. The inner wheel housing geometry was scaled proportionally to the radius. The radius was either increased or decreased by 50 mm.

It was chosen to keep the variation small in order to not interfere with the wheel housing opening when the volume was reduced. There was good agreement with the trends presented by Cogotti [9] as the drag coefficient was reduced with a smaller wheel housing volume, and increased when the inner wheel housing radius was larger. These configurations were tested without the influence of crosswinds, and thus a wind averaged drag coefficient was not determined.

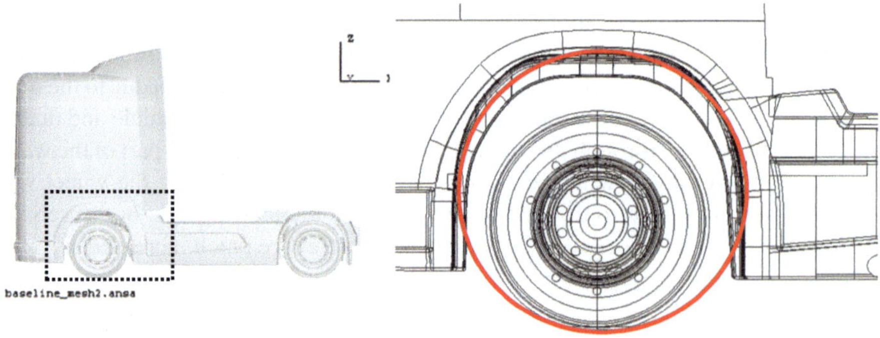

Fig. 5 Definition of the inner wheel housing radius when the wheel housing volume was altered

3.2 Wheel Housing Opening Shape

A parameter that was investigated thoroughly in this project was the shape of the wheel housing openings. In an early stage of this project the influence of the wheel housing opening size was investigated [12]. Recently several different wheel housing opening shapes were surveyed to evaluate the relation between the design and the aerodynamic properties of the vehicle [13], these configurations are summarised in Table 1. In the first alteration of the wheel housing opening shape (Case 2) the general shape was kept, but the size of the opening was reduced in order to leave room for the required modifications. The design of the wheel housing opening was taken to its two extremes, either circular (Case 3) or rectangular (Case 4) as shown in Fig. 6. Furthermore, different combinations of these shapes were also investigated, that is, circular in the front part of the opening and rectangular in the rear (Case 5), as well as the shapes reversed (Case 6).

It was seen that the drag coefficient was reduced with a decreased wheel housing opening. To evaluate the influence of the opening shape, the circular and rectangular

Table 1 Summary of the investigated wheel housing design configurations

Name	Description
Case 1	Baseline geometry
Case 2	Reduced opening size
Case 3	Circular wheel housing openings
Case 4	Rectangular wheel housing openings
Case 5	Circular front/Rectangular rear
Case 6	Rectangular front/Circular rear
Case 7	Outboard duct design I
Case 8	Outboard duct design II
Case 9	Inboard ventilation duct

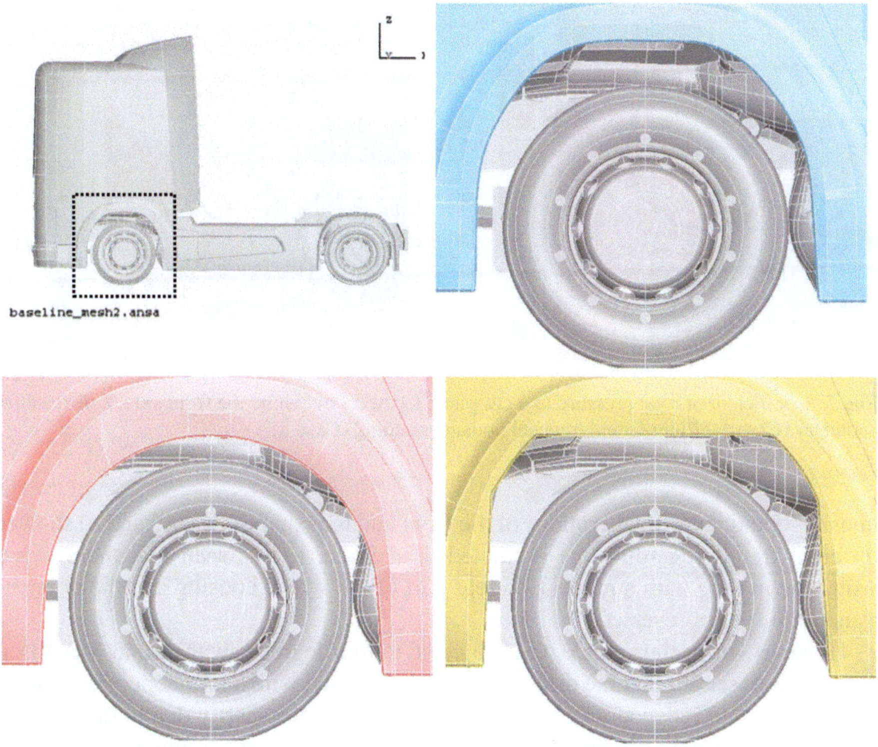

Fig. 6 Wheel housing opening shape. *Top right* Baseline geometry (Case 1). *Bottom left* Circular wheel housing opening (Case 3). *Bottom right* Rectangular wheel housing opening (Case 4)

openings were compared to the case with a reduced baseline opening. With reduced opening size the drag force distribution between the tractor and the trailer was altered. This was due to the reduced outflow from the wheel housing openings that had to exit underneath the vehicle and interfered with the trailer underbody. However, the net force in this case was a reduction in drag compared to the baseline when the openings were reduced in size. The circular opening (Case 3) showed a reduction in outflow from the upper part of the wheel housing which can be observed comparing the upper wheel wake structures between the different cases shown in Fig. 7.

However, the wheel wakes grew in size for the rectangular opening (Case 4) which indicated an increased amount of outflow through the openings. When the rectangular openings were introduced the front wheel wakes became longer as well and stretched so far downstream that there was a clear interference with the rear wheel wakes of the tractor. On the other hand, no significant difference in drag force distribution for the area around the rear wheels was noted.

Considering the cases with no influence of crosswind, it was noted that the circular openings resulted in a drag reduction whereas the rectangular openings increased the predicted drag coefficient as compared to the 'reduced' baseline opening. The wheel

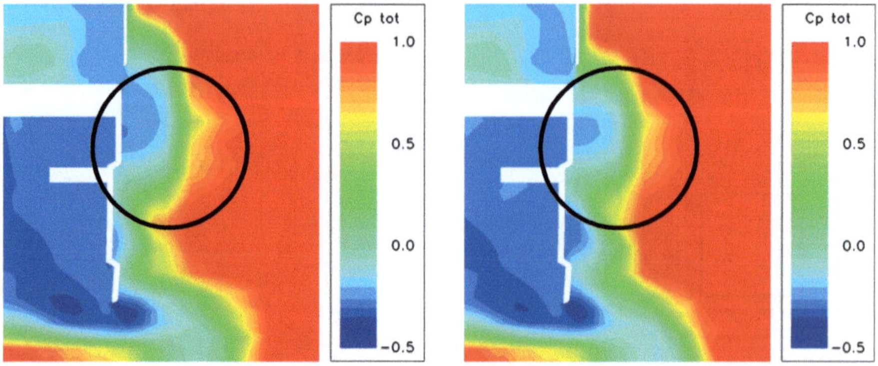

Fig. 7 Coefficient of total pressure in a cut-plane located 1 m behind the front axle. *Left* Baseline geometry (Case 1). *Right* Circular wheel housing opening (Case 3)

housing openings with alterations of the front and rear shape (Cases 5 and 6) showed different trends regarding the predicted drag force. It was seen that there was a reduction in drag with a more open rear part of the wheel housing and the opposite trend with shapes reversed.

However, when the vehicle was subjected to five degree yaw the trend was different. It was observed that the drag coefficient increased with a more open rear part of the opening and reduced when shapes reversed. This was partly due to the external flow interfering with the interior of the front wheel housings on the windward side of the vehicle when the openings were increased in size. There was also a change in the predicted drag coefficient for the circular and rectangular wheel housing openings compared to the reduced baseline opening (Case 2). In yawed conditions the aerodynamic drag increased for both the circular and rectangular configurations.

3.3 Wheel Housing Ventilation

The final parameter that was investigated in this study was the implementation of wheel housing ventilation. This parameter relied on the fact that the local pressure in the rear part of the wheel housing was relatively high compared to the pressure along the side of the vehicle and by the addition of a duct this pressure difference would drive the flow through the duct instead of exiting through the wheel housing opening perpendicular to the free stream flow. Three different designs were investigated. Two of them had an outboard duct located near the wheel housing opening (Cases 7 and 8), and the third design (Case 9) had a channel stretching from the inner part of the wheel housing and exiting in the foremost part of the recess on the side skirt similarly to the other designs.

A reduction in drag was noted for all three wheel housing ventilation designs, and the two cases with an outboard duct (Cases 7 and 8) showed a reduction in

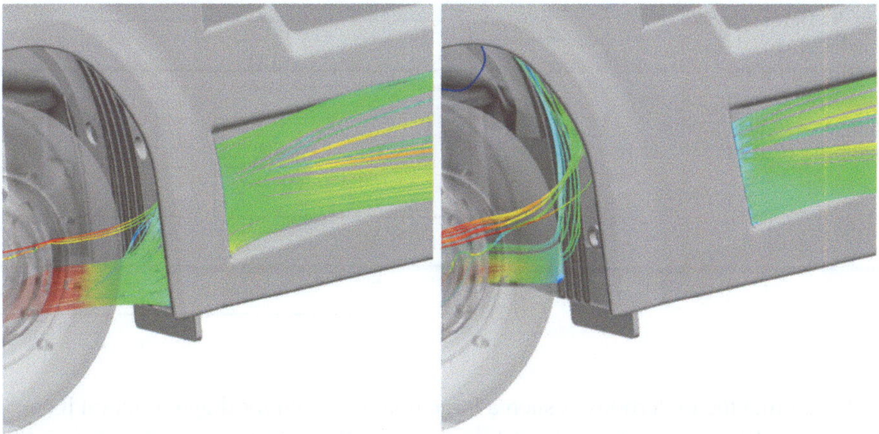

Fig. 8 Particle paths visualising the local flow through the wheel housing ventilation ducts. *Left* Outboard duct design II (Case 8). *Right* Inboard ventilation duct (Case 9)

both zero yaw and five degrees yaw conditions. For the case with the inboard duct (Case 9) there was a drag reduction at zero yaw but no change in the predicted drag coefficient at crosswind conditions. The major reductions in drag on the first two designs were noted on side skirts, rear wheels, rear suspension and the rear part of the inner wheel housings. In the rearmost part of the wheel housings the pressure relief through the ventilation duct contributed to the drag reduction, and it was also seen that the intermediate part of the wheel wake structure was reduced due to the addition of the wheel housing ventilation, which indicated that the duct operated as intended. In Fig. 8 the principle of the wheel housing ventilation is shown by particle paths coloured by velocity magnitude for two of the investigated designs.

In yawed flow conditions the drag was reduced for all but one configuration, which was the case with the inboard duct (Case 9). A drag reduction was noted in zero yaw for this configuration but due to the poor performance in yawed conditions this led to an increase of the wind averaged drag coefficient. However, larger reductions in drag are expected if the design of the ducts were to be optimised and by the addition of guiding vanes that direct the flow into the ducts. This is believed to be especially beneficial for the inboard duct design.

The delta values for the wind averaged drag coefficient for the investigated configurations in the recent study are presented in Fig. 9. It can be noted that there was a drag reduction for all but one configuration, and this configuration showed a reduction in drag at zero yaw but the penalties in yawed conditions led to an increased drag when the wind averaged drag coefficient was considered. Even though the drag reduction was small, there is still potential to improve the investigated designs further as well as combining them in order to take advantage of their individual performance and to find advantageous interference effects.

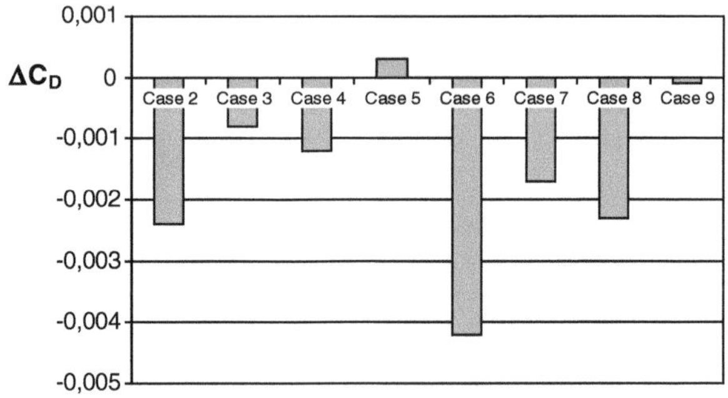

Fig. 9 Delta values of the wind averaged drag coefficient for the investigated configurations compared to the baseline geometry (Case 1)

Given that the underbody is such a large potential area for drag reduction it could be expected to see greater drag reductions than those that were presented in this study. Several causes may explain why the reduction was so small. For example, the focus in the later part of this study was put on the wheel housing opening shape and by adding wheel housing ventilation ducts in the rear part of the front wheel housings, whereas an addition of front wheel deflectors or altering the shape of the bumper skin would have influenced the inflow significantly. Influencing the inflow to the wheel housing may have a larger effect on the calculated drag coefficient, and a combination of the investigated parameters and influencing the inflow to the wheel housing might be beneficial and is of most interest in the continuation of this study. It was also chosen to perform modifications to the geometry that would be possible to implement on a production vehicle without interfering the steering ability and suspension travel. This was also expected to have an influence on the magnitude of the predicted drag reduction.

4 Conclusions

In this project the flow field around the front wheels of a heavy truck was investigated under the influence of ground simulation. The in- and outflows to the wheel housing was located and it was also seen that the flow approaching the front wheel was highly deflected as compared to the free stream flow. Several vortex structures originating from the flow past the rotating wheel were also identified.

The front wheel wake structures were investigated and the regions of flow losses were located. This was done in both zero yaw conditions and under the influence of crosswinds. The knowledge about the flow field was then used as input when different wheel housing designs were evaluated.

It was shown that by changing the design of the wheel housing a reduction in aerodynamic drag can be achieved. Several design parameters were investigated, and the general trend was that a reduced wheel housing opening resulted in a decrease of the predicted drag coefficient. It was noted that there was an alteration in the force distribution between the tractor and the trailer with reduced wheel housing openings. This was due to a redistribution of the outflow from the wheel housings, and that a greater amount of the outflow had to pass underneath the vehicle, interfering with the underbody, instead of exiting through the wheel housing openings.

There was also an interference between the front and rear wheel wakes for several of the wheel housing opening designs, but no change in force distribution on the rear wheels was observed. Furthermore, by implementing so-called wheel housing ventilation the drag coefficient was also reduced. It is believed that combining the investigated design parameters as well as influencing the inflow to the wheel housing can result in a significant reduction of the drag coefficient of heavy trucks.

Acknowledgments The authors would like to acknowledge Vinnova and Energimyndigheten for funding the project, and Chalmers Computational Center for Science and Engineering (C3SE) and Gridcore AB for providing computational resources. Furthermore, Stephen Conway, Guillaume Mercier and Kamran Noghabai at Scania and the Sebastien Gautier at EXA GmbH should be acknowledged for software support and providing guidelines for the case setup.

References

1. Lögdberg, O.: Turbulent boundary layer separation and control, Ph.D. thesis, KTH, Stockholm (2008)
2. Hjelm, L., Bergqvist, B.: European truck aerodynamics—a comparison between conventional and CoE truck aerodynamics and a look into future trends and possibilities. In: ECI—The Aerodynamics of Heavy Vehicles II: Trucks, Buses and Trains (2007)
3. Wiedemann, J.: The influence of ground simulation and wheel rotation on aerodynamic drag optimization—potential for reducing fuel consumption. In: SAE Paper 960672 (1996)
4. Elofsson, P., Bannister, M.: Drag reduction mechanisms due to moving ground and wheel rotation in passenger cars. In: SAE Paper 2002-01-0531 (2002)
5. Wäschle, A.: Numerische und experimentelle Untersuchung des Einflusses von drehenden Rädern auf die Fahrzeugaerodynamik, Dissertation, Expert Verlag, Stuttgart (2006)
6. Camosy, M.: Advanced experimental methods for the analysis and aerodynamic design of heavy vehicles. In: ECI—Aerodynamics of Heavy Vehicles II: Trucks, Buses and Trains, Conference Presentation (2007)
7. Sebben, S.: Numerical simulations of a car underbody: effect of front-wheel deflector. In: SAE Paper 2004-01-1307 (2004)
8. Modlinger, F., Demuth, R., Adams, N.: New directions in the optimization of the flow around wheels and wheel arches. In: 7th MIRA International Vehicle Aerodynamics Conference (2008)
9. Cogotti, A.: Aerodynamic characteristics of car wheels, International Journal of Vehicle Design, Technological Advances in Vehicle Design Series, SP3, Impact of Aerodynamics on Vehicle Design, pp. 173–196 (1983)
10. Fabijanic, J.: An experimental investigation of wheel-well flows. In: SAE Paper 960901 (1996)
11. Morelli, A.: A new aerodynamic approach to advanced automobile basic shapes. In: SAE Paper 2000-01-0491 (2000)
12. Söderblom, D., Elofsson, P., Hjelm, L., Löfdahl, L.: Heavy vehicle wheel housing flows—a parametric study. In: SAE Paper 2009-01-1169 (2009)
13. Söderblom, D., Elofsson, P., Hjelm, L., Löfdahl, L.: Flow field investigation of the wheel housing opening shape on a heavy truck using CFD. In: 8th MIRA International Vehicle Conference (2010)

Part VI
Truck Aerodynamics: Drag Reduction

Passive Devices for Reducing Base Pressure Drag in Class 8 Trucks

Ramesh Pankajakshan, C. Bruce Hilbert and David L. Whitfield

Abstract Full scale CFD simulations of the Generic Conventional Model (GCM), a simplified model of a Class 8 truck, were used to explore passive devices for improving the drag performance of the trailer base. Significant improvements over conventional straight base flaps were achieved using an Extended Bent (EB) flap that stays within the length limits imposed by US federal law. An additional boat tail device for the cab bogie base was also found to yield improvements in the base drag in that region. This device in combination with the EB flap leads to a wind-averaged drag reduction of 21 % over the non-modified GCM model. An under-trailer scoop to generate air for pressurizing the trailer base or for use in active flow control devices was found to add too much drag to be effective.

1 Introduction

The trailer base region of a modern Class 8 truck accounts for a significant portion of the aerodynamic drag experienced by the vehicle at freeway speeds. A large body of research aimed at improving the aerodynamics of this region has been conducted by academia, national labs and private companies. Of the passive devices which have been developed, the base flap, a set of flat plates approximately 25 % of the trailer width in length and attached to the edges of the trailer at slant angles ranging from 10–20°, have been the most effective. Extensive discussions on the design and effectiveness of various passive base flow devices including boat-tails, base plates and base flaps can be found in [1–4, 8].

The fundamental problem of allowing easy access to the trailer doors at loading docks while fitted with a base drag reduction device has been an engineering challenge which has dogged the widespread adoption of such devices. Recently, a number of devices with ingenious deployment mechanisms requiring minimal effort on the part

R. Pankajakshan (✉) · C.B. Hilbert · D.L. Whitfield
SimCenter: National Center for Computational Engineering, University of Tennessee at Chattanooga, Chattanooga, USA
e-mail: ramesh-pankajakshan@utc.edu

A. Dillmann and A. Orellano (eds.), *The Aerodynamics of Heavy Vehicles III*,
Lecture Notes in Applied and Computational Mechanics 79,
DOI 10.1007/978-3-319-20122-1_14

of the truck operator have become available. Concerns among fleet operators as to their efficacy, additional weight and robustness under rough usage in the field have prevented large scale adoption of such devices. The opportunity exists to look at improving the efficiency of these devices through simulation and/or testing so that a large number of effective concepts are available to be engineered into viable devices. Continuous improvements in effectiveness can only help in speeding up adoption of such devices.

2 Simulations

2.1 Flow Solver

The incompressible Unsteady Reynolds Averaged Navier Stokes (unRANS)/Discrete Eddy Simulation (DES) module of the UT Tenasi solver suite was used for all the simulations presented in this paper. The module uses Chorin's artificial compressibility approach [5] for low speed flows with a finite volume node centered Roe averaged flux formulation on unstructured grids. The 1-equation Menter SAS model in DES mode [6] was used for all the runs presented in this study. Integration was performed to the wall using grids with off-wall y+ values less than unity for full-scale Reynolds numbers. The specific solver parameters and grid generation approaches used for this set of simulations were validated using data from the Ground Transportation System (GTS) [7] and the Generic Conventional Model (GCM) experiments [8]. In the validation studies [9, 10], the pressure distributions in the regions of attached flow were well predicted with significant discrepancies in the base regions where the flow was massively separated. The predicted forces were found to be in a good agreement with measured values.

2.2 Generic Conventional Model (GCM)

The simplified truck model (Fig. 1) used for the current study was derived by making a number of modifications to the model originally used for the wind tunnel experiments.

Fig. 1 Generic Conventional Model (GCM) with full size trailer

The geometry was lowered to the ground to allow the wheels to contact the ground plane along with the removal of the support struts used for mounting the geometry in the wind tunnel. Fillets were introduced at the contact line between the wheel and the ground plane for ease of grid generation. Cab extenders where added to the sides of the tractor and the trailer length was extended to match the dimensions of a full-size trailer (53 ft) as opposed to the smaller scale version (42 ft) used in the original GCM model.

The initial surface mesh for the geometry as well as the inviscid tetrahedral mesh generation was done using the commercial package Pointwise. Modifications to the geometry were made either using the built-in manipulations tools of Pointwise or generated separately using SolidWorks and then integrated into the existing model. Viscous layers were inserted into the initial tetrahedral mesh using an in-house parallel viscous layer insertion code [11]. The off-wall spacing, growth parameters and the locations and spacing of transparent surfaces used for spacing control in the field near wake structures were based on the results of validation studies on truck models and other bluff bodies. The final grids have approximately 21 million control volumes with 1.2 million surface elements. Wall clock time for each case was approximately 6 days on ~200 Intel Xeon 3.0 Hz EM64T processor cores.

2.3 Flow Conditions

A Reynolds number of 5.15 million based on a trailer width of 8.5 ft (2.59 m) and a road speed of 65mph (29.05 m/s) was used for all simulations. Appropriate slip boundary conditions were imposed on the ground plane and the rear trailer wheels to simulate the effect of the truck moving relative to the ground and rotating wheels respectively. Simulations were conducted with a 7 mph (3.13 m/s) crosswind at six orientations to the horizontal axis of the truck were used to compute a wind-averaged drag number which was computed as a non-dimensional axial force with the road speed and trailer width as the reference speed and reference length respectively.

3 Results

3.1 Drag Components on the GCM

Table 1 shows components that contribute significantly to the total computed drag force on the GCM model. The trailer base and trailer bogie account for 32 % and 26 % of the total drag. The contribution of the trailer as a whole (55 %) is less than the sum of the two largest components due to the negative contribution (−9 %) from the front of the trailer caused by the low pressure in the cab-trailer gap. The gap flow is a net contributor to the drag (3.4 %) due its effect on the cab base. The relatively

Table 1 Truck components and their drag contributions

Truck component	Drag contribution (%)
Trailer front	−9.3
Cab base	12.7
Trailer wheels	21.1
Trailer bogie	26.4
Trailer base	32.0
Trailer	55.0
Cab	45.0
Truck	100.0

large contributions of trailer base and bogie to the total truck drag make them obvious targets for drag reduction efforts.

3.2 Short Base Flaps

As shown in Fig. 2, the original base flap design consisted of four straight flaps, 2.1 ft in length, each inclined to the corresponding surface of the trailer by an angle of 15°. These flaps reduce the drag on the model by 11 %. The effect of base flaps on the pressure is significant in that the drag due to the base decreases by 69 % while contribution of the base to the total drag drops from 30 to 12 %. However some of the gains are offset by the additional drag on the flaps themselves and the drag contribution of the trailer base including flaps accounts for 22 % of the total (reduced) drag.

3.3 Extended Base Flaps

The extended base flaps shown in Fig. 3, consist of additional flaps attached to the original design which extend to a total length of 5 ft behind the trailer. The new piece

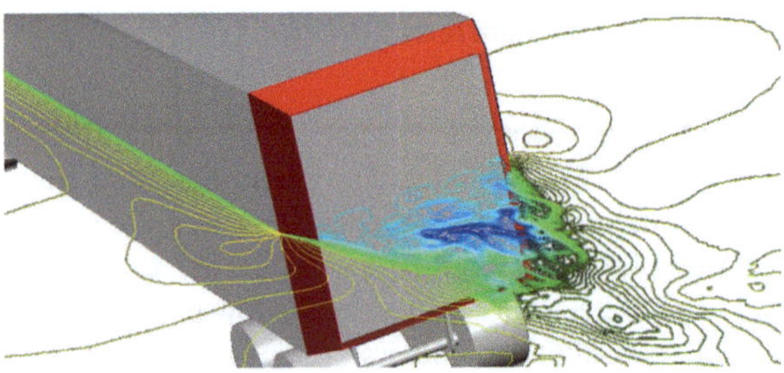

Fig. 2 GCM with short base flaps

Fig. 3 Extended base flaps with iso-surfaces of non-dimensional axial velocity ($V_{axial} = -0.001$)

is slanted an additional 15° such that the flaps are at an angle of 30° to the horizontal axis. It was found that the extended flaps produce no significant improvement in drag values compared to the short flaps. The lengthening of the flaps increased base pressure to the point where the trailer base drag was reduced by 91 %. The improvement was however completely offset by the added drag on the flaps due to separation on the extended section of the flaps.

3.4 Extended Bent (EB) Flaps

The EB flaps (Fig. 4) were an attempt to obtain the same base pressure increase as the extended base flaps while avoiding the drag cost of separated flow by making the aft section of the flap horizontal. The angles and lengths of the flaps were adjusted on the basis of the separation patterns that were observed on the extended flaps. In the final design, the slanted section extends 4.05 ft from the base of the trailer (perpendicular length) with the horizontal section of 0.95 making up the total extension length of 5 ft. The slant angles for the top and bottom flaps are 18 and 8° respectively with both the slide flaps at 13°. The base pressure was almost as high as that for the extended base flaps while the drag contribution from the EB flaps was slightly higher than that on the short flaps with the net effect that the total drag was reduced by about 17 %.

Fig. 4 Extended bent flaps with iso-surfaces of non-dimensional axial velocity ($V_{axial} = -0.001$)

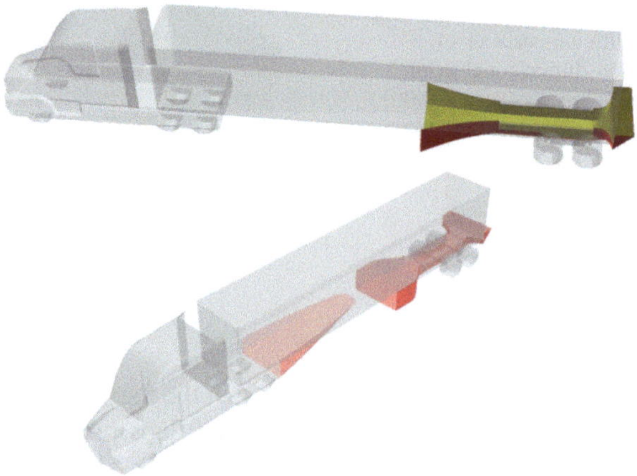

Fig. 5 Scoop designs with cab bogie boat tail

3.5 Under-Trailer Scoop

The under-trailer scoop was an attempt to craft a single solution to tackling the two large sources of drag on the trailer. The idea is to use the scoop to reduce drag on the trailer bogie and then use the airflow to reduce the drag on the trailer base either directly or though flow-turning using some kind of actuator device. The two scoops tried (Fig. 5) were unsuccessful in that the additional drag of the device far exceeded the drag reduction effects. Part of the problem is that the flow in the region might lack sufficient dynamic pressure to be effective. The problem is further exacerbated by the restrictions imposed on possible scoop designs by the existing bogie structures.

In analyzing the first scoop solution, it was found that the massively separated flow behind the base of the cab bogie not only accounted for almost 10 % of the total drag, but was also disturbing the flow into the scoop intake. The addition of boat tail to the cab bogie base (lower Fig. 5) cleaned up the flow while reducing drag by 11 %. The singular effectiveness of the boat tail in the presence of the scoop naturally leads to the question of whether the effect can be replicated with the best trailer base treatment, viz. the EB flap in the absence of the scoop.

3.6 EB Flap with Cab Bogie Boat Tail

This combination of devices (Fig. 6) was found to reduce the wind averaged drag by 21 % compared to the base GCM with no base treatments of any kind. The boat tail was found to be less effective (8 % drag reduction) in the absence of the scoop which was increasing the pressure on the device. Moreover, it increased the average flow velocity under the trailer and thereby increased the drag on the trailer bogie by 20 % with the net effect that total drag (wind-averaged) improvement over the EB

Table 2 Drag contributing components for the base GCM and design with EB flaps

Truck component	GCM drag contribution (%)	
	Base	EB flaps + boattail
Trailer front	−9.3	−12.5
Base flaps	0.0	8.3
Cab base	12.7	20.9
Trailer wheels	21.2	28.0
Trailer bogie	26.4	35.2
Trailer base	32.0	18.2
Trailer	55.0	47.6
Cab	45.0	51.4
Truck	100.0	100.0

Fig. 6 EB flaps with cab bogie boat tail and instantaneous streaklines

flap alone was about 5 %. The drag contributions of the truck components with and without drag reduction devices are show in Table 2.

It is interesting to note that with the improvement in the drag profile of the trailer, the cab now accounts for more than half of the (reduced) drag while the trailer bogie now accounts for more than a third. The cab-trailer gap has also become more significant as it now accounts for 8.5 % of the drag.

3.7 Eliminating the Bottom Flap

One question that was raised during the conference was the effect of the lower flap on the performance of the drag devices. Cooper [3] noted that the addition of a horizontal bottom flap to a three-sided design consisting of straight flaps (non-dimensional length = 0.18) at a 15° inclination increased the drag reduction from 25 to 28 %. The improvement was also found to increase with increasing flap length. Some commercial designs consist of only the top and side flaps but the relative importance of the bottom flap is still an unresolved question. In order to study this problem, two additional geometries were created by removing the bottom flap from both the short base flap and the EB flap. A zero yaw case was run for both geometries and the resulting drag compared with the corresponding four-flap cases. For the short flap design the absence of the bottom flap increases the total drag by 8.6 % while the increase for the EB flap was 7.3 %. The zero yaw drag reduction for the four cases when compared to a flap-less trailer base is shown in Table 3.

Table 3 Effect of removing the bottom flap on drag reduction

	Four flaps (%)	Three flaps (%)
Short flap variant	12.5	4.9
EB flaps	17.5	11.5

4 Discussion

The trailer base flap is an effective device for reducing the drag contribution from the massive flow separation region behind the truck. The design of the device involves finding the shape and orientation that maximizes the base pressure while minimizing the drag contribution of the flaps themselves such that the total effect is maximized. Since the geometry of the optimal device is likely to be highly dependant on the details of the actual tractor-trailer combination, this proof-of-concept effort is an attempt to show that there might be sufficient efficiency rewards to exploring the design space afforded by current regulations. An optimal design is also a good starting point to begin the inevitable process of making engineering tradeoffs needed for usability, robustness, etc. that will lead to a sub-optimal but deployable device.

The tractor bogie is a solid structure in the GCM model while it is a complex structure with considerable flow through in a real truck. Thus, the drag produced by the base of that region in a real truck is probably less than the 10 % of the total as seen in the GCM. If the drag contribution from flow separation in that region is significant, a boat-tail would be a highly effective device. Due to the need for trailer rotation about the kingpin and to avoid interference with the trailer landing gear, a boat tail would have to be a two piece device with a circular interface with one piece each attached to the bogie base and the underside of the trailer. A simulation with detailed tractor geometry and under-hood flow will be required to evaluate any devices in this region of the truck.

The simplified trailer bogie on the GCM offers far less blockage to the flow than the bogie in a real truck. This factor must be kept in mind while evaluating the effects of removing the bottom flap on the effectiveness of the base flap devices.

5 Conclusions

Variations in the length and shape can be used to considerably improve the effectiveness of base flaps particularly in comparison to the short straight flaps which have been studied extensively. They may however offer considerable challenges in transitioning to deployable devices in the field. Efforts to simultaneously tackle two of the largest drag contributors were stymied by the quality of the flow and design limitations imposed by the trailer bogie geometry. Improvements in one region of the truck may have a detrimental impact on downstream components as was observed in the case of the cab boat-tail and the trailer bogie. This implies that upstream devices have to be designed and evaluated first or under a systems approach, the design has to be done in tandem. The bottom flap was found to be a critical part of two of the base flap devices simulated in this work, at least for the GCM model. With the addition of the EB flap and boat tail, the cab and trailer bogie hold out the best areas for improvement and will be targeted in ongoing and future studies.

Acknowledgments This works was funded in part by the Tennessee Higher Education Commission Center of Excellence in Applied Computational Science and Engineering with Dr. Harry McDonald as technical monitor. This support is gratefully acknowledged.

References

1. Browand, F., Radovich, C., Boivin, M.: Fuel saving by means of flaps attached to the base of a trailer: field test results. SAE Paper 2005-01-1016 (2005)
2. Coon, J.D., Visser, K.D.: Drag reduction of a tractor-trailer using planar boat-tail plates. In: McCallen, R., Browand, F., Ross, J. (eds.) The Aerodynamics of Heavy Vehicles: Trucks, Buses and Trains. Springer, Berlin (2004)
3. Cooper, K.R.: The Effect of Front-Edge Rounding and Rear-Edge Shaping on the Aerodynamic Drag of Bluff Vehicles in Ground Proximity. SAE Paper No. 850288, Detroit, Michigan (1985)
4. Lanser, W.R., Ross, J.C., Kaufman, A.E.: Aerodynamic performance of a drag reduction device on a full-sclae tractor/trailer. SAE Paper No. 912195 (1991)
5. Chorin, A.J.: A numerical method of solving incompressible viscous flow problems. J. Comput. Phys. **2**, 12–26 (1967)
6. Nichols, D.S., Hyams, D.G., Sreenivas, K., Mitchell, B.J., Taylor, L.K., Whitfield, D.L.: Aerosol propagation in an urban environment. AIAA Paper 2006-3726 (2006)
7. Croll, R.H., Gutierrez, W.T., Hassan, B., Suazo, J.E., Riggans, A.J.: Experimental investigations of the ground transportation systems (GTS) project for heavy vehicle drag reduction. SAE Paper 96-0907 (1996)
8. Storms, B.L., Satran, D.R., Heineck, J.T., Walker, S.T.: A study of reynolds number effects and drag-reduction concepts on a generic tractor-trailer. AIAA Paper 2004-2251 (2004)
9. Sreenivas, K., Nichols, D.S., Hyams, D.G., Mitchell, B.J., Sawyer, S., Taylor, L.K., Whitfield, D.L.: Computational simulation of heavy trucks. AIAA Paper 2007-1087 (2007)
10. Sreenivas, K., Mitchell, B.J., Nichols, D.S., Hyams, D.G., Whitfield, D.L.: Computational simulation of the GCM tractor-trailer configuration. In: McCallen, R.C., Browand, F., Ross, J. (eds.) The Aerodynamics of Heavy Vehicles II: Trucks, Buses and Trains. Springer, Berlin (2007)
11. Karman, S.: Unstructured viscous layer insertion using linear-elastic smoothing. AIAA Paper 2006-0531 (2006)

Numerical and Wind Tunnel Analysis Together with Road Test of Aerodynamic Add-Ons for Trailers

G.M.R. van Raemdonck and M.J.L. van Tooren

Abstract Due to increasing environmental constraints it is crucial for the heavy duty transport sector to find solutions to stay sustainable in this aggressive and fast changing market. Improving aerodynamic quality, i.e. reducing drag, will contribute largely to the solution for this issue. With numerical analysis of a standard tractor semi-trailer combination a general overview of the flow has been obtained: a highly complex flow and large separated regions can be observed. Wind tunnel experiments with a 1:14 scaled truck model are performed to experimentally analyze the effect of aerodynamic aids to reduce the drag of the tractor with trailer. Aerodynamically shaped skirts improve the drag reductions more then 14 %. Mounting a boat tail reduces the drag with 12 %. Full-scale prototypes are built and road tests are conducted. The aerodynamic skirts, called SideWings showed an averaged fuel economy increase 1.6 l/100 km when tested on the circuit and the road. During one year four different boat tail configurations are tested. A fuel saving of 2 l/100 km for the 2 m tail is obtained.

1 Introduction

In this article the development of two different aerodynamic add-ons for trailers will be presented. Two solutions will be discussed: aerodynamic shaped skirts, called SideWings, and the boat tail concept. Both aerodynamic fuel saving techniques are developed with the aid of numerical simulations, wind tunnel experiments and extensive road tests.

G.M.R. van Raemdonck · M.J.L. van Tooren (✉)
Faculty of Aerospace Engineering (TU Delft), Kluyverweg 1, 2629 HS Delft, The Netherlands
email: m.j.l.vantooren@tudelft.nl

G.M.R. van Raemdonck
email: g.m.r.vanraemdonck@tudelft.nl

A. Dillmann and A. Orellano (eds.), *The Aerodynamics of Heavy Vehicles III*,
Lecture Notes in Applied and Computational Mechanics 79,
DOI 10.1007/978-3-319-20122-1_15

1.1 Background

The expected road transport demand in the next decades and the increasing environmental constraints together with the rising fuel prices has renewed the interest in aerodynamic truck and trailer design in the last decade; any reduction in truck fuel consumption can be associated with large annual fuel cost reductions and considerable emission savings for the transport sector.

The heavy duty transport sector covers a very large business area. It is crucial, however, to find solutions to stay sustainable in this aggressive and fast changing market. Cargo transport by road using tractor semi-trailer combinations is still one of the most efficient transport solutions within the European context. This fact manifests itself in the large amount of trucks on the road and the associated total volume of fuel consumed annually. The environmental concerns and the harsh competition force the transport companies to reduce fuel consumption in an economically and environmentally sustainable way.

Generally there are two ways to reduce the fuel consumption of a vehicle. One can improve the efficiency of the engine delivering power: improvements on the side of the available power. Or one can lower the different forces acting on a truck traveling over the road: the required power side. The latter can be achieved by reducing the weight of the vehicle, reducing its aerodynamic drag and by reducing the friction resistance of the tires.

1.2 Past Research

Improving the fuel economy of trucks, and more specifically tractors, by aerodynamic means has become an accepted practice in the last decades. Tractor manufacturers improved the aerodynamic performance of their tractor by applying roof deflectors, side fenders and corner vanes. Several aerodynamic solutions have been developed for the front and top of the tractor and for the gap between the tractor and semi-trailer, [2, 7, 11, 16].

Beside the development of aerodynamic add-ons for the cabin, modifications of the general shape of the vehicle are introduced to improve the aerodynamic efficiency through, for instance, rounded cabin corners and sloping front window. Futuristic aerodynamic designs concepts have been presented by the European truck manufactures. Up until this moment the main focus of the manufacturers was on the development of the tractor and not particularly on the trailer and the vehicle as a whole.

1.3 Present Research

In 2005 a research program was initiated at the faculty of Aerospace Engineering to reduce the fuel consumption of heavy duty vehicles by lowering the aerodynamic drag of articulated trucks. These types of vehicle are mostly used for long haul road transport and have a high aerodynamic drag contribution due the bluff shape of the trailer and its high average traveling velocity.

The presented research has three components: numerical analysis, wind tunnel experiment and road testing. A tractor is equipped with a data acquisition system to monitor the performance of a standard tractor with semi-trailer during operational activities. With the aid of computational fluid dynamics (CFD) numerical simulations of a tractor semi-trailer combination is conducted to identify the critical regions that are responsible for the high aerodynamic drag. Several aerodynamic add-ons for a trailer are designed and tested in the wind tunnel with a scaled truck model. Finally full-scale prototypes are built to perform several road tests in order to indicate the fuel savings of the final designs of the aerodynamic devices.

2 Pre-design Phase

Before one can start designing aerodynamic add-ons the boundary conditions of the problem has to be defined. This is achieved with the aid of monitoring the performance of a long-haul vehicle and initial numerical analysis of the flow around an articulated truck.

2.1 Data Acquisition and Monitoring

In preparation of the numerical simulations and the wind tunnel experiments, the performance of a general tractor semi-trailer combination is monitored during a testing period of two months throughout Europe. A data acquisition system was connected to the on-board computer of the tractor through a CAN bus network connection. The obtained results are used as input and as design parameters for the design process of the aerodynamic add-ons.

Based on the obtained data more insight in the aerodynamic performance is obtained through calculation: an estimated C_D value of 0.430; mechanical performances: the power required broken down into 47 % rolling resistance, 39 % of aerodynamic drag and 15 % mechanical losses; an average fuel consumption of 30 l/100 km and the majority of the velocities higher than 80 km/h, see Fig. 1. More than 88 % of all yaw angles measured, illustrated in Fig. 1, represent the yaw angle range $\pm 10°$, van Raemdonck [15].

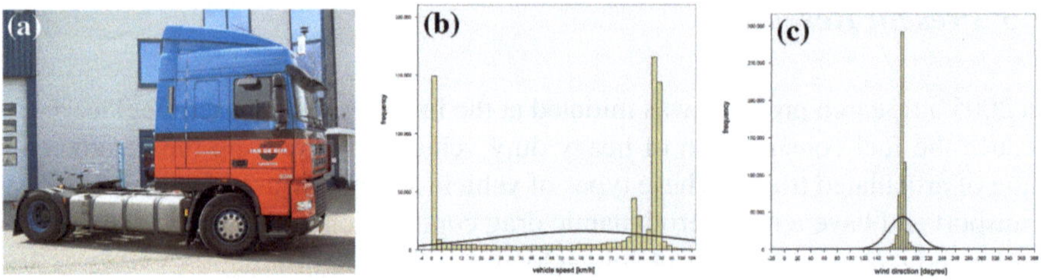

Fig. 1 Tractor with data acquisition system and performance monitoring results. **a** Data monitoring system on tractor, **b** vehicle speed histogram, **c** wind direction histogram

2.2 Numerical Analysis of a Long Haul Truck

Numerical analysis of a generalized tractor-trailer combination with the aid of Computational Fluid Dynamics (CFD) is conducted, van Raemdonck [14], to obtain insight in the flow behavior and its characteristics.

A surface model is generated by laser scanning a full-scale tractor. A long range scanner (accuracy of ± 3 mm) and a Konica Minolta VI-9i laser (accuracy of ± 1 mm) are used. The model of the standard trailer is based on technical information provided by a trailer manufacturer.

The discretization of the volume around the surface model is conducted by introducing blocks. The surface mesh consists of triangles that are extruded normally to create a six-layered boundary layer made of prisms ($y^+ = 50$, height of first cell = 1 mm, growth rate = 1.15). The volume around the model (length: 127.5 m, width: 30 m; height: 30 m) is filled with tetrahedrals. A half model is used for the simulations in order to reduce the amount of cells and the corresponding processing time.

The Navier-Stokes flow equations together with the Realizable $\kappa - \varepsilon$ turbulence model are solved with the commercial package Fluent. As near-wall treatment for the boundary layer, the non-equilibrium wall function, is chosen. The operating pressure of 101325 Pa is defined in the outer corner of the computational domain.

The upstream inlet has the velocity inlet boundary condition: an absolute magnitude of 25 m/s in the direction of the flow. The outlet is set as pressure outlet. The actual symmetry plane as well as the other two outer planes (side and upper plane) have the symmetry boundary condition. The floor is defined as a moving wall with a translational velocity of 25 m/s.

All the surfaces of the tractor and trailer are solid stationary walls. The wheels of the tractor and trailer have a moving wall boundary condition. In this truck model no boundary condition is set for the radiator.

The three dimensional streamlines around the vehicle together with the pressure coefficient are shown in Fig. 2. At the front surface the oncoming flow stays attached around the cabin corners and the roof deflector. The recirculation zone with the two counter rotating vortices at the rear are clearly visible. The streamlines at the bottom of the vehicle, especially at the underside of the trailer, have a highly separated

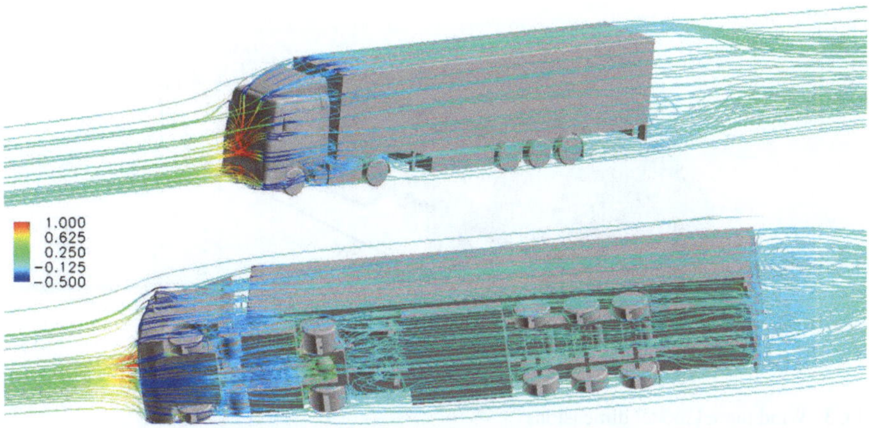

Fig. 2 Streamlines at underside of tractor semi-trailer combination

behavior. Figure 2 indicates that two regions, the underside and the back of the trailer, have highly separated flows, which cause a lot of pressure drag.

Considering the total drag of the vehicle, the frontal area and rear trailer surfaces has a drag contribution of 0.23 and 35 %, respectively. The region in between, including the underside of the vehicle, delivers 42 % of the total drag.

3 Aerodynamically Shaped Skirts

The drag caused by the underside of semi-trailer is significant, therefore it is interesting to improve the aerodynamics towards a lower total vehicle drag. In this section the development of the aerodynamic skirt, called SideWing, will be discussed.

3.1 Wind Tunnel Experiments: Skirts

A series of wind tunnel experiments is conducted in the Low Turbulence Tunnel. This closed circuit wind tunnel has an octagonal test section with a cross sectional area of 2.07 m^2 (width: 1.8 m; height: 1.25 m) and a maximum velocity of 120 m/s. The wind tunnel does not have a moving belt to simulate the road. According to Cooper [5, 6] a fixed elevated ground board with a thinned boundary layer is sufficient to conduct wind tunnel experiments.

A 1:14 scaled truck model (TAMIYA Mercedes Benz 1838LS truck and TAMIYA container-trailer) is used. The initial truck model had sharp corners which will result in front flow separation and therefore would make any aerodynamic add-on inef-

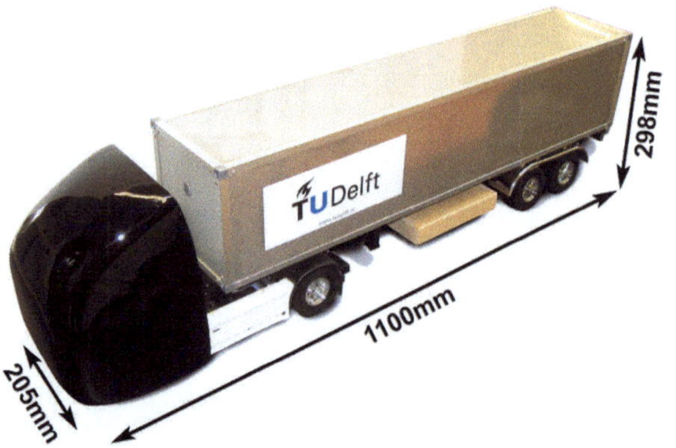

Fig. 3 Wind tunnel model dimensions

fective. A new cover, see Fig. 3, with properly defined round-off edges, based on consideration given in Hoerner [8] and Cooper [6], is made which prevents flow separation at the front cabin corners. The model is not equipped with a fan system and is mounted belly-up in the test section due to the location of the balance system.

The Reynolds number of a full-scale truck, based on a square root of the frontal area $A = 10.34\,\text{m}^2$ and a driving velocity $V = 25$ m/s, becomes 5.4×10^6. The scaled wind tunnel model has a Reynolds number (tunnel speed of 60 m/s) of 0.98×10^6 which is high enough for bluff bodies, SAE J1252 [10]. After testing a Reynolds number range, no Reynolds effects are deducted for the drag coefficient, illustrated in Fig. 4. No wind tunnel corrections methods are applied on the dynamic pressure.

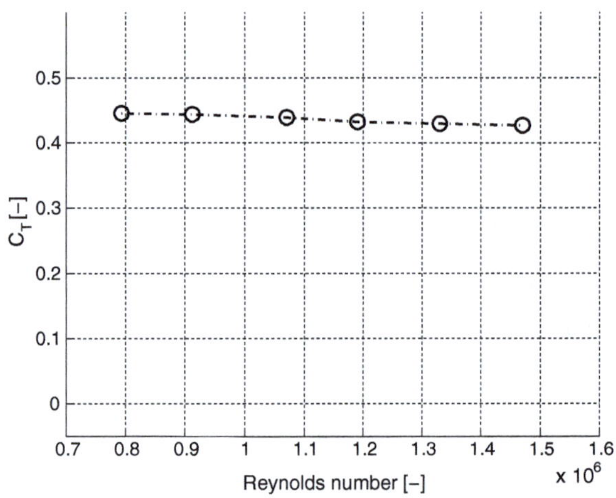

Fig. 4 Drag coefficient over a Reynolds number sweep

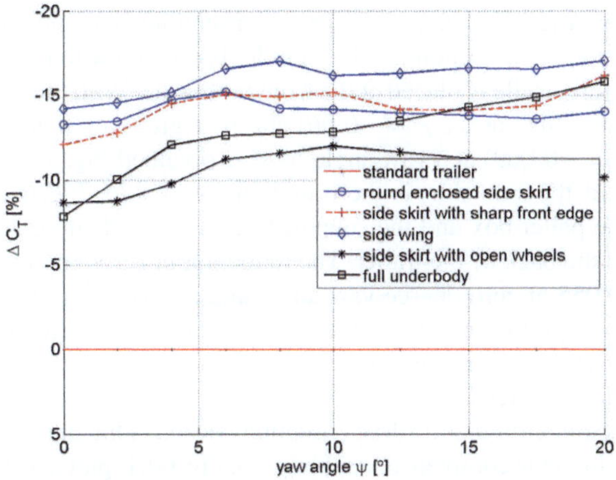

Fig. 5 Drag reduction results

Fig. 6 Aerodynamic add-ons tested in the wind tunnel. **a** Skirt with uncovered wheels, **b** skirt with sharp front edge, **c** round enclosed skirt

During the experimental measurements more than 100 different aerodynamic devices are designed and tested. Only a short selection is being discussed here. A more profound overview of the results can be found in van Raemdonck [12, 14]. All the drag coefficients C_T in longitudinal direction of the vehicle with the different aerodynamic devices are compared with the standard tractor-trailer combination ($C_T = 0.443$ at zero yaw angle) equipped with a pallet box, landing legs, mudflaps and a rear bumper.

The most well know side skirt with uncovered wheels, see in Fig. 6a, reduces the drag with 8 % as indicated in Fig. 5. The same skirt, but with covered wheels, see Fig. 6b, reduces the drag coefficient with 12 % or $\triangle C_T = -0.049$ at a zero yaw angle and stays almost the same with increasing yaw angle, as can be seen in Fig. 5. Uncovered wheels introduce significant drag that results in a lower drag reduction: 4 % at zero yaw angle, or an increase of C_T with 0.018 compared to the covered wheel configuration.

With increasing yaw angles the flow separates at the sharp front vertical edge of the standard skirt. In order to prevent flow separation several rounding configurations are applied. The configuration with the rounding at the outside of the skirt performed not better. A better result is obtained when the rounding is applied at the inner side of the skirt. The applied rounding, a half circle, is not an ideal aerodynamic shape. Therefor an airfoil profile is placed at the front inner side. With the optimized profile a drag reduction of 14 % or $\triangle C_T = -0.061$ at zero degree yaw angle and a maximum of 17 % ($\triangle C_T = -0.069$) at a yaw angle of 8° is obtained Fig. 5.

A variation of the side skirt is the round enclosed skirt, Fig. 6c. This variation encapsulates the pallet box and the support legs in a round curve and covering the wheels towards the back of the trailer. The reduction in aerodynamic drag is 12.5 % or $\triangle C_T = -0.055$ at zero degree yaw angle and a maximum of 14.5 % ($\triangle C_T = -0.064$) at 6°, as plotted in Fig. 5. Modifiations of this configuration (i.e. with a closed bottom) are tested as well and discussed in detail by van Raemdonck [12], but did not perform better.

Another type of aerodynamic drag reduction add-on is the under body. The measured drag benefit of the complete under body, see Fig. 6d, is plot in Fig. 5: it generates at zero yaw angle 7 % less drag ($\triangle C_T = -0.030$) and 12 % or $\triangle C_T = -0.052$ at 6° yaw angle. A variation of the under body has covered wheels to illustrate its influence on the drag coefficient C_T. Again better results are achieved with the wheels covered: a decrease of 3 % ($\triangle C_T = -0.013$) of the drag coefficient with respect to the similar configuration with the open wheels.

Out of this series of wind tunnel experiments the aerodynamic shaped skirt is selected as the best performing aerodynamic add-on for the underside of the trailer. The measured drag reduction of 14 % can be translated in a fuel economy reduction of 7 %, Rose [9], when the vehicle is driving 90 km/h.

3.2 Road Tests: SideWings

Full-scale prototypes of the SideWings are designed, manufactured from an engineering plastic and mounted with state-of-the-art techniques at the underside of a semi-trailer. Thirteen vehicles are equipped with the SideWings and subjected to an extensive testing program.

Controlled fuel economy tests are conducted on a testing circuit (RDW test circuit, Lelystad, The Netherlands) together with a reference vehicle. During the test day the vehicle combinations are swopped in order to analyze the fuel saving of the SideWings for both tractors. The vehicles together with the SideWings are also used during operational activities in corporation with several international transport companies. The vehicles are equipped with a data monitoring system that registers the fuel consumption through a CAN bus connection with the board computer of the corresponding vehicle.

The circuit tests are conducted at a constant speed (≥85 km/h) with cruise control activated. This way the influence of the driver is canceled. With six different vehicles, see Fig. 7, on four different days a fuel consumption reduction between almost 1.5 and 2 l/100 km is measured, see Table 1. This corresponds with fuel savings of 4.1–7.5 %, depending on the weight of the empty vehicle, its cargo weight and the weather

Fig. 7 Circuit tests with different tractor-trailer combinations

Table 1 Overview circuit tests results

Test partner	abs. consump. [l/100 km]	Savings [l/100 km]	Savings [%]
1	36.86	1.53	4.14
2	20.86	1.58	7.50
3	28.15	1.33	4.73
4	27.20	1.96	7.22

conditions of that particular day. One has to remark that test partner 1 had a total weight of 40 ton. All other test partners were not carrying any cargo.

During several months of operational testing the SideWings proved to have an aerodynamic impact on the fuel economy of a tractor semi-trailer combination. The fuel consumption of the vehicle with the SideWings is compared with the averaged fuel consumption of a selection of the fleet for the same period. A fuel economy increase of average 1.6 l/100 km is measured when cruise control is activated amd driving maximum velocities.

4 Boat Tail

Besides the underside of the trailer, the rear-end has a large share in the total drag built-up of a heavy duty vehicle. To reduce the drag, four trapezoidal shaped plates are placed at the back of a truck, such that an open box is created with the shape of a pyramid called a boat tail, see Fig. 8. If the flow is able to follow the structure, the static pressure at the rear is increased reducing the drag.

4.1 Wind Tunnel Experiments: Boat Tail

The wind tunnel results of the boat tail presented here are based on the research executed in the Low Turbulence Tunnel, Assen [1]. Again a 1:14 scaled tractor-trailer combination was used for the experiments. During this wind tunnel research different slant angels of the boat tail are tested to analyze its influence on the wind average drag $\bar{C}_T$. The average drag coefficient of the clear configuration was $\bar{C}_T = 0.5122$.

Fig. 8 Standard semi-trailer equipped with a boat tail

As can be observed in Fig. 9 a boat tail with a zero slant angle reduces the drag by 2.5 % on average, which means a $\triangle\bar{C}_T = -0.0132$. Increasing the slant angle to 5° positively influences the drag reduction. The optimal slant angle is 10°, resulting in an average drag reduction of 12 %, corresponding with a $\triangle\bar{C}_T = -0.0635$. Increasing the slant angle further, up to 15°, is not beneficial for the aerodynamic drag, as can be seen in Fig. 9. Removing the bottom plate reduces the drag reduction results: only a $\triangle\bar{C}_T = -0.0418$ is measured.

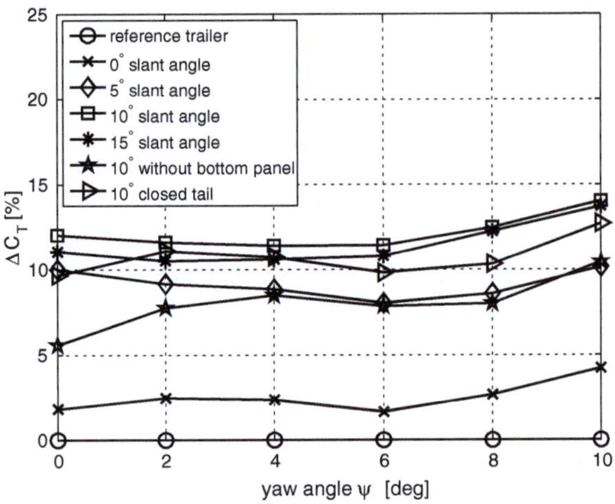

Fig. 9 Drag coefficient of different boat tail configurations

In Fig. 10 the drag results for a simplified truck model (van Raemdonck [13]) are illustrated when the offset of the bottom plate is varied. A complete boat tail, configuration (1) in Fig. 10, gives a drag reduction of 40 % at zero yaw angle, while

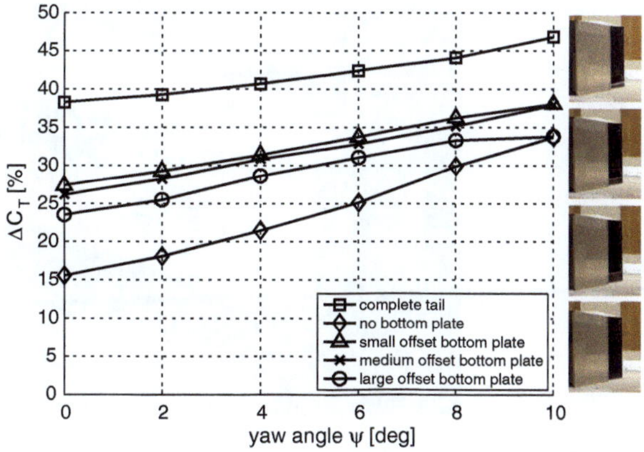

Fig. 10 Bottom plate offset variation on a simplified truck

the bottom plate is removed, i.e. configuration (5), only drag reduction of 15 % is measured. Increasing the offset of the bottom plate, configuratinos (2), (3) and (4), reduces the drag reduction significanly.

It is evident that the presence of the bottom panel is a prerequisite for obtaining the largest drag reduction possible, since the drag reductions achieved without the bottom panel is less then half of that of the versions with bottom panel for the simplified truck model. Tests performed by Coon and Visser [4] show similar results.

4.2 Road Tests: Boat Tail

As the wind tunnel results seemed promising, extensive road test have been performed with a full-scale prototype of the boat tail, see Fig. 11. For more than two years, a truck equipped with a data monitoring system has been driving around The Netherlands with and without a boat tail. In total three different boat tail lengths are tested: 1, 1.5 and 2 m. The 2 m boat tail is also tested with an extra rear bumper which was required for safety reasons.

The vehicle combination without a boat tail has a fuel consumption of 27.5 l/100 km, driving a constant velocity of 85 km/h and with a total vehicle weight of 40 ton. Analysis of the tests show, see Fig. 12, that the 2 m boat tail generates a fuel economy improvement of 7.5 %, which results in a reduction of 2 l/100 km. Adding the extra rear bumper costs 2 % in fuel efficiency: only a fuel consumption of 26 l/100 km is measured. As expected, shorter boat tails go together with smaller fuel savings. A boat tail of 1.5 m has a fuel saving of 6 % or 1.7 l/100 km, while the 1 m boat tail results in a 3 % or 0.8 l/100 km reduction in fuel consumption compared to the truck without a boat tail. As indicated with the trend line in Fig. 12 shorter boat tails then 1 m will give drastically lower or even no fuel savings.

Fig. 11 Road test with the boat tail

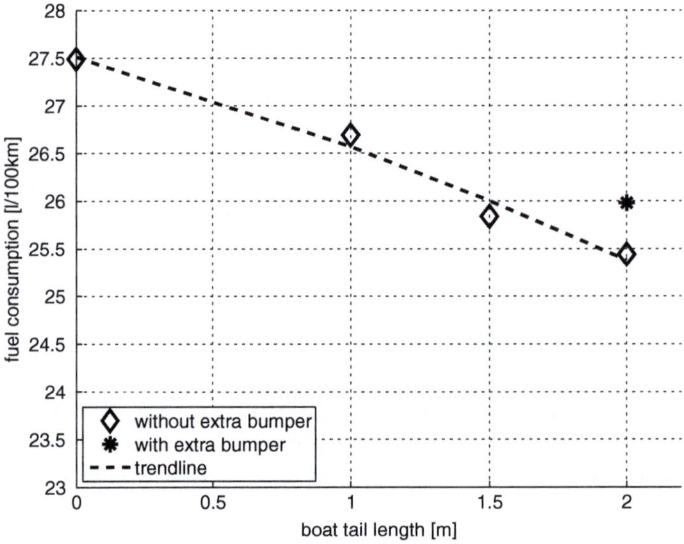

Fig. 12 Results of road test with boat tail

5 Combining SideWings and a Boat Tail

In the above sections the aerodynamic shaped skirt and the boat tail are tested in the wind tunnel and on the road on individual vehicles. When both aerodynamic add-ons are combined on one vehicle higher drag reductions are expected.

5.1 Semi-trailer

Figure 13 shows a combination of two aerodynamic add-ons for a semi-trailer: SideWings and a boat tail. An averaged drag reduction of 24 % is achieved during wind tunnel experiments, Assen [1]. This 24 % is almost the summation of the individual drag reductions of both aerodynamic aids. It is unclear what and how large the effect is of the shaped skirt on the boat tail and vice versa. Also what the contribution is of each aerodynamic add-on on the total drag reduction is unknown.

5.2 Combi-Vehicle

Within the same research program wind tunnel experiments of a combi-vehicle, see Fig. 14, are performed in the combination with skirts and with a boat tail, Buijs [3]. As can be seen from Fig. 15 the boat tail is able to reduce the drag coefficient of the combination by 11 %, with a maximum of 12 % for a yaw angle of 10°. When using both the boat tail and skirts, a drag coefficient reduction of 15 % is reached in head wind conditions, and the wind averaged drag coefficient is reduced by 23.6 %. These experiments with the combi-vehicle illustrate that a combination of both add-ons have a positive effect on the total drag of this type of vehicle. The drag reduction results of the aerodynamic add-on also indicate that the skirts improve the performance of the vehicle with the boat tail especially for increasing yaw angles.

Fig. 13 Semi-trailer equipped with SideWings and a boat tail

Fig. 14 Wind tunnel model of a combi-vehicle

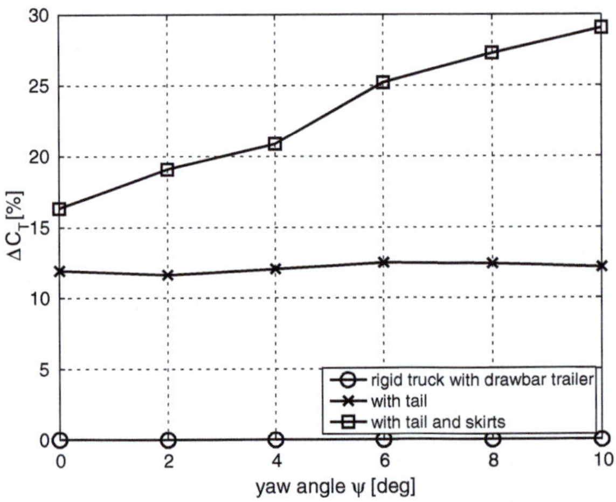

Fig. 15 Drag reduction results of a combi-vehicle equipped with skirt and boat tail

6 Concluding Remarks

Initial road tests defined the design boundary conditions while a full-scale numerical analysis indicated the problem areas of an articulated truck: the underside and rear side of the semi-trailer. For both area's two aerodynamics drag reduction devices are designed and tested.

An aerodynamic add-ons for the underside, called SideWings, is successfully developed. Scaled wind tunnel experiments showed drag reductions up to 17 % for a yaw angle of 8°. Several full-scaled prototypes are built to conduct an extensive test plan. Circuit and operational road tests during daily activities are executed with different testing partners. An average fuel consumption reduction of 1.6 l/100 km is measured when driving the maximum velocity.

For the rear-end of the semi-trailer the boat tail concept is extensively tested in the wind tunnel. With an optimal slant angle of 10° a drag reduction of 12 % is measured. Again a full-scale prototype is measured and tested on the public road for one year. The highest fuel economy improvement of 2 l/100 km is achieved with a two meter boat tail. An extra bumper reduces the fuel efficiency. When shorter boat tails are applied the fuel savings will reduce drastically.

A combination of the SideWings and the boat tail is tested in the wind tunnel on a semi-trailer and a combi-vehicle. Averaged drag reductions of more then 20 % are measured on both vehicles types. The next step in this development is to test the combination of the aerodynamic devices on the road.

Acknowledgments The manufacturing of the prototype and the execution of the road tests were made possible and successfully conducted thanks to the close cooperation with PART, Platform for Aerodynamic Road Transport (www.part20.eu). PART is a public-private partnership that initiates aerodynamic projects to improve the fuel efficiency of trucks. The spin-off company Ephicas, www.ephicas.eu, is further developing the SideWings into a commercial product to introduce this aerodynamic add-on in the international transport market.

References

1. Assen, R.C.A.G.: Base Drag Reduction of Articulated Trucks by Means of Boat-Tailing. Master's thesis, Delft University of Technology, Faculty of Aerospace Engineering, The Netherlands. (2007)
2. Browand, F., McCallen, R., Ross, J.: Aerodynamics of Heavy Duty Vehicles II: Trucks Busses and Trains. Springer, Berlin (2008)
3. Buijs, L.J.: Numerical and Experimental Analysis on Aerodynamic Solutions for Drag Reduction on Truck-Trailer Combinations. Master's thesis, Delft University of Technology, Faculty of Aerospace Engineering, The Netherlands (2010)
4. Coon, J.D., Visser, K.D.: Drag Reduction of a Tractor-Trailer Using Planar Boat Tail Plates. Clarkson University, Potsdam (2004)
5. Cooper, K.: The wind tunnel testing of heavy duty trucks to reduce fuel consumption. SAE Paper 821285 (1982)
6. Cooper, K.: Bluff-body aerodynamics as applied to vehicle. J. Wind Eng. Ind. Aerodyn. **49**, 1–21 (1993)
7. Freight Best Practice: Aerodynamics of Efficient Road Transport. Department of Transport, United Kingdom (2003)
8. Hoerner, S.: Fluid-Dynamic Drag. Hoerner Fluid Dynamics, Vancouver (1965)
9. Rose, M.J.: Commercial vehicle fuel economy—the correlation between aerodynamic drag and fuel consumption of a typical truck. J. Wind Eng. Ind. Aerodyn. **9**, 89–100 (1981)
10. Society of Automotive Engineers. SAE Wind Tunnel Test Procedure for Truck and Buses. SAE J1252 (1981)

11. van Raemdonck, G.M.R., van Tooren, M.J.L., Boermans, L.: Aerodynamics and Trucks. Platform for Aerodynamic Road Transport, Faculty of Aerospace Engineering—TU Delft, The Netherlands (2009)
12. van Raemdonck, G.M.R., van Tooren, M.J.L.: Design of an aerodynamic aid for the underbody of a trailer within a tractor-trailer combination. In: Bluff Body Aerodynamics and Application Conference, Milan, Italy (2008)
13. van Raemdonck, G.M.R., van Tooren, M.J.L.: Time-averaged phenomenological investigation of a wake behind a bluff bofy. In: Bluff Body Aerodynamics and Application VI Conference, Milan, Italy (2008)
14. van Raemdonck, G.M.R.: Design of an Aerodynamic Aid for a Tractor-Trailer Combination. Master's thesis, Delft University of Technology, Faculty of Aerospace Engineering, The Netherlands (2006)
15. van Raemdonck, G.M.R., van Tooren, M.J.L.: Data Acquisition System of a Tractor-Trailer Combination to Register Aerodynamic Performances, pp. 299–310. Aerodynamic of Heavy Duty Vehicles II: Trucks, Buses and Trains. Springer (2008)
16. Wood, R., Bauer, S.: Simple and low-cost aerodynamic drag reduction devices, SAE Paper 2003-01-337 (2003)

Computations and Full-Scale Tests of Active Flow Control Applied on a VOLVO Truck-Trailer

Mohammad El-Alti, Valery Chernoray, Per Kjellgren, Linus Hjelm and Lars Davidson

Abstract Large-eddy simulations and full-scale investigations were carried out that aimed to reduce the aerodynamic drag and thus the fuel consumption of truck-trailers. The computational model is a relevant generic truck-trailer combination, and the full-scale is a corresponding Volvo prototype vehicle. Passive and active flow control (AFC) approaches were adopted in this work and applied at the rear end of the trailer. Flaps were mounted at an angle that induces separation, and synthetic jet actuators were placed close to the corner of the rear end and the flaps. The drag reduction obtained is in the order of 30 %. The flow was analyzed by comparing the phase-averaged and time-averaged flow field of the unforced and the forced cases. The full-scale prototype is a Volvo truck-trailer. The trailer is mounted by three flaps at the rear sides and top end. The actuators consist of loudspeakers in sealed cavities, connected to amplifiers that are supplied with a frequency generator controlled by LabVIEW. The full-scale test includes passive and active flow control investigations by varying the flap angle, with and without AFC, investigating different frequency and slot angle configurations. The fuel flux was measured during the full-scale test. The test shows a fuel reduction of about 4 % in a comparison of two flap angles. The test of active flow control shows a reduction of 5.3 % compared to the corresponding unforced case. Compared with the baseline case, the passive flow control fails to reduce the total fuel consumption.

1 Introduction

Many efforts are made by vehicle manufacturers and research institutions around the world to reduce the drag of vehicles. The objective of drag reduction was earlier related to the increase in vehicle speed. Today, efforts to reduce drag are driven by

M. El-Alti (✉) · V. Chernoray · P. Kjellgren · L. Hjelm · L. Davidson
Department of Applied Mechanics, Division of Fluid Dynamics,
Chalmers University of Technology, 41258 Göteborg, Sweden
e-mail: mohammad.el-alti@chalmers.se

A. Dillmann and A. Orellano (eds.), *The Aerodynamics of Heavy Vehicles III*,
Lecture Notes in Applied and Computational Mechanics 79,
DOI 10.1007/978-3-319-20122-1_16

economic and environmental requirements. In the aerodynamic shaping of truck-trailers, the major advances have been implemented in the truck and fewer improvements have been considered for the trailer. The reason for this is both practical and economic. Every cargo company strives to maximize its cargo in order to increase income. The trailer design is therefore almost untreated by aerodynamics engineers. By studying the different contributions to the drag of a truck-trailer, interest has increased in analyzing the wake and altering its intensity by several add-on devices, sush as, splitters, vanes, cavities and boat tails. However, the gain in drag reduction and in fuel economy achieved by these rear-end add-on devices has never exceeded 10 and 5 %, respectively. This work investigates the opportunities for using an active flow control approach consisting of add-on flaps together with synthetic jet actuators to reduce the drag of truck-trailers. The idea is to increase the base pressure by narrowing the wake size. The efficiency of the AFC system is not considered in this work; the aim is to study the possible effect of the AFC system and the possibility to reduce the overall fuel consumption of truck-trailers. The design cycle for this work starts with the simulations. Large-eddy simulations of a simplified truck-trailer geometry with AFC ON and AFC OFF were carried out to find proper values of the different design parameters, test a simple actuator, employ add-on flaps with integrated actuators and finally to test the prototype vehicle at the proving ground. The idea of using flaps and reattaching the separated flow is in line with research showing successful results in tilt-rotor aircrafts carried out by the authors in [1, 2] that was a continuation of previous research [3–5]. Several active flow control efforts have been reported in the literature, both with and without an extended flap. The case without a flap was investigated in [6, 7], where the authors in [6] used a circular cylinder that had a built-in actuator and achieved a drag reduction of 20 %. In [7] they used closed-loop control on a blunt trailing edge and achieved a 10 % drag reduction. The authors in [8, 9] used base flaps with AFC and achieved a drag reduction below 10 %. One successful effort was reported in [10], where a drag reduction of about 30 % was achieved using steady blowing. This is a promising result, but the use of steady blowing is energy consuming and the net drag reduction is not particularly high.

The simulations in this work were carried out using a finite element code called FlowPhys v2.0a. This code has previously succeeded in simulating active flow control in several cases [3, 4]. The details of the code are given in [11, 12].

2 Computational Model

The computational model is a bluff body comparable to a truck with a width of ($w = 2.6$ m) and length of ($l = 13.0$ m), a typical size for today's truck trailers. Assuming a sufficiently large height of the truck, an appropriate large slice in the middle was isolated. The domain is shown in (Fig. 1a). The trailer was also supplied with additional flaps on the rear end. The oscillating synthetic jet actuators were placed immediately downstream of the corner between the rear end of the trailer and the flaps (Fig. 1b). The actuators were modelled as a time varying boundary condition

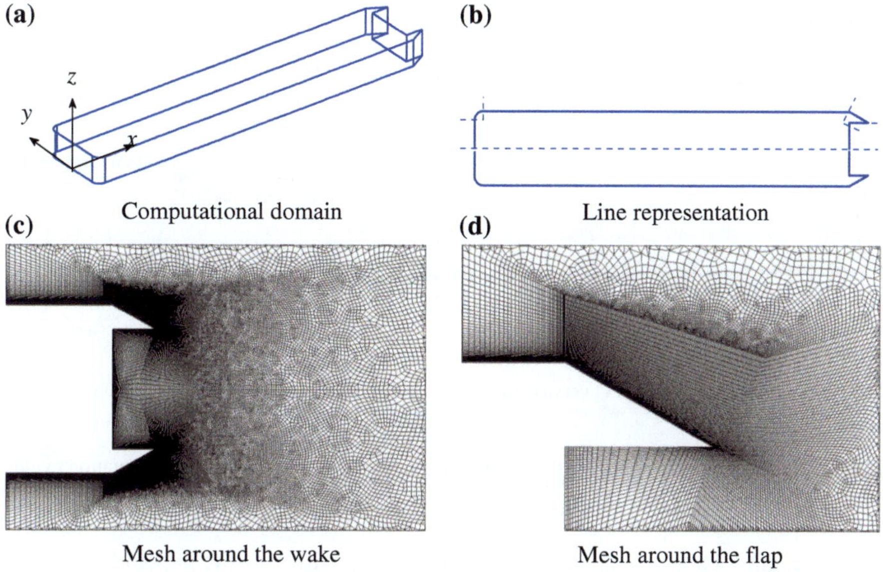

Fig. 1 **a** Simplified truck model: inlet at $x = -12w$; outlet at $x = 30w$. **b** *Line* representation of the truck. **c** 2D slice of the mesh, wake region and **d** flap region

(slot) and extended over the entire height (z) of the truck. The inlet flow was modelled as a constant free stream velocity, $U_\infty = 25$ m/s. The Reynolds number was reduced from 4300000 to 200000 by increasing the viscosity to $\mu = 3.25 \times 10^{-4}$ kg/ms, and the density was set to $\rho = 1.0$ kg/m^3. A part of the computational mesh is shown in Fig. 1c, d. There is a much larger concentration of nodes in the wake and flap regions as compared to the free stream region. The mesh size is about 1.4–3.3 million nodes and consists of a structured O-grid around the truck-trailer model, a free mesh of unstructured quad elements in the outer region in the x–y plane and structured quad elements in the span-wise direction. The mesh quality (Table 1) is very high in the wall and flap regions. A thorough dependence study of span-wise mesh resolution and domain size is reported in [13].

2.1 Results

The results of the numerical simulations are presented in this section. It starts with the analysis of the phase locked flow. Some results are presented for a smaller slot at

Table 1 Mesh quality values prior to the flap

Max y^+	Mean y^+	Max Δx^+	Mean Δx^+	Max Δz^+	Mean Δz^+
<2	∼1	<30	∼20	∼130	∼100

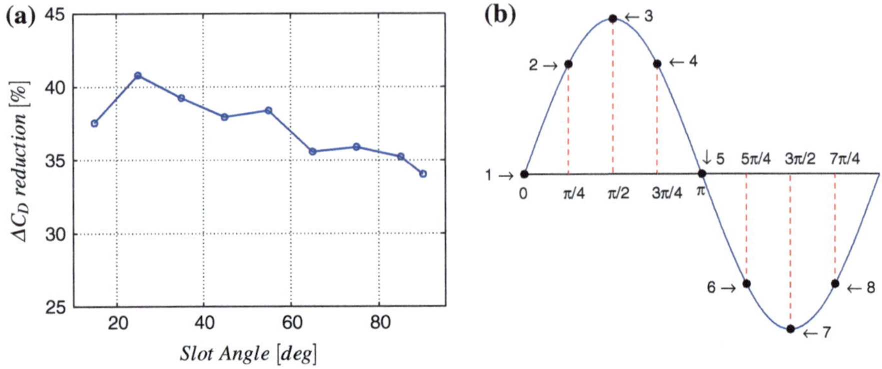

Drag reduction using AFC versus slot angle configuration.

Phase averaging along the actuation cycle.

Fig. 2 Dependence on slot angle of drag (**a**) and phase lock positions (**b**) of the actuation cycle

two different locations, one at the most upstream position of the flap and one 12 cm downstream of the rear end corner. A detailed analysis of the unforced case is given in [13, 14].

2.1.1 Phase Averaging

The phase averaged velocity is analyzed in this section to study the mechanism of AFC. Two cases are studied with different slot angles, a small one of $SA = 25°$ (Case A) and a larger one of $SA = 90°$ (Case B). All other parameters are identical in both cases. These two angles are shown to give the highest and lowest drag reduction, according to Fig. 2a. The phase averaging procedure is presented in Fig. 2b. The actuation cycle is phase locked at eight positions, labelled 1–8, i.e. at each $\pi/4$.

Flow Control Dynamics Around the Flap

The dynamics of the flow is visualized and analyzed in detail in the flap region (Fig. 3). *Phase* 1, plotted in Fig. 3a, b, is the phase following *Phase* 8 in Fig. 3o, p, i.e. the flow is already affected by the actuation cycle when the first phase is analyzed. It can be observed that the flow is fully attached and a shear layer is present at the trailing edge of the flap in both cases. In the second phase, Fig. 3c, d, the ejection stroke is started with just a small amount of blowing; vortices are created immediately downstream of the slot location in both cases. However, the vortex in Case B (Fig. 3d) is significantly larger than the one in Case A (Fig. 3c), and the shear layers move away from the flap surface. In *Phase* 3 (Fig. 3e, f), the size of the vortices increases and the major difference between the cases is the creation of two vortices in Case A as opposed to one vortex in Case B. Analyzing the ejection stroke, the possible

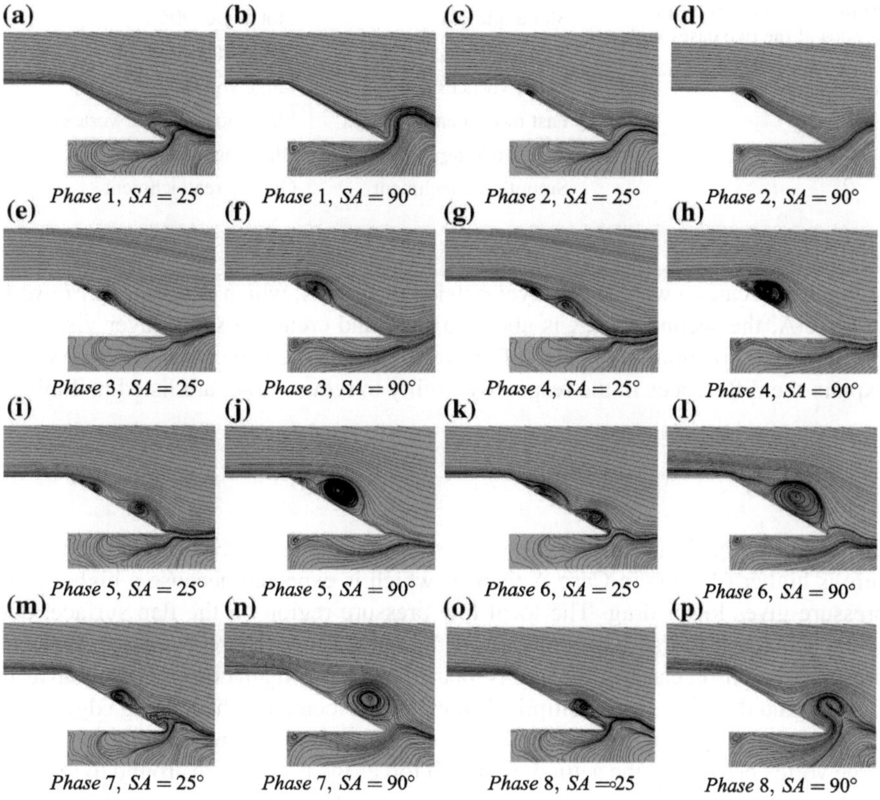

Fig. 3 Phase locked velocity streamlines for two different slot angles, 25° (Case A) and 90° (Case B): *Phases1–8*

reason for the creation of two vortices is the faster movement of the first vortex in Case A as compared to Case B. This is expected because the velocity component of the synthetic jet parallel to the flap surface is larger in Case A than Case B. We can also observe in both *Phase* 1 and 2 that the vortex in Case B is blocking the streamlines upwards of the slot, a possible explanation for the higher drag in this case. This blocking effect is a large scale virtual surface change of the flap induced by the small scale synthetic jet, as described in [15]. In *Phase* 4 (Fig. 3g, h), the blocking effect in Case B is still noticeable and the difference in the magnitude of the velocity of the moving vortex in Case B is significantly slower than that in Case A. The size of the Case B vortex is still increasing in the following *Phase* 5 (Fig. 3i, j) and 6 (Fig. 3k, l). In Case A the first vortex is circular and the second has an elliptical shape. In *Phase* 7 (Fig. 3m, n) the first vortex in Case A leaves the flap region and creates a shear layer and the vortex in Case B is at its largest, reaching to the trailing edge position of the flap. The flow upstream of the vortex in both cases is fully attached. Finally, in *Phase* 8 (Fig. 3o, p), the vortex in Case B starts to shed the flap

Table 2 Comparison of the physics of the two cases with different slot angle configurations

Slot angle: 25°	Slot angle: 90°
Two vortices per cycle	One vortex per cycle
Small vortices	Large vortex
Fast movement of vortex	Slow movement of vortex
Non-blocking of flow	Blocking of flow
Smooth reattachment	Gauche reattachment

region and creates a wide shear layer outside the vortex, which is visible in *Phase* 1. In Case A, the second vortex is about to shed and creates a shear layer visible in *Phase* 1. A short summary of the differences between the two cases, which possibly explain the differences in the respective ability to reduce drag, are listed in Table 2.

Wake Region

The time averaged pressure in the wake region is plotted in Fig. 4. The base pressure is higher (darker) in Case A than B, which is expected because a higher base pressure gives lower drag. The local low pressure region on the flap surface, due to the actuation, is also larger in Case B than in Case A. There are in fact two low pressure regions in Case B: the first is located at the upstream corner of the flap as in Case A, and the other has an elliptical shape and is located at the leading edge of the flap, immediately after the first region. The second low pressure region is due to the large vortex created by the actuation, which forces the flow to turn around to reattach on the flap surface. This is in agreement with the above discussion of the flow. The

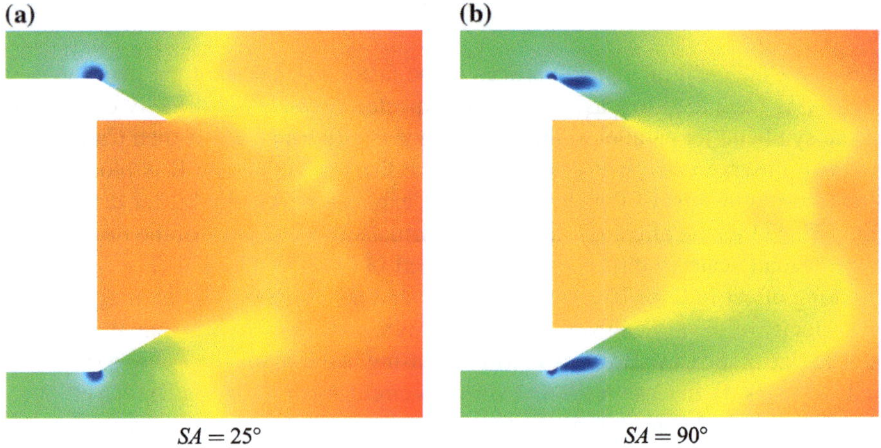

Fig. 4 Time-averaged pressure comparing Case A and Case B

existence of a large low pressure region on the flap is not beneficial to reducing drag, but the small regions are important because of the flow reattachment mechanism.

2.1.2 Small Slot

This section describes the actuation using a smaller slot width of $h = 2$ cm instead of $h = 6$ cm at two different locations and slot angles. The aim is to use a configuration as close as possible to the full-scale test configuration. The full-scale slot width is 0.5–0.7 mm and is located 12.2 cm downstream of the leading edge of the flap. The simulations compare two different locations and slot angles at different momentum strength, C_μ. C_μ is the amount of momentum out from the slot and is defined as

$$C_\mu = \frac{u_{rms}^2 \cdot h}{\frac{1}{2} w \cdot u_\infty^2} \tag{1}$$

where h is the slot width, u_{rms} is the RMS velocity from the jet and w a characteristic length used in the model in which AFC is applied. In this case, w is the width of a truck [16]. The location of the slot is either immediately downstream of the rear corner ($loc1$) or 12 cm downstream of the corner ($loc2$). The second location is very close to the full-scale configuration. The slot angles investigated are 45° and 90°, which correspond to the slot angles investigated in the full-scale tests.

Figure 5a shows the drag reduction at slot angle $SA = 45°$. It is shown that the reduction in drag improves with high C_μ values for both slot locations. However, the difference between the two locations in the ability to reduce drag becomes smaller with increased C_μ. For $SA = 90°$ the difference between the locations is large for $C_\mu > 0.5$; $loc2$ is even better for very low C_μ values and $loc1$ is better for high C_μ values.

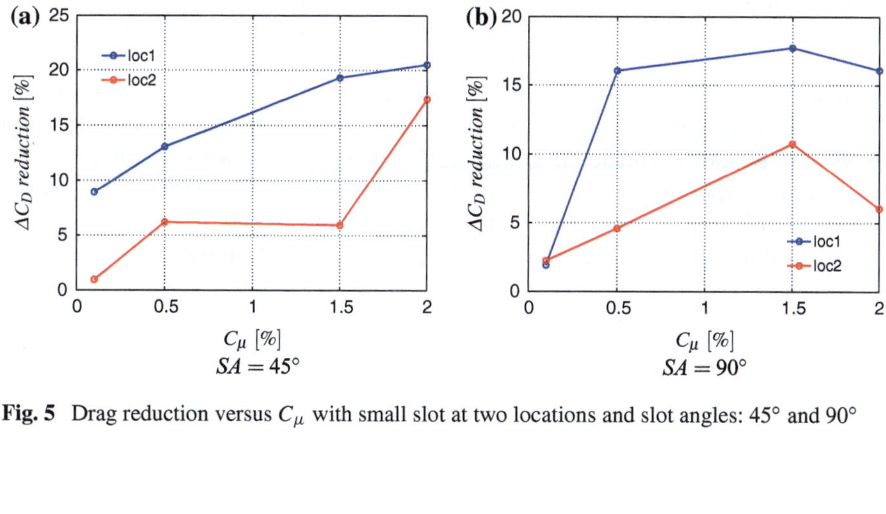

Fig. 5 Drag reduction versus C_μ with small slot at two locations and slot angles: 45° and 90°

3 Synthetic Jet Actuator

To make use of the AFC technique, a reliable and sufficiently powerful synthetic jet (also known as zero net-mass flux) actuator is evaluated for the experimental full-scale tests. Synthetic jets can be produced by imposing an alternating pressure drop over an orifice, either by a vibrating membrane or a piston motion. To investigate the momentum output from a simple actuator, a prototype with a vibrating membrane (loudspeaker) was manufactured and tested as a first step. With the experience gained from this, another prototype, which was integrated with the mounted flaps, was manufactured. This section deals with an isolated synthetic jet but without cross flow. The actuator is active when the loudspeaker membrane is set in vibration; the air will be compressed out from the slot and produce trains of vortices. The actuator creates a zero net mass flux sinusoidal synthetic jet [15]. The aim is not to have an optimized and effective actuator, considering size, weight, effectiveness (power input) and powerfulness (maximum momentum output) but rather to create a simple test model that can produce enough power and that is cheap and easy to manufacture for testing purposes. This actuator has to be sufficiently powerful in the low frequency domain. Our frequency domain is between 10 and 30 Hz, and the goal is to reach a momentum coefficient of $0.5\,\% \leq C_\mu \leq 1\,\%$.

3.1 Configuration

The prototype actuator consists of a rectangular wood cavity ($V_1 = W \times L \times H = 28.3 \times 28.4 \times 5$ [cm]), a loudspeaker and an aluminum plate with an orifice (slot) in the middle. The cavity volume can be adjusted by moving the loudspeaker, i.e. different H, see Fig. 6a. Three volumes were investigated: V_1, $2V_1$ and $4V_1$. There are also several aluminum plates with different slots (h). The slot length is 28 cm and the different slot widths (h) are 0.5, 1.0, 2.0 and 5.0 mm. One slot (2.0 mm) was cut with an angle of 25° to the surface. The actuator is shown in Fig. 6. The loudspeaker mainly used is AUDAX PR 240 Z0 with a 24 cm diameter [17]; other loudspeakers were also investigated. The running voltage was 22.0 V RMS, giving a power of 80 W. The frequency was set to $f = 16.67$ Hz, which is a frequency typically used in AFC simulations. The actuator was well sealed inside in order to maximize the momentum output. The full-scale flap (Fig. 6c) has 5–6 integrated actuators, each having a configuration similar to the test actuator (Fig. 6b) but different designs and materials in order to fit the flaps with respect to weight and strength.

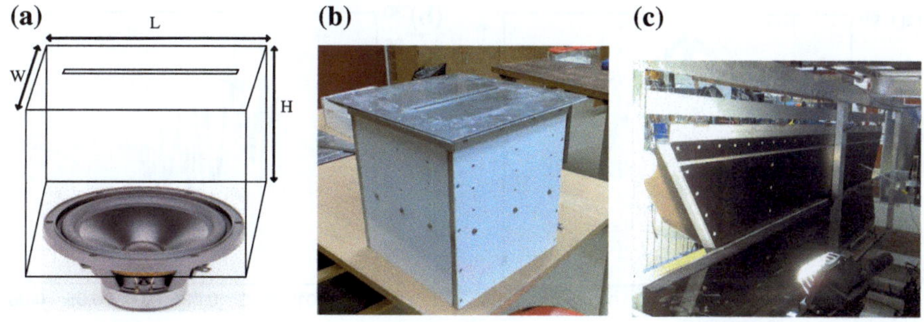

Schematic view of the actuator. Test actuator Full-scale flap with actuator

Fig. 6 The actuator investigated for test purposes

3.2 Measurements

Phase averaging is adopted in order to separate the turbulence and the periodic motion imposed by the actuator. The sampling rate is $f_s = 50{,}000\,\text{Hz}$. Considering the fact that a single hot wire is unable to measure the direction of the flow, the signal will consist of two half periods with a positive amplitude: one for the ejection and the other for the suction. A trigger signal is used for the phase averaging and is imposed at the beginning of each ejection. When measuring the velocity profile from the integrated actuator, the sample time at each position is 6 s and the spatial resolution is 0.1 mm, with an adaptive spatial range that covers more than the whole slot width. Only one part of the final integrated actuator is measured, as similar behavior, is assumed in all other parts.

3.3 Results

This section gives an analysis of the integrated actuator. Results of the model actuator are given in [13]. The output velocity, the velocity profile out from the slot and the span-wise ejection/suction continuity along the flap are presented below. The instantaneous velocity signal and the phase averaged one are plotted in Fig. 7. Figure 7a shows the signal at $z = 760$ mm and Fig. 7b shows the signal at $z = 1100$ mm. Figure 8 shows the velocity line along the flap for two different frequencies, $f = 16.67\,\text{Hz}$ in Fig. 8a and $f = 10.00\,\text{Hz}$ in Fig. 8b. The velocity profiles along two slots are shown (recall that the entire integrated configuration includes 5–6 slots, i.e. separated actuator cavities). The maximum velocity from the slot is shown. The discontinuities at $z = 700$ mm and $z = 1100$ mm are due to the separation between the actuator cavities, which is needed to create enough pressurization inside each actuator cavity. As can be seen in Fig. 8a, b, the maximum velocity out from the slot is strongly frequency dependent. The frequency response is further analyzed in Fig. 8c, where the output maximum velocity is plotted versus frequency. The behavior is similar to the electrical characteristics of a dynamic loudspeaker but with a

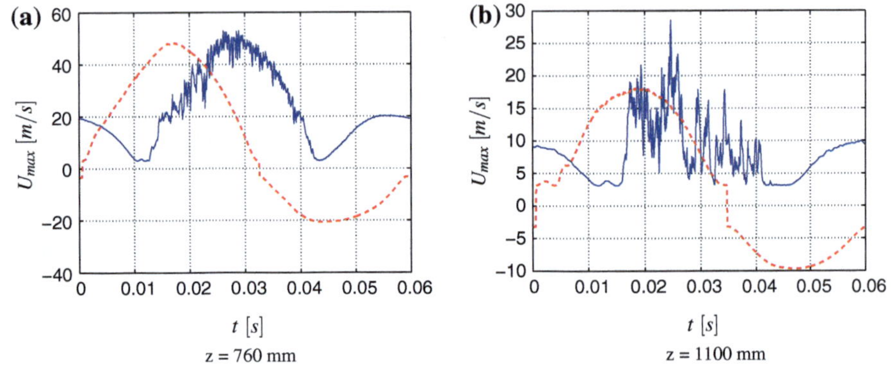

Fig. 7 The instantaneous velocity signal (——) and with phase-locked (- - -) at two positions along the slot span

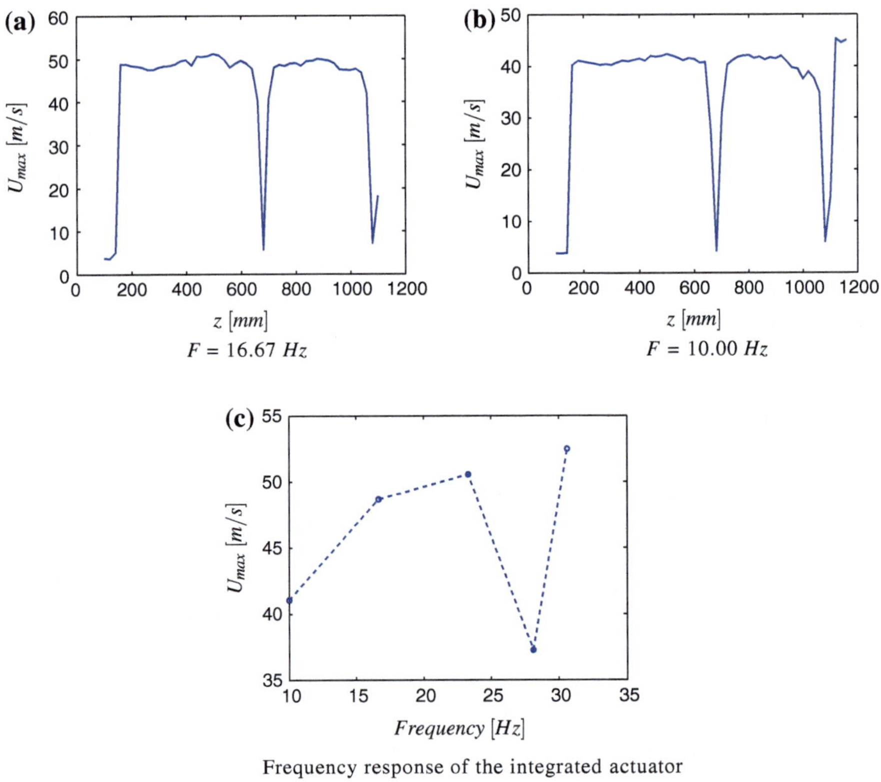

Fig. 8 Maximum velocity along the slot and actuator frequency response

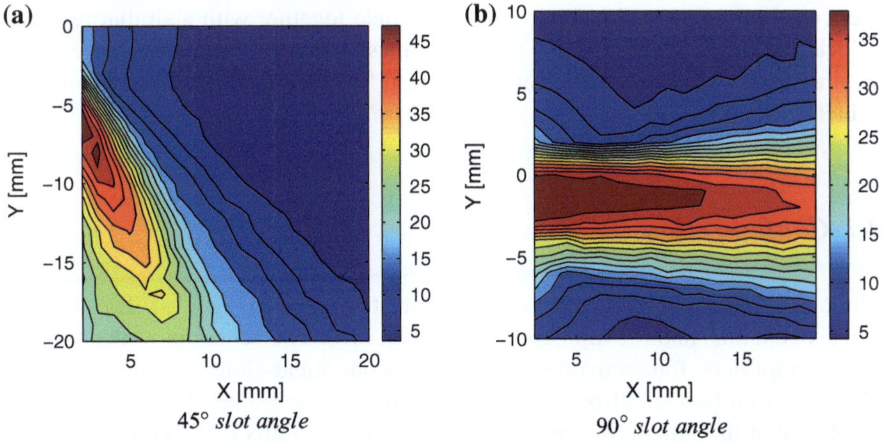

Fig. 9 Contour plots of phase averaged velocities at maximum velocity during the ejection stroke at two slot angles: 45° and 90°

modified resonance frequency due to cavity and load combinations. The free-space resonance frequency of the actual loudspeaker is $F_c = 28.10\,\text{Hz}$.

The two-dimensional velocity profile is plotted in Fig. 9. It is noticeable that the flow in the 45° slot exits at a much smaller angle and deflects towards the surface of the flap. This is probably due to the Coanda effect. This phenomenon shows that we should not use slot angles less than 45° in the full-scale test.

4 Full-Scale Test

The full-scale tests were conducted at the Hällered proving ground in Sweden (Fig. 10). The fuel consumption was measured and compared for different cases. The "Joint ATA/SAE Fuel Consumption Test Procedure, Type II" was adopted to assure consistency in the tests. This procedure is a standard method for full-scale

Fig. 10 Hällered proving ground in Sweden

fuel consumption investigations. The test vehicle together with a similar reference vehicle are run at the road track repeated times in order to minimize the effect of weather and other sources of error. In this test, five runs were carried out with both vehicles for each case.

4.1 Configuration

Three flaps were mounted at the rear end of the test vehicle (Fig. 11a). The side flaps had six actuators and the top flap had five (Fig. 11b). The actuators were connected to five amplifiers, four actuators/amplifier plus one stand-alone, and the power was obtained from two small power stations located in the trailer. The frequency and amplitude of the actuation signal were controlled by a LabVIEW program from the truck via WIFI network. The actuation was active only in the two straight paths of the proving ground.

(a) **(b)**

The mounted flaps — The integrated actuators

Fig. 11 The VOLVO full-scale test vehicle

Table 3 Full-scale test cases specifications and results, colored rows by *light blue* unforced and *light red* forced

Case	AFC	Flap angle	Slot angle	Frequency	Ratio	Difference
1	ON	30	45	16.7	101.1%	+1.1%
2	OFF	30	–	–	105.0%	+5.0%
3	OFF	20	–	–	100.7%	0.7%
4	ON	30	45	23.3	99.6%	-0.4%
5	ON	30	90	16.7	102.7%	+2.7%
6	ON	30	90	23.3	102.3%	+2.3%
7	ON	20	90	16.7	99.5%	-0.5%
8	ON	20	90	23.3	99.5%	-0.5%
9	ON	30	Large slot	16.7	101.9%	+1.9%
10	Baseline	–	–	–	100.0%	0.0%

4.2 Results

The ratios of total consumed fuel by the test vehicle to the baseline case (case 10) are shown in Table 3. The table includes full specifications of the test cases. A comparison between the unforced (no AFC) Case 2 with a flap angle of 30° and the corresponding forced (AFC) Case 4 shows a reduction of 5.4 %. We also observe that, by using a smaller flap angle of 20° and no AFC (Case 3), the reduction is 4.3 %. The reduction as compared to Case 2 is almost as large as the reduction gained in the forced case (Case 4). Comparing all cases with regard to frequencies, one can conclude that the highest frequency gives a slight improvement in reduction of drag, e.g. comparing Cases 1 and 4. The baseline case 10 has better fuel consumption than both passive Cases 2 and 3. The reason is probably that there is no flap on the lower edge of the trailer. This will be further investigated to get a better understanding of this unexpected result.

5 Conclusions

Computational and full-scale investigations using AFC were carried out to improve the drag of truck-trailers. An experimental investigation was also made of a prototype actuator. The computational work investigates different parameters of the AFC.

It is shown that the influence of slot angle is important for drag reduction. Smaller slot angles are preferable. The phase-locked flow at two different slot angles is analyzed, and it is shown that the improved ability to reduce drag of small slot angles has to do with the dynamics of the AFC around the flap region. Two vortices per cycle were created by the actuation for the smaller slot angle case (Case A) compared to one vortex per cycle for the larger slot angle case (Case B). The flow in

Case A smoothly reattached upstream of the vortices that were created, which were significantly smaller and had a faster movement than the vortex in Case B. The time averaged pressure was also investigated for both cases, and it is shown that the base pressure in Case A is larger than that in Case B, as expected.

Using a smaller slot width and investigating two different slot locations and angles, the reduction in drag at slot angle $SA = 45°$ is improved with high C_μ values for both slot locations. At low C_μ values, $loc1$ gives a significantly better reduction as compared to $loc2$, while the difference becomes smaller with increasing C_μ. The difference between the locations for $SA = 90°$ is large for $C_\mu > 0.5$.

This experimental investigation of the integrated actuator shows that the output velocity is highly frequency dependent. The maximum output velocity is around 50 m/s. The two-dimensional velocity profile shows that the flow in the 45° slot exists at a much smaller angle. The flow is deflected towards the surface of the flap due to the Coanda effect.

Results of the full-scale tests show that, for a particular unforced case (no AFC), as compared to its corresponding forced (AFC) case, a reduction of 5.4 % in fuel consumption is achieved. It is also shown that, by using a smaller flap angle of 20° and no AFC, the reduction is 4.3 %. The baseline Case 10 shows better fuel consumption than both passive cases.

This work is supported by the **Swedish Agency of Innovation Systems (VINNOVA)**, **Volvo 3P**, **SKAB** and **CD-ADAPCO**. The financial support of SNIC (the Swedish National Infrastructure for Computing) for computer time at C3SE (Chalmers Centre for Computational Science and Engineering) is gratefully acknowledged.

References

1. El-Alti, M., Kjellgren, P., Davidson, L.: On the download alleviation for the XV-15 wing by active flow control using large-eddy simulation. In: ERCOFTAC Workshop: Direct and Large-Eddy Simulation, vol. 7, Trieste, Italy, 8–10 Sept 2008
2. Kjellgren, P., El-Alti, M., Davidson, L.: Download alleviation of a tilt-rotor wing by active flow control strategies. In: KATnet II Conference: Key Aerodynamic Technologies, Bremen, Germany, 12–14 May 2009
3. Kjellgren, P., Anderberg, N., Wyganski, I.: Download alleviation by periodic excitation on a typical tilt-rotor configuration—computation and experiment. In: Fluids 2000 Conference and Exhibit, Denver, CO, 19–22 June 2000
4. Kjellgren, P., Hassan, A., Sivasubramanian, J., Cullen, L., Cerchie D., Wyganski, I.: Download alleviation for the XV-15: computations and experiments of flows around the wing. In: Biennial International Powered Lift Conference and Exhibit, Williamsburg, Virginia, 5–7 Nov 2002
5. Darabi, A., Wygnanski, I.: Active management of naturally separated flow over a solid surface. part 1. The forced reattachment process. J. Fluid Mech. **510**, 105–129 (2004)
6. Seifert, A., Stalnov, O., Sperber, D., Arwatz, G., Palei, V., David, S., Dayan, I., Fono, I.: Large trucks drag reduction using active flow control. In: The Aerodynamics of Heavy Vehicles II: Trucks, Buses and Trains (2007)

7. Henning, L., King, R.: Drag reduction by closed-loop control of a separated flow over a bluff body with a blunt trailing edge. In: 44th IEEE Conference on Decision and Control and European Control Conference ECC 2005, pp. 494–499, Seville, Spain (2005)
8. Taubert, L., Wygnanski, I.: Preliminary experiments applying active flow control to a 1/24th scale model of a semi-trailer truck. In: The Aerodynamics of Heavy Vehicles II: Trucks, Buses and Trains (2007)
9. Nayeri, C.N., Haff, J., Greenblatt, D., Loefdahl, L., Pashereit, C.O.: Drag reduction on a generic tractor-trailer using active flow control in combination with solid flaps. In: The Aerodynamics of Heavy Vehicles II: Trucks, Buses and Trains (2007)
10. Englar, R.J.: Improved pneumatic aerodynamics for drag reduction, fuel economy, safety and stability increase for heavy vehicles. In: SAE 2005 Commercial Vehicle Engineering, Congress and Exhibition, SAE Paper 2005-01-3627, Chicago, Illinois, USA (2005)
11. Kjellgren, P.: A semi-implicit fractional step finite element method for viscous incompressible flows. Comput. Mech. **20**, 541–550 (1997)
12. Kjellgren, P.: An arbitrary lagrangian-eulerian finite element method. Comput. Mech. **21**, 81–90 (1998)
13. El-Alti, M.: Active flow control for aircrafts and heavy vehicles. Thesis for licentiate of engineering no. 2009:011, Division of Fluid Dynamics, Department of Applied Mechanics, Chalmers University of Technology, Göteborg, Sweden (2009)
14. El-Alti, M., Kjellgren, P., Davidson, L.: Drag Reduction for trucks by active flow control of the wake behind the trailer. In: 6th International Symposium on Turbulence, Heat and Mass Transfer, Rome, Italy, 14–18 Sept 2009
15. Glezer, A., Amitay, M.: Synthetic jets. Ann. Rev. Fluid Mech. **34**, 503–529 (2002)
16. El-Alti, M., Kjellgren, P., Davidson, L.: Drag reduction of trucks by active flow control of the wake behind the trailer. In: 6th International Symposium on Turbulence, Heat and Mass Transfer, Rome, Italy, 14–18 Sept 2009
17. AUDAX Industries. Ttieftonlautsprecher, PR 240Z0. http://www.audax-speaker.de/ibase/module/medienarchiv/dateien/PDF-DOWNLOAD/AUDAX-PR240Z0.pdf (2009)

Experimental and Full Scale Investigation of Base Cavity Drag Reduction Devices for Use on Ground Transport Vehicles

J. Kehs, K. Visser, J. Grossmann, C. Horrell and A. Smith

Abstract Comparison studies have been conducted on a 1:16th scale model and a full scale tractor trailer of a variety of sealed aft cavity devices as a means to develop or enhance commercial drag reduction technology for class 8 vehicles. Eight base cavity geometries with pressure taps were created for the scale model. Drag data were acquired on the models using a 6-axis internal force balance for a range of yaw sweeps and at three Reynolds numbers for each base cavity. Pressure surveys for selected base cavities were also completed for the same yaw angles and Reynolds numbers to quantify the change in base pressure. The scale model force data indicated a marked decrease in drag at up to 12 % for two base cavity shapes, however most base cavities reduced drag by 3–10 % and a few did not decrease drag any significant amount. Pressure data indicated the base cavities increased the base pressure over baseline with a ΔC_P of up to 0.3. Moreover, drag computed from pressure data implied that although the base cavities decreased the base drag due to a pressure increase, the drag may have increased elsewhere on the model. Full-scale tests were also completed using SAE Type II testing procedures. Full-scale tests on the same geometry indicated a fuel savings of over 6.5 %. Overall, the use of these devices shows to be a viable, effective and economical way to reduce fuel consumption on ground transport vehicles.

J. Kehs (✉) · K. Visser
Clarkson University, Potsdam, NY, USA
e-mail: kehsjp@clarkson.edu

K. Visser
e-mail: visser@clarkson.edu

J. Grossmann · C. Horrell · A. Smith
ATDynamics, San Francisco, CA, USA
e-mail: jgrossman@atdynamics.com

A. Smith
e-mail: asmith@atdynamics.com

A. Dillmann and A. Orellano (eds.), *The Aerodynamics of Heavy Vehicles III*,
Lecture Notes in Applied and Computational Mechanics 79,
DOI 10.1007/978-3-319-20122-1_17

1 Introduction

The US Department of Transportation reported that over 2.2 million combination trucks were registered in the United States, driving a total of 143 billion miles (230 billion km) and burning a total of 132 billion gallons (500 billion liters) of fuel in 2008 [8]. At these levels even a modest 5–10 % reduction in fuel usage could save billions of liters of fuel each year, result in billions of dollars of savings for the transportation business, reduce the demand of domestic and foreign oil, have a significant environmental impact, and reduce global oil dependency.

Drag reduction measurements on tractor-trailers configurations, specifically using aft geometrical modifications, has been the focus of many scale and full scale studies in the past [1, 3–5, 9–12, 14, 17]. Cooper, for example, at the National Research Council in Canada, conducted wind tunnel experiments on scale models with rounded and beveled panels attached to the rear end of the trailer. A reduction in drag for such devices of 7–10 % was reported.

At Clarkson University, experimental scale model and full-scale studies have been on-going for the past decade, focusing on the optimization of sealed aft cavities [6, 7, 13]. Coon reported in a numerical and experimental study of 0° angle, variable inset aft cavities, that a 4 sided, 48 in. deep cavity with a 6 in. inset on the sides and top was the best case at zero yaw with a drag reduction of $\%\Delta C_D = 9\,\%$ or $\Delta C_D = 0.075$. Figure 1 illustrates the full scale protoype driven cross country in the year 2000.

Maragno investigated the effectiveness of boat tail angle on the side plates, inset from the side edges, length of the boat tail and the number of plates, at a Reynolds number of 2.3e5, and found that the best configuration was a sealed 4 sided aft cavity with a 10° side boat-tail angle, no inset, and a 48 in. (full-scale) length. A ΔC_D of approximately 0.12 and a $\%\Delta C_D$ of approximately 9.5 % were reported. Full-scale

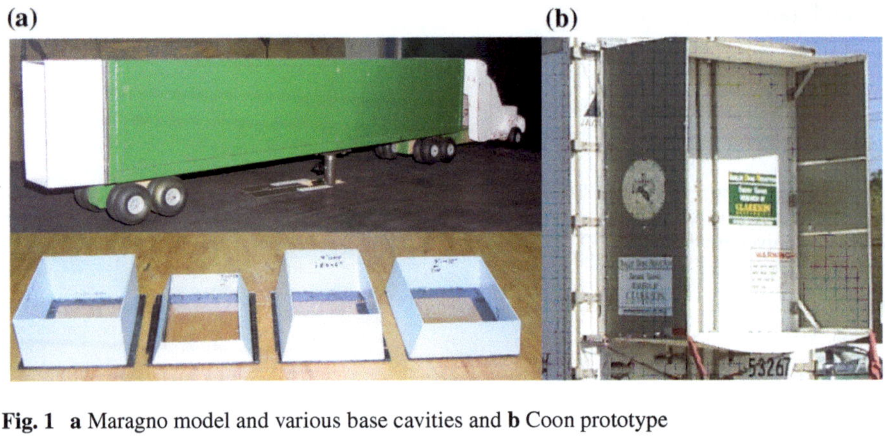

Fig. 1 **a** Maragno model and various base cavities and **b** Coon prototype

Fig. 2 MAKA innovations rear end Configuration

over-the-road tests by Grover on a class 8 tractor trailer in 2006 indicated an optimal sealed aft cavity geometry of 15° on the side and top panels, and a 7° inset on the bottom. An improvement of 0.63 mpg was observed in the back to back runs, which would result in a fuel savings of about 10 %.

Attempts at commercialization and patents of aftermarket base cavities have been many over the past 30 years. MAKA Innovations, now Transtex [18] of Quebec, Canada have marketed 3-sided base plate design (Fig. 2). MAKA's base cavities were relatively short (under 24 in./0.61 m) and were angled inward. Boivin and Roberge [2] filed a patent of this type of configuration with 3 inward angled plates attached to the rear of a trailer. However, to date, there has not been a commercial product that has been widely adopted.

The purpose of the current study was to conduct a joint study to compare wind tunnel tests of passive aft cavity, drag reduction devices and to compare the results to test data acquired on a full-scale configuration. A partnership between Clarkson University and ATDynamics, Inc. was formed to conduct this testing. ATDynamics, Inc is a transport aerodynamics company based in South San Francisco, California that markets a sealed base cavity drag reduction device under the brand name TrailerTail®. In this study, drag and pressure measurements were acquired on a 1:16th scale model of a tractor-trailer with 8 interchangeable rear end configurations. Full-scale SAE-J1321 Type II tests were then completed on ATDynamics' production TrailerTail®.

2 Experimental Setup

The following section describes the setup for the scale model tests conducted by Clarkson University in the wind tunnel and the setup for the full-scale SAE J1321 Type II truck tests completed by ATDynamics, Inc.

3 Scale Model Wind Tunnel Setup

Wind tunnel tests were conducted in the Clarkson University high-speed wind tunnel, an open circuit tunnel equipped with a 150 hp motor driving a 6-foot diameter fan. A 9:1 contraction lead to a test section that is 1.22 m wide, 0.91 m tall and 1.65 m long. The tunnel was capable of 60 m/s (134 mph). A pitot-static probe was used to measure free-stream total and static pressure. Installed into the floor of the tunnel was model-positioning system with controlled yaw and pitch. Depending on the measurements being taken, either a 6-axis force balance or a rigid tube was mounted to the model positioning system.

A 1:16th scale model of a generic tractor-trailer was fabricated from a Nylon and aluminum powder matrix using selective laser sintering. The model was designed to resemble a modern aerodynamic cab and standard 53 ft trailer. The total length of the model is approximately 1.4 m. The model included a mounting location for the stings and a method for mounting the different rear end configurations. The truck model is shown in Fig. 3.

Eight rear end configurations, including one baseline configuration (flat rear end) and seven aft cavity configurations, were tested. All of the configurations had an inset of 0.0 m on the top and sides but the inset on the bottom changes between configurations. The shape of the cavity walls also varied for different configurations. Table 1 contains illustrations of the rear end configurations as well as the differing specifications between them.

A 0.75 in. (19.05 mm) thick splitter plate was fitted to the wind tunnel to simulate a road surface. The combination of the truck at 0° and the splitter plate contributed to a 5.67 % blockage in the tunnel.

6-axis force measurement data were acquired at Reynolds numbers of 2.0e5 (20 m/s), 3.0e5 (30 m/s), and 4.5e5 (45 m/s) based on the width of the trailer. At each Reynolds number, data were taken at yaw angles from −9° to 9° in increments of 1° for a total of 19 data acquisition points at each Reynolds number. At each yaw angle, 8000 data points were collected over 8 s to average out any oscillations.

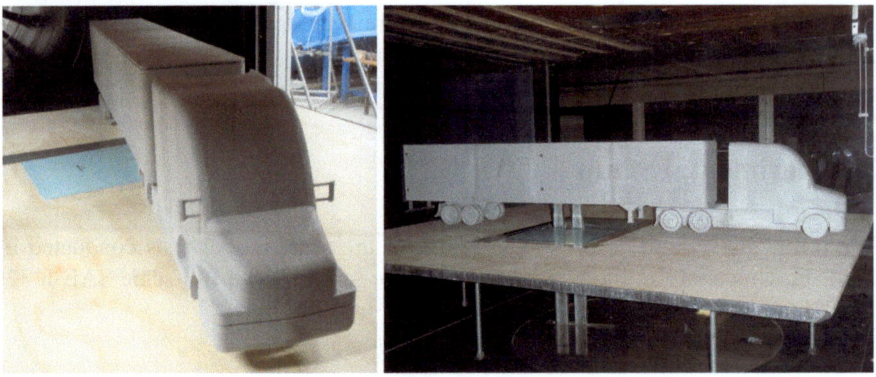

Fig. 3 Tractor trailer scale model in Clarkson University wind tunnel

Table 1 Table of rear end configurations

TT1.0	TT2.0	TT3.0sym	TT3.0nonsym
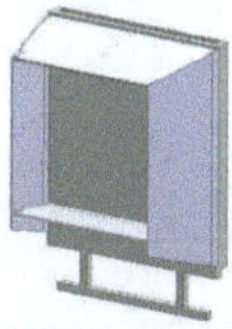	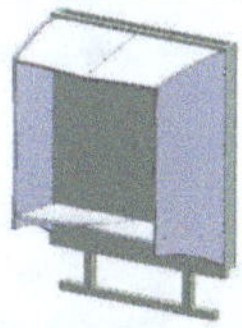	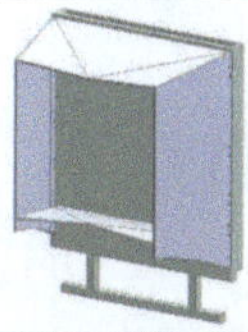	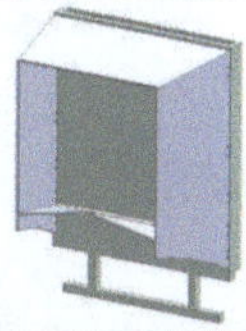
Straight angled walls	Curved Walls	Straight Side Walls with Origami Folding Top and Bottom Walls	Straight Side and Top Walls with Origami Folding Bottom Wall

TT3.0reverse above LR	TT3.0reverse below LR	TT3.0reverse below LR w/cutouts
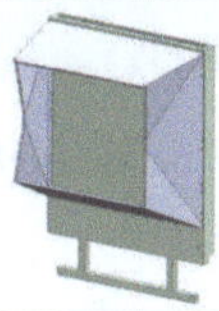	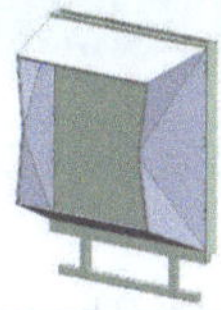	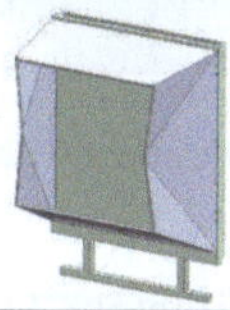
Straight Top and Bottom Walls with Origami Side Walls, Bottom Wall Placed Above Lock-Rods as all previous	Straight Top and Bottom Walls with Origami Side Walls, Bottom Wall Placed Below Lock-Rods	Straight Top and Bottom Walls with Origami Side Walls, Bottom Wall Placed Below Lock-Rods with Cutouts to Reach Lock-Rods

Pressure surveys were performed on 4 of the 8 rear end configurations that showed promising results. Pressure measurements were taken at a range of yaw angles from $-9°$ to $9°$ in increments of $1°$ at the same 3 Reynolds numbers as the drag experiments. Each pressure survey consisted of measurements taken at 31 taps on the downwind side of each rear end configuration. The pressures were measured using two 16-channel pressure transducers with an accuracy of $\pm 0.10\,\%$ of the full-scale range of ± 10 in. of water (2490 Pa).

4 Road Tests Setup

The Goodyear Tire Test Division in collaboration with ATDynamics, Inc completed full-scale road tests at the Goodyear Proving Grounds in San Angelo, Texas in April 2010. The tests followed the SAE J1321 type II protocols [16]. The tractors used in the test were low mileage International ProStars with Cummins ISX400 ST engines and Fuller 10 speed transmissions. The trailers were Vanguard 53' trailers loaded to a total weight of 46000 lbs. Tire inflation was set to 110 psi. The test was run with TT3.0sym only (See Table 1) as this was the current production TrailerTail®. Three baseline runs and three test runs were completed on an oval track with a total

Fig. 4 Test trailers with baseline configuration on the *Left* and 'TT3.0sym' on the *Right*

of approximately 41 miles for each run. The test measured percent fuel saved and percent improvement over baseline. A photo of the test trucks is shown in Fig. 4.

5 Results

The following section presents and discusses the data acquired from the scale and full-scale experiments.

5.1 Scale Model Wind Tunnel Results

The drag coefficient, C_D, and pressure coefficient, C_P, were calculated from the data collected and were based on the frontal area of the truck using:

$$C_D = \frac{Drag\ Force}{1/2\rho V_\infty^2 A_F} \quad (1)$$

$$C_P = \frac{\overline{P}_{s,Tap} - \overline{P}_{s,\infty}}{1/2\rho V_\infty^2} \quad (2)$$

where 'Drag Force' is the measured drag in [N], ρ is the density of the air in [kg/m^3], V_∞ is the free-stream velocity in [m/s], A_F is the frontal area of the model [m^2] and $P_{S,Tap}$ and $P_{S,\infty}$ are the average static pressure at the tap and in the free-stream respectively [Pa].

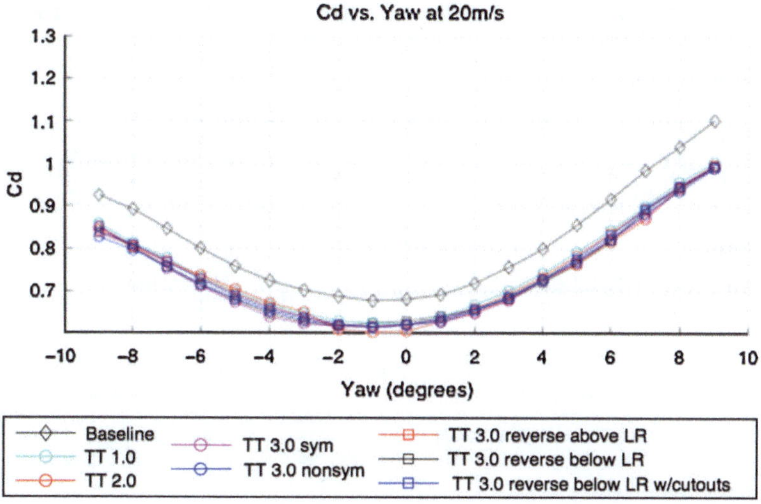

Fig. 5 Example plot of C_D versus yaw at Re = 2.0e5 for all rear end configurations

Figure 5 presents an example of the C_D curves for the rear ends at Re = 2.0e5. The shape of these curves is roughly a parabola, which is typical for such an experiment. Drag coefficients for the baseline configuration at 0° of yaw ranged from 0.68 at the lowest Reynolds numbers to 0.75 at the highest Reynolds numbers. This is within the range of modern full-scale tractor-trailers without drag reduction devices.

The change in drag coefficient, ΔC_D, and the percentage drop in drag coefficient, $\%\Delta C_D$, between the baseline and each rear end configuration were used to compare the drag reducing capabilities of each rear end configuration where:

$$\Delta C_D = C_{D_{Base}} - C_{D_{device}} \tag{3}$$

$$\%\Delta C_D = \frac{\Delta C_D}{C_{D_{Base}}} * 100 \tag{4}$$

At a Reynolds number of 2.0e5, 'TT2.0' performed best out of all the rear end configurations at 0° with a ΔC_D greater than 0.07. TT2.0 also performed relatively well at high yaw angles at this Reynolds number but performed relatively poor between −6° and −3° and between 1° and 5°. The jumps in the data for that configuration suggest that hysteresis may be present for this Reynolds number but is only present at the lowest Reynolds number. TT3.0nonsym posted a very consistent $\%\Delta C_D$, never straying outside 8–10 %. Although other configurations showed a $\%\Delta C_D$ greater than this range, none remained above 8 % for all yaw angles except TT3.0nonsym. The drop in C_D for this rear end ranged from 0.06 to approximately 0.11. The ΔC_D and $\%\Delta C_D$ for a Reynolds number of 2.0e5 (20 m/s wind speed), along with the others of the following discussion, are shown below in Fig. 6.

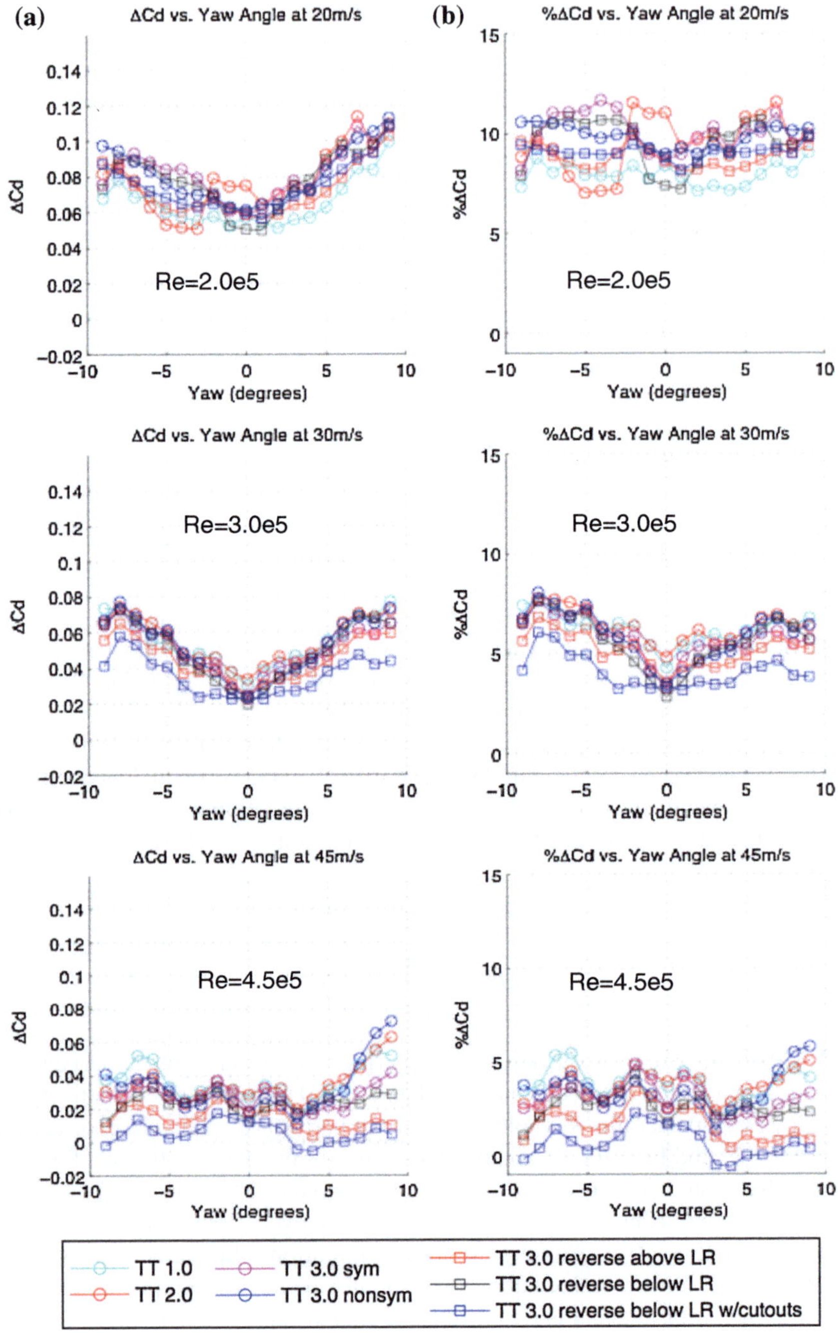

Fig. 6 **a** ΔC_D versus yaw angle and **b** $\%\Delta C_D$ versus yaw angle for all rear end configurations

The data from Re = 3.0e5 indicated again that all rear end configurations reduced the total drag. 'TT3.0nonsym' showed the highest reduction in drag at this Reynolds Number with a maximum ΔC_D of approximately 0.08 at $-8°$ and $+9°$ yaw and a ΔC_D of 0.025 at 0° of yaw. 'TT3.0nonsym' had a maximum $\%\Delta C_D$ of approximately 8 %. 'TT2.0' showed the best reduction in drag at 0° yaw with a ΔC_D of 0.035 or 5 %. The remaining rear end configurations posted a range of ΔC_D values from 0.02 to 0.075 with 'TT3.0 reverse below LR w/cutouts' performing the worst with a $\%\Delta C_D$ at a low of approximately 3 % drag reduction from baseline at 0° of yaw.

At Re = 4.5e5, 'TT1.0', 'TT2.0', 'TT3.0nonsym', and 'TT3.0sym' were all at the top of the range with a ΔC_D of 0.02 to 0.07 depending on the angle of yaw. This corresponded to a $\%\Delta C_D$ of approximately 2 % to 6 %. All rear end configurations posted a decrease in drag but 'TT3.0 reverse below LR w/cutouts'. This particular rear end configuration showed an increase in drag at $-9°$, 3°, and 4°, and showed minimal reduction in drag on all of the other yaw angles.

The progression of increasing Reynolds number in Fig. 6 suggests that the performance of the drag reduction devices is reduced with increasing wind speed at a rate of approximately $\Delta C_D = -0.016$ per Re = 1e5. It is unclear why exactly this would occur at this relatively small range Reynolds numbers, but it is suggested that the separation point on the rear end moved forward with increasing tunnel velocity, causing effectiveness to be reduced. Unfortunately, no flow visualization was conducted to verify this. Cooper's [4] findings suggest that this decline in performance levels out with increasing Reynolds number. The suggested minimum Reynolds number for wind tunnel tests on trucks and buses according to SAE [16] is 7.0e5, however the model and data acquisition system used in the present study limited the Reynolds numbers to <5.0e5.

6 Pressure Experiment Results

The following section presents the data from pressure experiments conducted on 4 of the 8 rear end configurations including the baseline configuration. The pressures presented represent the base pressure on the aft end of the tractor-trailer. In this respect, an increase in base pressure is an increase in forward force on the rear end and therefore a reduction in the pressure drag. An example of the coefficient of pressure map on a rear end is shown in Fig. 7 for 'TT3.0nonsym' and the baseline configuration. Other configurations indicated similar pressure distributions.

Pressures on the rear end configuration were compared using the change in coefficient of pressure and percent change in coefficient of pressure averaged spatially and temporally for each configuration at each yaw angle using:

$$\overline{\Delta C_P} = \overline{C}_P - \overline{C}_{P,Baseline} \tag{5}$$

$$\overline{\%\Delta C_P} = \frac{\overline{C}_P - \overline{C}_{P,Baseline}}{\overline{C}_{P,Baseline}} \tag{6}$$

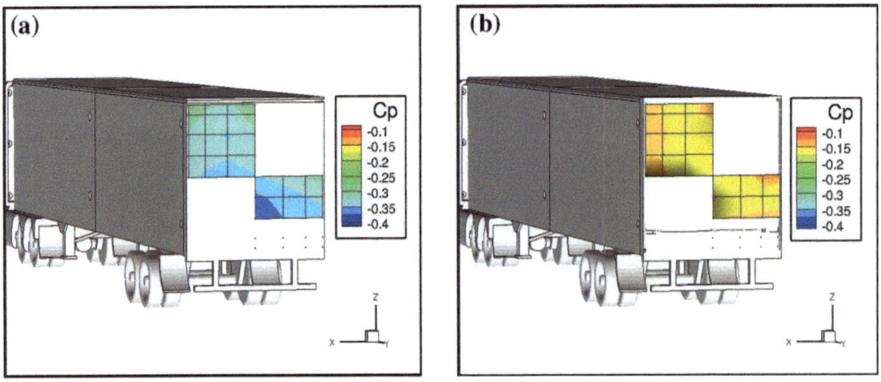

Fig. 7 Coefficient of pressure at all measured pressure taps on the rear face of the truck with **a** Baseline configuration and **b** TT3.0nonsym configuration

Figure 8 displays the spatially and temporally averaged coefficient of pressure and the averaged change in C_P from baseline at Re = 2.0e5, 3.0e5, and 4.5e5 for the four selected rear end configurations at each yaw angle. The pressure results at Re = 2.0e5 (20 m/s) show a significant increase in base pressure from the baseline, up to an average of $Cp = -0.2$ for 'TT2.0' and 'TT3.0nonsym' from the baseline at $Cp = -0.4$ at 0° of yaw. The curves for ΔC_P are curved upward indicating that the effectiveness of the devices increases as the yaw angle increases. The best performing rear end configuration with respect to pressure was 'TT3.0nonsym' which posted a maximum drop in C_P of approximately 0.32 at −9° yaw and a minimum drop in C_P of approximately 0.17 at 0° yaw. 'TT2.0' was very close to that in performance. The remaining configuration ('TT3.0 reverse below LR w/cutouts') indicated a maximum drop in C_P of approximately 0.24 at −9° yaw and a minimum drop in C_P of approximately 0.12 at 0° yaw.

The pressure results at Re = 3.0e5 showed similar results to the results at Re = 2.0e5. 'TT2.0' and 'TT3.0nonsym' had very similar pressure results and both performed better than 'TT3.0 reverse below LR w/cutouts'. The maximum Cp of the best devices remained nearly the same at $C_P = -0.2$ for but the baseline C_P increased slightly up to $C_P = -0.38$ from the lower Reynolds number case. The change in C_P from baseline for 'TT2.0' and 'TT3.0nonsym' had a maximum of approximately $\Delta C_P = 0.32$ at −9° yaw and a minimum at 0° yaw of $\Delta C_P = 0.17$.

The coefficient of pressure curves for the three modified rear end configurations at Re = 4.5e5 indicated a significant increase in base pressure and were similar to the two previous Reynolds number cases. The maximum C_P occurred on 'TT2.0' at −1° yaw and showed a C_P of −0.20. TT3.0nonsym posted the highest change in drag at −8° yaw with $\Delta C_P \approx 0.30$. An interesting observation to note is that at Re = 4.5e5, ΔC_P dropped at the extreme yaw angles due to the baseline C_P curve flattening out at those yaw angles to a much greater degree than at lower Reynolds numbers.

The increase in base pressure shown in Fig. 8 correlates to a decrease in drag and the upward facing parabola shaped Average ΔC_P curves show that the devices

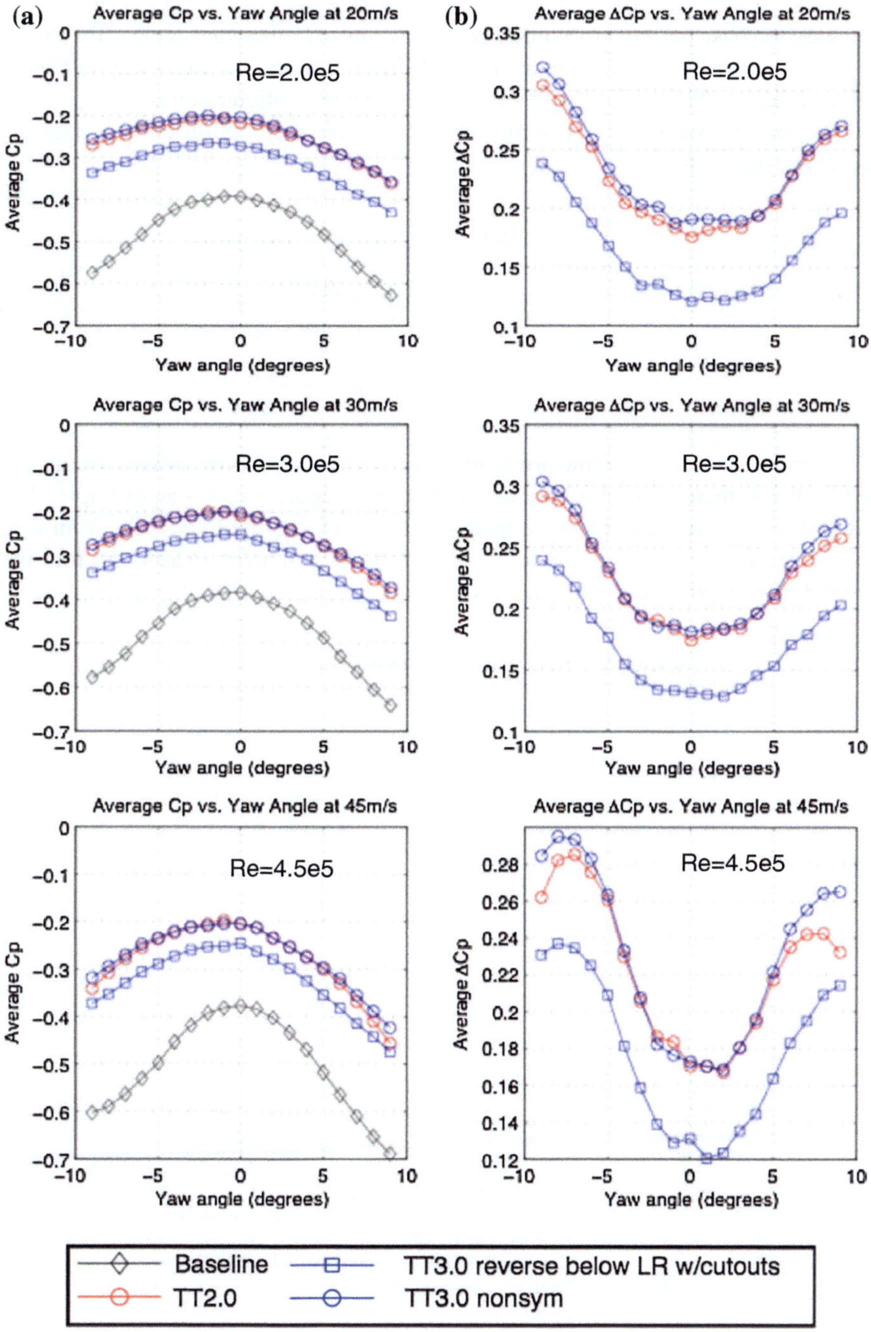

Fig. 8 **a** Average C_p versus yaw angle and **b** Average ΔC_p verus yaw angle at Re = 4.5e5 for 4 selected rear end configurations

increase the base pressure to a much greater degree at higher yaw angles. This agrees with the upward facing parabolas in the ΔC_D curves and confirms the increase in effectiveness at higher yaw angles for these rear end configurations.

The base pressure difference measurements were then converted to a change in C_D using Eqs. 7, 8, 9:

$$\Delta F = \Delta C_P(\frac{1}{2}\rho V_{\infty}^2)A_{base} \tag{7}$$

$$\Delta C_{D,base\ pressure} = \frac{\Delta F}{(\frac{1}{2}\rho V_{\infty}^2)A_{frontal}} \tag{8}$$

$$\Delta C_{D,base\ pressure} = \Delta C_p \frac{A_{base}}{A_{frontal}} \tag{9}$$

The change in C_D from the pressure data can be viewed as the portion of the drop in C_D due to increase of pressure on the rear end. Figure 9 shows an example of the ΔC_D due to increase in pressure from the rear end and the total ΔC_D from force measurements. From this figure it is seen that the drop in pressure accounts for much more drag reduction than actually was measured indicating that although this device

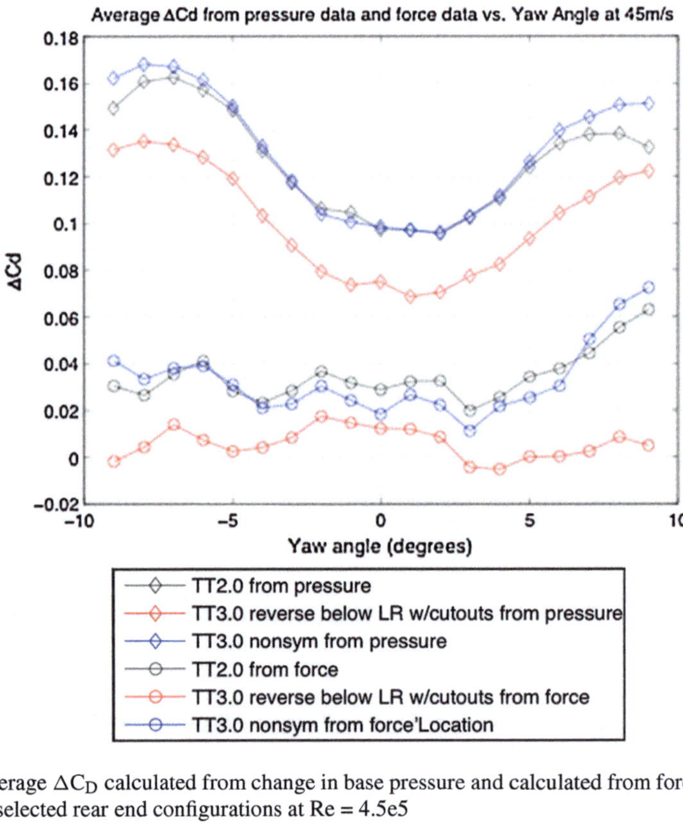

Fig. 9 Average ΔC_D calculated from change in base pressure and calculated from force measurements for selected rear end configurations at Re = 4.5e5

creates a much higher pressure in the back, there was an increase drag elsewhere on the vehicle.

Increased skin friction from the plates would be one component of this increase. The cavity pressure would also exert a pressure on the inside surface of the boat tail plates which would decrease the effectiveness and increase the overall drag, and these pressures were not measured. Further flow visualization and flow measurements will be needed to investigate these results.

7 Road Tests Results

The following section presents the results from the SAE J1321 Type II [15] road tests. Figure 10a, b shows the amount of fuel used on the baseline and test runs respectively for the control truck and the test truck. To compare the fuel usage over these runs, the following Eqs (10), (11), (12) were used to calculate the percentage of fuel saved using the TT3.0sym device configuration and the percentage improvement over the baseline configuration.

$$\overline{T:Cratio} = mean\left(\frac{FuelUsed_{test\,truck}}{FuelUsed_{control\,truck}}\right) \tag{10}$$

$$\%FuelSaved = \frac{\left(\overline{T:Cratio}_{baseline} - \overline{T:Cratio}_{test}\right)}{\overline{T:Cratio}_{baseline}} \tag{11}$$

$$\%Improvement = \frac{\left(\overline{T:Cratio}_{baseline} - \overline{T:Cratio}_{test}\right)}{\overline{T:Cratio}_{test}} \tag{12}$$

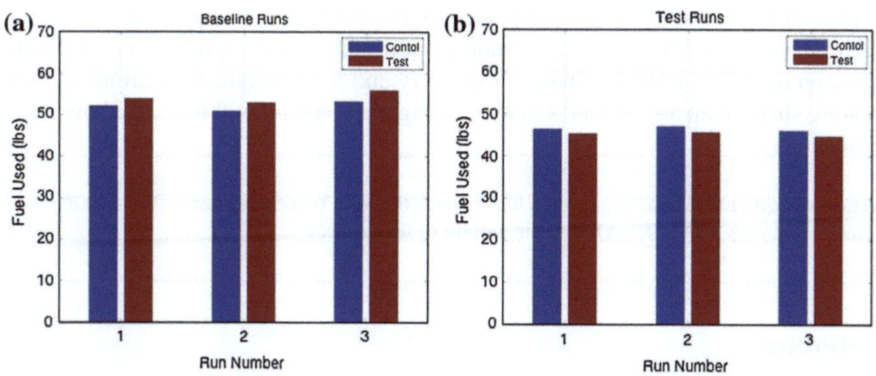

Fig. 10 **a** Fuel used on test and control trucks on baseline runs (No device on either truck) and **b** Fuel used on test and control trucks on test runs (TT3.0sym device on test truck)

From Eqs. 10, 11, and 12, the TT3.0sym rear end configuration provided a savings in fuel of 6.58 % and a percent improvement over baseline configuration of 7.04 %.

Using current "rule of thumb" estimates from the industry, the 6.58 % fuel savings saved translates into approximately a 13 % reduction in ΔC_D, comparable to the $\%\Delta C_D = 8–13\%$ from the scale model experiments at Re = 2.0e5 but is more than double the $\%\Delta C_D$ of approximately 3–5% that is seen at Re = 4.5e5. The reason for this inconsistency is not clear and is currently under investigation. The 6.58 % fuel savings seen in the full-scale road tests is estimated save over 4500 l (1200 gallons) of fuel on a truck that travels 74,600 km (120,000 miles) per year.

8 Conclusions

The wind tunnel data presented here shows that the use of rear mounted sealed aft cavity devices can be effective in reducing the drag on tractor-trailers by a significant amount. Many of the aft end geometries indicated reductions in C_D of 0.04 to 0.12, depending on the Reynolds number and the yaw angle. The highest reductions in C_D from the wind tunnel were at lower speeds (Re = 2.0e5) and at higher yaw angles. The best performing rear end configurations were TT1.0, TT2.0, TT3.0sym, and TT3.0nonsym and had similar results with $\%\Delta C_D$ ranging from 5 to 13 %. The pressure data from the wind tunnel confirmed the reduction in drag through an increase in base pressure with average ΔC_p in excess of 0.3 higher than baseline. When converted to a drop in C_D the pressure measurements account for more of a reduction in drag than was actually measured by the force balance, which implies an increase in drag elsewhere on the vehicle. Full-scale SAE J1321 Type II road tests on the same configuration revealed a reduction in drag resulting in a 6.58 % savings in fuel use. A savings of 4500 l (1200 gallons) of fuel per year, per vehicle, could be realized through normal use.

The results of this study have prompted a more generalized study of the aerodynamics of the sealed aft cavities. A greater understanding of how fluids move around such devices can lead to a better understanding of how to optimize them. Future research involving particle image velocimetry and flow visualization around a more generic shape with a sealed aft cavity is being planned and will be carried out in the near future.

Acknowledgments The authors would like to thank the New York State Energy Research & Development Authority (NYSERDA) for their support of this project.

References

1. Bilanin, A.J.: Vehicle drag reducer. US Patent # 4,682,808, 1987
2. Boivin, M., Roberge, K.: Air drag reducing apparatus. US Patent # 6,257,654, 2001
3. Cooper, K.: The wind tunnel testing of heavy trucks to reduce fuel consumption. In: SAE Technical Series. 8–11 Nov 1982. SAE Paper # 821285

4. Cooper, K.: The effect of front-edge rounding and rear-edge shaping on the aerodynamic drag of bluff vehicles in ground proximity. In: International Congress and Exposition, Detroit, Michigan. 25 Feb–1 March, 1985. SAE Paper # 850288
5. Cooper, K.: Bluff-body aerodynamics as applied to vehicles. J. Wind Eng. Ind. Aerodyn. **49**, 1–21 (1993)
6. Coon, J.D., Visser, K.D.: Drag reduction of a tractor-trailer using planar boat tail plates. In: McCallen, R., Browand, F., Ross, J. (eds.) The Aerodynamics of Heavy Vehicles: Trucks, Buses, and Trains, Lecture Notes in Applied and Computational Mechanics, vol. 19, pp. 249–265, illus (2004) With CD-ROM., Hardcover, ISBN:3-540-22088-7
7. Grover, K., Visser, K.D.: Over-the-road tests of sealed aft cavities on tractor trailers. In: SAE 2006 Transactions Journal of Commercial Vehicles, No. 2006-01-3529, p. 170, March 2007
8. Highway Statistics (Washington, DC: Annual issues), table VM-1; www.fhwa.dot.gov/policy/ohpi (2010)
9. Hsu, T.Y., Hammache, M., Browand, F.: Base flaps and oscillatory perturbations to decrease base drag. In: McCallen, R., Browand, F., Ross, J. (eds.) The Aerodynamics of Heavy Vehicles: Trucks, Buses, and Trains, Lecture Notes in Applied and Computational Mechanics, vol. 19, pp. 303–316, illus (2004). With CD-ROM., Hardcover, ISBN: 3-540-22088-7
10. Lanser, W.R., Ross, J.C., Kaufman, A.E.: Aerodynamic performance of a drag reduction device on a full-scale tractor-trailer. In: Aerospace Technology Conference and Exposition, pp. 1–9, Long Beach CA, 23–26 Sept 1991
11. Leuschen, J., Cooper, K.R.: Summary of Full scale wind tunnel test of aerodynamic drag-reducing devices for tractor trailers. In: Browand, F., McCallen R., Rose J. (eds.) The Aerodynamics of Heavy Vehicles: Trucks, Buses, and Trains, Lecture Notes in Applied and Computational Mechanics, vol. 41, pp. 451–462, illus (2009). With CD-ROM., Hardcover, ISBN: 978-3-85069-4
12. Mair, W.A.: Reduction of base drag by boat-tailed afterbodies in low speed flow. Aeronaut. Q. **20**, 307–320 (1969)
13. Maragno, D., Visser, K.D.: Experimental optimization of non-ventilated aft cavities for tractor trailer drag reduction. In: 2003 SAE International Truck and Bus Meeting and Exhibition, Fort Worth, Texas, 10–12 Nov 2003
14. McCallen, R., Flowers, D., Dunn, T., Owens, J., Leonard, A., Brady, M., Browand, F., Hammache, M., Salari, K., Rutledge, W., Ross, J., Storms, B., Heineck, J.T., Driver, D., Bell, J., Zilliac, G., Walker, S.: Aerodynamic Drag of Heavy Vehicles (Class 7–8): Simulation and Benchmarking, 2000-01-2209, pp. 19–21. SAE Gov/Industry Meeting, Washington, D.C. (2000)
15. SAE J1321: Joint ATA/SAE Fuel Consumption Test Procedure, Type II. SAE International (1981)
16. SAE J1252: SAE Wind Tunnel Test Procedure for Trucks and Buses. SAE International (1981)
17. Saltzman, E.J., Meyer, R.R.: A Reassessment of Heavy Duty Truck Aerodynamic Design Features and Priorities. NASA/TP-1999-206574, June 1999
18. Transtex Composites Inc. http://www.transtexcomposite.com/home.html, 201

Part VII
Truck Aerodynamics: Experimental Techniques

In Depth Cd/Fuel Economy Study Comparing SAE Type II Results with Scale Model Rolling Road and Non-rolling Road Wind Tunnel Results

Adrian Reynard, Mike Camosy, Fritz Marinko, Henri Kowalczyk and Tim Jennings

Abstract The trucking industry is moving into a new era of development brought on by governmental concerns over energy independence as well as the realities of increasing fuel costs. This has renewed interest in optimizing the aerodynamics of Class 8 tractor-trailer trucks. However, many of the large aerodynamic gains have already been developed, for example trailer skirts and boat tails that give fuel economy improvements of approximately 5 %. Research continues in order to better understand the aerodynamics of these vehicles and further improve their efficiency. Scale model rolling road testing has been around for several decades. In fact, the earliest rolling road wind tunnel test of a Class-8 truck that the authors are aware of occurred in the late 1980s [1]. In order to define the performance of a heavy duty truck, it has been well established that the use of wind averaged drag coefficients are required. To achieve wind averaged drag coefficients, it is necessary to measure data with a model in yaw for either a static floor or rolling road tunnel. The authors previously published results comparing a generic truck model tested using both a moving ground plane and a static ground plane. This paper builds on the work using a more detailed and modern truck model. The improvements in aerodynamic drag by fitting trailer skirts are discussed. These drag reductions are converted into fuel efficiency improvements and are compared to SAE type II testing. The reader will be able to appreciate the difficulty associated with attempting to correlate wind tunnel results to SAE Type II results. Some of the issues related to scale model rolling road testing such as Reynolds number dependency and yaw over a rolling road will be explored. Due to the size and nature of heavy duty trucks, cross winds must be taken into account. In addition to track or on-road testing, static tunnels have been

A. Reynard · M. Camosy · F. Marinko · H. Kowalczyk (✉) · T. Jennings
Auto Research Center LLC, Indianapolis, IN, USA
e-mail: hkowalczyk@arcindy.com

A. Reynard · M. Camosy · F. Marinko · H. Kowalczyk · T. Jennings
Auto Research Center LLC, Detroit, MI, USA

A. Reynard · M. Camosy · F. Marinko · H. Kowalczyk · T. Jennings
Auto Research Center LLC, Brackley, UK

A. Reynard · M. Camosy · F. Marinko · H. Kowalczyk · T. Jennings
Auto Research Center LLC, Mooresville, NC, USA

A. Dillmann and A. Orellano (eds.), *The Aerodynamics of Heavy Vehicles III*,
Lecture Notes in Applied and Computational Mechanics 79,
DOI 10.1007/978-3-319-20122-1_18

the primary experimental tool for developing heavy duty trucks. Using the static floor tunnel method, heavy duty truck models have fixed (non-rotating tires) and are yawed via a turn table mounted in the floor of a tunnel. Multiple studies have been published which illustrate the necessity for rotating the tires in order to achieve improved correlation to real-world results, for examples see [1, 3–6].

1 Introduction

In recent years, the Auto Research Center (ARC) has seen an increased interest in testing of class 8 tractor-trailer aerodynamic characteristics. The ARC has worked with truck, trailer, after-market manufacturers and fleets using scale model, rolling road testing to advance the heavy trucking industry's ability to improve fuel economy and reduce green house gases by improving the aerodynamic efficiency of their vehicles. This testing allows accurate, quick and cost efficient development which is important since over 21 percent of a heavy truck's power requirement is needed to overcome aerodynamic drag, [2].

It is a common belief that a moving ground plane is only required for vehicles that are low to the ground and would not be required for class 8 trucks, however, Sardou [1] found otherwise. He states: "We just wanted to analyse [*sic*] whether or not high speed moving ground plane must be used for truck development. The answer seems to be yes…" [1]). Subsequent research also showed that the interaction between the moving ground plane and the vehicle being tested were very complicated and unpredictable. Research based on different types of vehicles ranging from trains to race cars showed "that it is just impossible to find any kind of correlation between stationary and moving ground results" [3]. These types of findings are consistent with the experience of the authors at the ARC tunnel. Typically passenger cars show a reduction in drag with a moving ground plane [4]. The authors have previously published results [5] that show an increase in drag with a moving ground plane for class 8 tractor trailers. These results are consistent with recently published studies using CFD to study the effects of yaw and a moving ground plane on semi trucks [6]. Clearly the effect of a moving ground on drag coefficients cannot be easily predicted a priori.

In order to understand the impact of aerodynamic improvements on the overall efficiency of a truck, it is important to consider the environment in which the truck will be used. When trucks travel in a road environment they will experience a relative wind from various directions depending on the road speed and wind speed, thus it is accepted practice that a drag coefficient should include the effects of yaw. SAE J1252 [7] procedure shows how to compute the wind averaged drag to account for the expected operating conditions in the USA. Results at ARC are usually reported for a speed range of 50–75 mph, for this paper the wind averaged drag was computed at the same speed as the SAE type II test that is shown. Figure 1 [8] shows what the probability of yaw angle versus speed is for a truck traveling on a highway. Based on this as well as the SAE1252 procedure for calculating wind averaged drag at greater

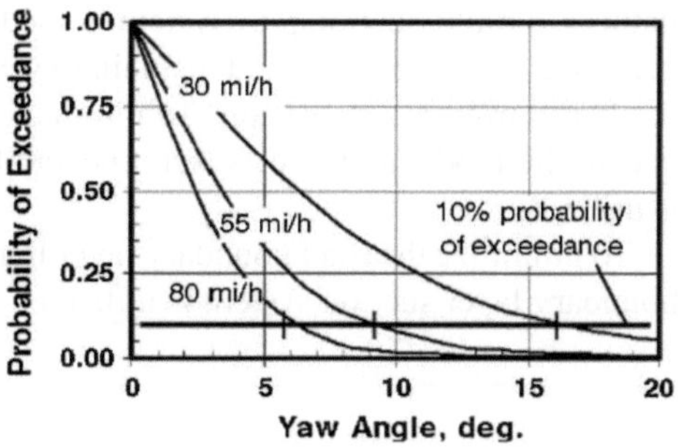

Fig. 1 Probability of exceeding a given yaw angle [8]

Table 1 Aerodynamic drag reduction required to increase fuel economy 1 % at given speed [9]

Vehicle speed (mph)	Aerodynamic drag reduction to increase fuel economy 1 % (%)
60	2
40	3
20	6

than 50 mph, it is felt that limiting testing to 9° of yaw is a reasonable time and cost savings measure.

A common rule-of-thumb used in the industry is that to achieve a 1 % fuel economy increase you must decrease aerodynamic drag by 2 % at 60 mph. Table 1 shows this relationship for a few different speeds [9]. It shows the difficulty in achieving meaningful fuel economy through aerodynamic improvements increases at lower speeds. In this paper the authors prefer to use the results from a more refined methodology [10] in order to present the fuel efficiency results to the end users. This is discussed in the data reduction section.

2 Description of Test Facility

The ARC tunnel is a scale model rolling road tunnel that tests models of vehicles. The model is mounted to a balance cradle that houses a 6-component load cell balance. The balance cradle is attached to a sting that is attached to a steel beam structure that is mounted on vibration pads 8 ft underground. This mounting allows for the model to be completely isolated from any building noise. The model's wheels run on top of the rolling belt that has a surface roughness to match the average surface roughness of highways and roads in the US. The belt speed matches the wind speed to ±0.01 m/s. The turbulence intensity is 0.24 % and the flow angularity is 0.24°. Boundary layer velocity is 99.8 % freestream at 1.0 mm height at center of the model.

The vehicle model motion system (VMMS) is mounted within the cradle that is mounted inside the model. The VMMS allows the vehicle to automatically yaw, pitch, roll, heave and front wheel steer during a test session. To maintain the road flatness, a multi-suction port system is used throughout the platen. To eliminate any belt edge curl, a pneumatic tensioning system is used to apply constant tension to the belt during a test session. To eliminate the static electricity build up before it gets to the model and affects test repeatability, a static electricity discharge system is utilized.

To minimize the road boundary layer that builds up in wind tunnels, a three stage boundary layer suction system is utilized.

3 Test Procedure

At the beginning of each testing session, the tunnel and its data acquisition systems are run for a minimum of 30 min to warm up the facility. The balance and laser systems are warmed up for a minimum of four hours prior to use. The model is inspected for any wheel bearing issues or suspect/damaged parts that could affect testing. The tunnel and road system are inspected. The yaw, heave, roll and pitch position are all measured and maintained.

Each test series began with a static weight tare of the model. Following the static weight tare, a rolling wheel tare was taken for each model position that data was to be recorded. The model position remained the same for all of the runs. The positions were standard road height at yaw equaling 0, 9, 6, 3, 0, −3, −6, −9, 0 degrees. There were three 0° points to allow for a "first to last" run match that is used to determine potential belt stretch, or in test model issue. Positive 9° is yawing the truck 9° to the driver's left. All rolling wheel tares were completed with the belt running 50 m/s without any air speed.

Following the rolling wheel tares, the data sampling process took place for all 9 yaw angles listed above. All tests were run at a constant dynamic pressure equivalent to 50 m/s. After the model data samples were measured, the tunnel was shut down and a change was made to the model. Following a model change, the general process would be repeated.

4 Data Reduction

All force and moment data was measured using an AEROTECH 6 component load cell balance, 124 static pressures and all associated tunnel parameters (i.e. DC motor information, belt temperature, wind speeds etc.) were recorded using a PI Mistral system. The data was organized through a program called PI AERO which utilizes an ACCESS database. PI AERO was used to display the reduced and raw data into Excel. Using Excel, the data was then summarized for reporting purposes. Once a full test run was completed, the measured drag coefficients for each yaw angle were imported into an Excel sheet that calculated the wind averaged drag coefficient from 22.3 m/s (50 mph) through 33.5 m/s (75 mph). Each wind averaged drag

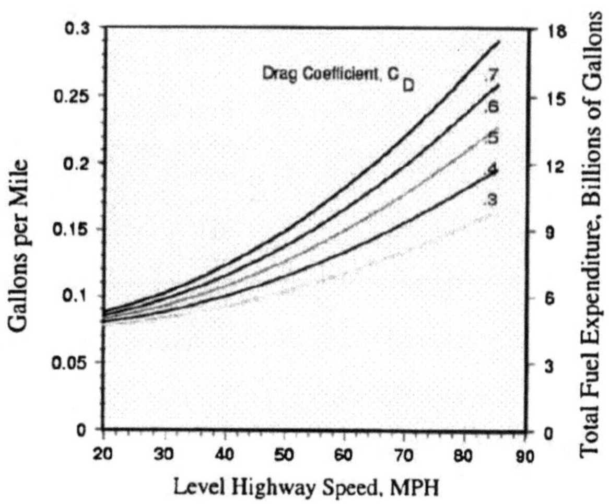

Fig. 2 Fuel consumption for typical Class 8 Truck [10]

coefficient was individually compared to its baseline's proper wind averaged drag coefficient and a percentage change calculated. Based on McCallen [10] we have created a spreadsheet which will convert from percent change of wind averaged drag coefficient to percentage savings of fuel. This percentage change only applies to a theoretical class 8 tractor-trailer, but it gives end users an understanding of the order of magnitude of fuel economy that they can expect, thus allowing them to explore the most economically viable aerodynamic solutions. Figure 2 shows the graph from [10] that the spreadsheet is based on.

A wind speed of 50 m/s, constant dynamic pressure, was used for these investigations using a 1/8 scale mode. A Reynolds number of Re = 1.1 Million results from the model's width as characteristic length. SAE J1252 recommended practice recommends that Re_{min} = 0.7 Million. Storms [11] studied Reynolds number effects on various generic configurations and found that for Reynolds numbers ranging from 1.1 to 7.0 Million very little change on the wind averaged drag coefficient was measured. In this test our speed sweep showed a minimum of 40 m/s will yield reasonable results for wind averaged drag as seen in Figs. 3 and 4. The appendix shows more detailed graphs of C_d at different yaw angles and Reynolds numbers. Figure 5 shows a comparison of drag coefficient for two different yaw angles at various speeds (Reynolds numbers). It shows that the higher yaw angles have a higher sensitivity to Reynolds number. This yaw angled Reynold's dependency shows that simply listing one minimum Reynolds number to define the testing of Class 8 Trucks is not appropriate and should be treated very carefully to ensure quality data is being recorded.

A source of measured drag that must be accounted for is the drag induced by rolling the wheels with a yaw angle. For much of the testing at ARC, the absolute value of this induced drag is of little interest, since much of the aerodynamic development is geared towards observing changes or deltas in drag coefficients. The observed

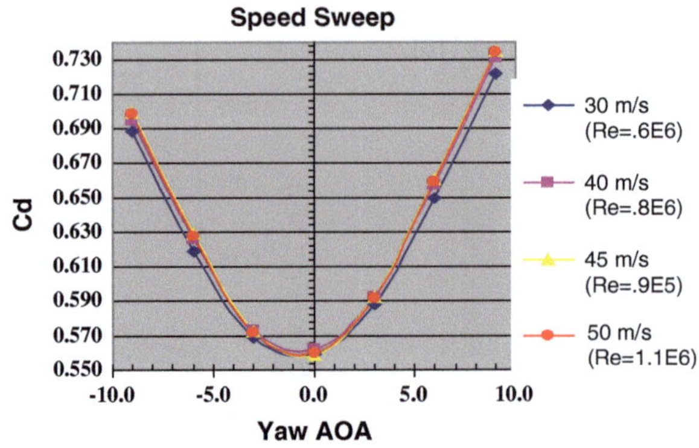

Fig. 3 C_d versus yaw at various test speeds

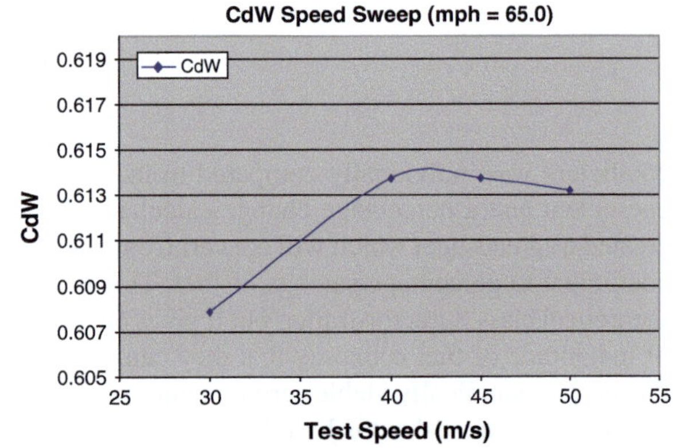

Fig. 4 Wind averaged drag versus test speed

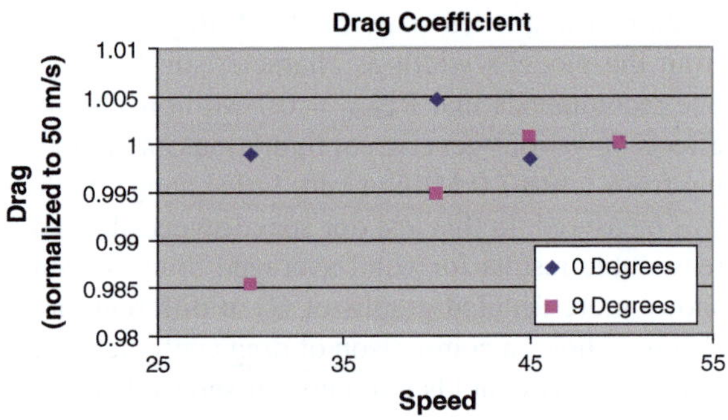

Fig. 5 Drag coefficient normalized to 50 m/s

repeatability of the ARC wind tunnel for this type of testing is ±0.3 %. For this test, however, we were also interested in quantifying the amount of induced mechanical drag that is created by yawing the model over the rolling belt in order to more accurately measure the total aerodynamic drag of the truck to compare to full scale road testing results. We were able to do this by running the model through the yaw sweep without the wind blowing and taring out the rolling drag. The end result is that 0.54 kg (1.2 lbs) is tared out in this initial process versus an overall drag force of 22.68 kg (50 lbs).

5 Discussion of Testing Limitations

As road and track tests are subject to variations in conditions between runs, the accounting for these variables becomes a large task that does induce errors into their final results. Wind tunnel testing can control a majority of the issues a full scale track test faces, but is also subject to smaller variations. Typically, a full scale track test has an error of 2 % (±1 % in fuel economy). Scale model wind tunnel testing has an error of 0.6 % (±0.3 % in fuel economy). We refer the reader to [12] which has a detailed discussion of some of the difficulties in comparing wind tunnel and on-road testing. Unfortunately the SAE type II [13] test is not designed to give an absolute level of drag so we cannot compare the wind tunnel results directly to SAE type II results. However, we present them here so that the reader can understand the impact aerodynamic changes have on fuel efficiency.

6 Approch

The test vehicle used was a 1/8 scale tractor and trailer. The tractor was a Navistar Prostar sleeper. The tractor was created by scanning a full size tractor and building a scale model. The trailer was a 53 foot Wabash trailer and was built with data provided by Wabash. The model had articulating suspension. It also included pressure taps that were not used for this test. Any part of the tractor/trailer that touched the air was accurately represented in this model. This model also included internal flow modeling from the tractor grill into the radiator and then into the engine bay. The bogey was adjustable for multiple positions. Pictures of the model used are shown on the following pages.

Navistar Prostar Sleeper Tractor and Wabash Trailer

Navistar Prostar Sleeper Tractor

Front of Wabash 53' Trailer

Wabash Trailer Landing Gear

6.1 Truck and Trailer Specification

- 2008 International Pro-Star Sleeper Premium 6 × 4
- 270" Wheelbase
- Full Height Roof Fairing
- Vertical Exhaust
- Full Aero Side Mirrors
- Hood Mounted Mirrors
- Full Length Fuel Tank Covers
- Full Length Fuel Tank Skirts
- Factory Cab Extenders
- 2008 Wabash 53' Dura Plate Dry Van
- 102" Wide
- Swing Doors
- 36" King Pin
- Hendrickson Air Ride Suspension
- 13.6' Height with 1" Taper
- Sliding Bogey Set to California Position

7 Configurations Tested

1. Ridge Corp 32" Greenwing Trailer Skirt
2. Ridge Corp 36" Greenwing Trailer Skirt
3. Ridge Corp 36" Greenwing Trailer Skirt integral struts
4. Wabash Trailer Skirt

To compare to all SAE Type II tests for these components, the component's baseline used during the SAE Type II test was also independently established [14–19]. Non-rolling road and non-rolling wheel (NRRNRW) data were measured at varying yaw angles to deliver a wind averaged Cd value. The same process was used to measure rolling road with rolling wheel (RRRW) Cd values.

8 Results

The authors were interested in determining the effects on drag of testing trailer skirts on a fully detailed 1/8 scale class 8 tractor trailer model in a moving ground plane tunnel. The following table shows the difference in wind averaged drag coefficient for both moving plane and fixed floor conditions. It is interesting to note that for these cases the road on condition results in more drag, but it is not simply a small offset. Also, notice the difference between the road on and road off cases for the two baselines A and B. Baseline A resulted in 1.28 % more drag with moving ground plane and baseline B resulted in 3.22 % more drag with moving ground plane. The only difference between these two cases was a 9" full scale difference in tractor to trailer gap and the mud flaps were moved to the trailer bumpers. Clearly the interaction of the truck and ground plane road is a complex one.

Wind averaged drag coefficient at 60 mph			
	Road on	Road off	% Difference (%)
Baseline A	0.623	0.615	1.28
Configuration 1	0.563	0.554	1.60
Configuration 2	0.557	0.545	2.15
Configuration 3	0.554	0.542	2.17
Baseline B	0.683	0.661	3.22
Configuration 4	0.620	0.595	4.03

The next table shows the change in drag from the baseline configuration. The results here are interesting because the fixed floor testing shows a greater difference between the configuration and the baseline. Thus the static floor testing tends to be overly optimistic in predicting aerodynamic improvements for these tests.

The Table 2 shows the fuel economy improvements gained by installing the different aerodynamic devices. Here we can see that the SAE Type II testing and the wind tunnel testing are in general agreement. Also we can see that the fixed ground plane testing overestimated the fuel savings compared to rolling road. For three configurations both rolling road and fixed ground plan tests overestimated the fuel savings and for one configuration both underestimated the savings comparing to SAE Type II testing. We should add here that other aerodynamic devices (aside from trailer skirts) yields a similar mix of the SAE Type II testing both over and underpredicting wind tunnel results based on the tractor/trailer starting configuration. Aside from the

Wind averaged drag coefficient at 60 mph	% Difference to baseline			
	Road on	Road off	Road on (%)	Road off (%)
Baseline A	0.623	0.615		
Configuration 1	0.563	0.554	−9.63	−9.92
Configuration 2	0.557	0.545	−10.59	−11.38
Configuration 3	0.554	0.542	−11.08	−11.87
Baseline B	0.683	0.661		
Configuration 4	0.620	0.595	−9.22	−9.98

Table 2 Summary of results

Device tested	SAE Type II (%)	WT (rolling road) (%)	WT (non-rolling road) (%)
Configuration 1	4.0	4.8	4.97
Configuration 2	5.2	5.31	5.69
Configuration 3	5.1	5.54	5.94
Configuration 4	5.6	4.62	4.99

tractor/trailer baseline sensitivity, the authors also note that significant discrepancies with respects to SAE Type II ambient testing conditions were identified. It is believed that these ambient condition discrepancies play a very large roll in making comparisons to wind tunnel data difficult.

Figure 6 shows the difference in drag between moving ground plane and static floor conditions of four configurations. It is interesting to note that the delta is not a simple offset but instead varies throughout the yaw range. This has important implications since a truck traveling on a highway will be influenced by significant cross wind components. What is especially interesting is the change in yaw dependence between the two baseline cases. For Baseline A we see a hysteresis effect throughout the yaw range that is not present for Baseline B.

Another example of this asymmetry in relation to yaw angle is the percent difference in drag coefficient versus yaw angle show in Fig. 7. This is interesting because it highlights a potential difficulty in correlating wind tunnel testing to road testing. When computing the wind averaged drag coefficient, the yaw dependence is subsumed into a single drag coefficient which assumes the wind direction is somewhat evenly distributed around 360°. However, it is very likely that a single over the road test such as an SAE type II test, will have wind coming from only a single direction or at most a single quadrant, this could skew the results and make it difficult to correlate to these types of tests.

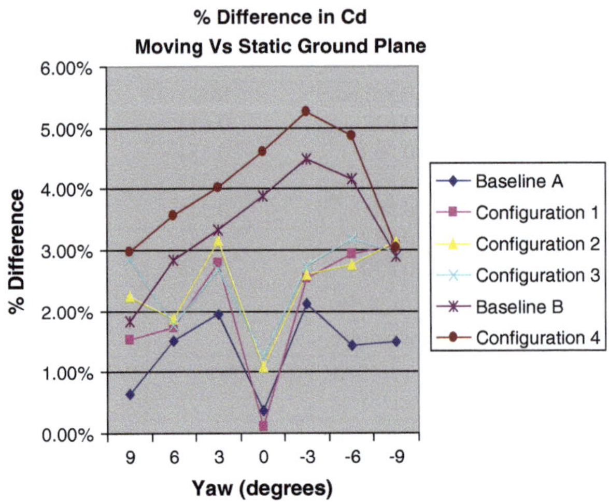

Fig. 6 Moving versus static ground plane

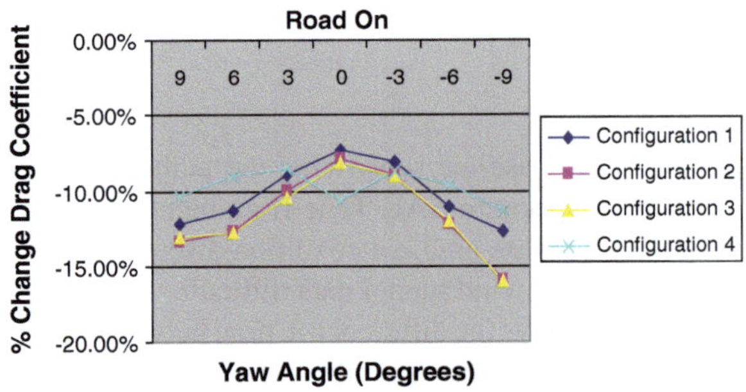

Fig. 7 Percentage change in drag coefficient

9 Conclusions

Because wind averaged drag coefficients are calculated based on wind tunnel testing and take into account the average of all cross flows a tractor/trailer could see (over 360°), wind tunnel comparisons to SAE Type II testing could be skewed since during a typical SAE Type II test the tractor/trailer is most likely not seeing the average estimated cross flows acting over 360°, but more likely acting from one direction or quadrant.

The interaction between the tractor/trailer and the ground plane is complex and it is difficult to estimate the impact of the rolling road on the aerodynamic drag. It must be tested for each case of interest.

For the cases studied, the static ground plane tends to overestimate aerodynamic improvements compared to the rolling road testing.

Depending on the tractor/trailer configuration, a hysteresis effect can be measured throughout the yaw range. Trailer skirts do not change this hysteresis effect, but simply will offset it.

Tractor/Trailer combinations have a Reynolds dependency that is a function of yaw angle (or cross wind). This yaw angled Reynold's dependency shows that simply listing one minimum Reynolds number to define the testing of Class 8 Trucks is not appropriate and should be treated very carefully to ensure quality data is being recorded.

10 Future Work

The authors are currently working to improve correlation between the wind averaged drag coefficient and on the road testing. As a part of this, the authors plan to improve on the correlation between aerodynamic drag and fuel economy.

This paper dealt with a limited subset of aerodynamic devices available. Further work will compare boat tails, trailer leading edge add-on devices, as well as comparing the position of mud flaps, bogeys and the tractor to trailer gap.

Further work is currently being carried out by the authors using CFD to understand the yawed flow conditions. This work will explore the Reynolds number effects on yawed flow and tractor to trailer gap, as well as understanding the localized angled flow around the rolling wheels.

Understanding why certain tractor/trailer configurations have a hysteresis effect over a standard yaw range could provide clues towards improving overall vehicle stability.

Appendix A—Reynolds Number Sweep Results

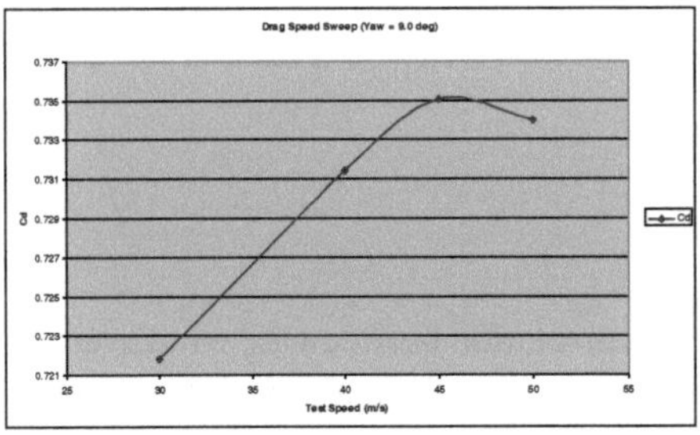

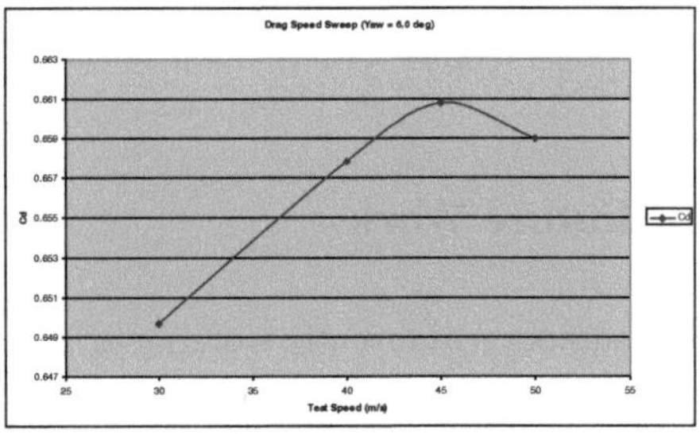

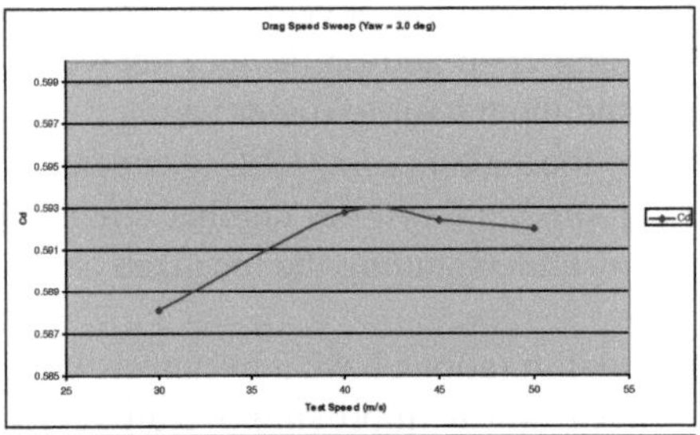

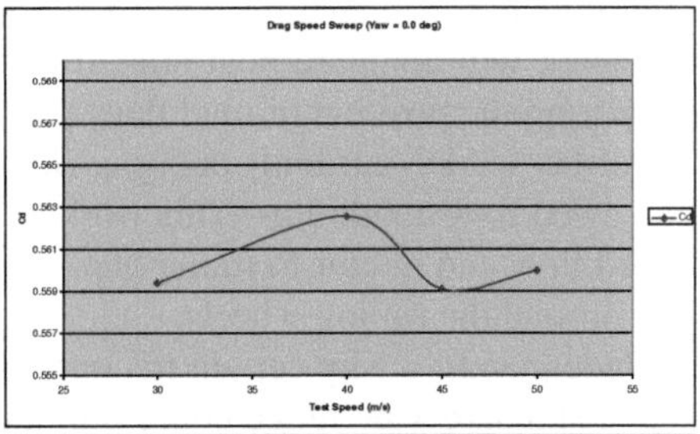

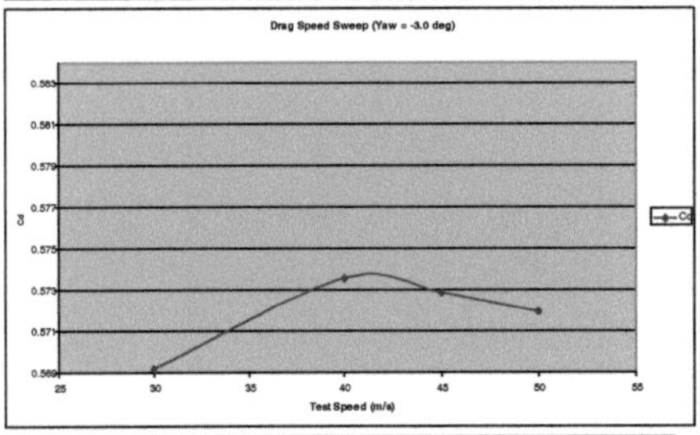

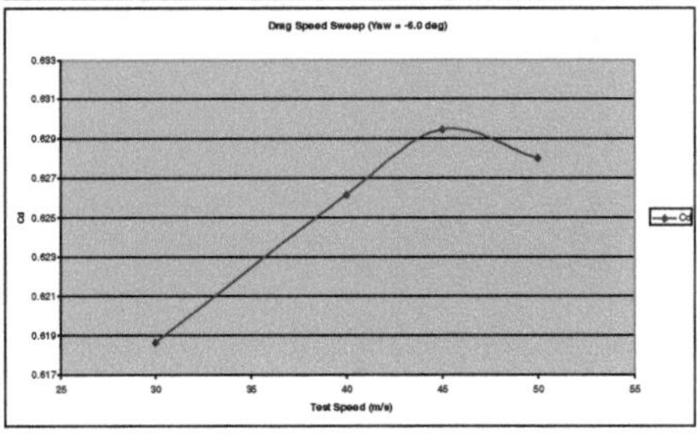

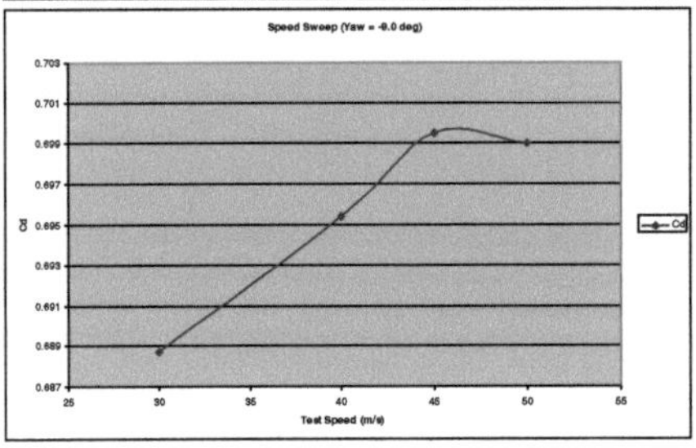

References

1. Sardou, M.: Moving ground and reynolds effect on tractor-trailer. SAE 870707 (1987)
2. U.S. Department of Energy. Technology Roadmap for the 21st Century Truck Program: A Government-Industry Research Partnership, Report 21CT-001, Office of Heavy Vehicle Technologies (2000)
3. Sardou, M.: The sensitivity of wind-tunnel data to high-speed moving ground for different types of road vehicles. SAE 880246 (1988)
4. Elofsson, P., Bannister, M.: Drag reduction mechanisms due to moving ground and wheel rotation in passenger cars. SAE2002-01-0531 (2002)
5. Camosy, M.: Advanced experimental methods for the analysis and aerodynamic design of heavy vehicles. In: ECI—The Aerodynamics of Heavy Vehicles II: Trucks, Buses and Trains, Conference presentation (2007)
6. Elofsson, P.: An investigation of the aerodynamic drag mechanisms due to ground simulation in yawed flow conditions for heavy trucks. In: Proceedings of the ASME 2009 Fluids Engineering Division Summer Meeting, FEDSM2009-78591 (2009)
7. SAE Wind Tunnel Test Procedure for Trucks and Buses. SAE J1252, July 1981, SAE Recommended Practice (1981)
8. Cooper, K.: Truck aerodynamics reborn—lessons from the past. SAE2003-01-3376 (2003)
9. Wood, R.M.: Simple and low-cost aerodynamic drag reduction devices for tractor-trailer trucks. SAE 2003–01-3377 (2003)
10. McCallen, R.: Progress in reducing aerodynamic drag for higher efficiency of heavy duty trucks. SAE1999-01-2238 (1999)
11. Storms, B., Satran, D., Heineck, J., Walker, S.: A study of reynolds number effects and drag-reduction concepts on a generic tractor-trailer. AIAA 2004–2251 (2004)
12. Kettinger, James N.: Tractor-trailer fuel savings with an aerodynamic device—a comparison of wind tunnel and on-road tests. SAE 820376 (1982)
13. Joint TMC/SAE Fuel Consumption Test Procedure—Type II. SAE J1321, SAE Recommended Practice (1981)
14. Wabash National Fuel Economy Test Report. Transportation Research Center Inc. (2009)
15. Freight Wing. Type II Fuel Economy Test Report, Transportation Research Center Inc. (2004)
16. Energotest 2009 Fall: Fuel Consumption Tests of the Ridge Corporation Aerodynamic Side Skirts, FPInnovations Inc. (2009)
17. Energotest 2008: Fuel Consumption Tests for Prototypes of the Freight Wing Trailer Belly Fairing, FPInnovations Inc. (2008)
18. Surcel, M.: Track-test evaluation of aerodynamic drag reducing measures for class 8 tractor-trailers. SAE 2008-01-2600 (2008)
19. Energotest 2009 Fall: Fuel Consumption Track Tests of Fuel-Savings Technologies for Tractor-Trailers, FPInnovations Inc. (2009)
20. Soderblom, D., Lofdahl, L., Elofsson, P., Hjelm, L.: Heavy vehicle wheel housing flows—a parametric study. SAE 2009-01-1169 (2009)
21. Landstrom, C., Lofdahl, L., Walker T.: Detailed flow studies in close proximity of rotating wheels on a passenger car. SAE2009-01-0778 (2009)
22. Leuschen, J., Cooper, K.: Full-scale wind tunnel tests of production and prototype, second-generation aerodynamic drag-reduction devices for tractor-trailers. SAE2006-06CV-222 (2006)

Towing Tank Experiments for Bluff Body Aerodynamics

C.N. Nayeri, J. Glas and C.O. Paschereit

Abstract In ongoing studies on drag reducing devices for heavy vehicles wind tunnel experiments have been carried out at the Hermann-Föttinger Institute of the TU-Berlin. However, these wind tunnel experiments lack a moving belt or a suction system to remove the boundary layer of the wind tunnel floor. Therefore, the use of a towing tank which is available at the TU-Berlin was considered and realized. In order to gain experience with the towing tank as a tool for vehicle aerodynamic studies, preliminary experiments with a bluff body with out any ground effect were carried out. The box-shaped body corresponds to a simplified European trailer at a scale of 1:10. Towing speeds of up to 4 m/s leading to a Reynolds number of $Re = 5.2 \times 10^6$ were used. Comparative measurements in the wind tunnel and the towing tank were carried out with the box. They showed good agreement, however maximum deviations of up to 2.5 % for the axial drag coeefficient were observed. The effect of passive drag reducing devices, so called base flaps, were measured and also their positive effect was well reproduced in the towing tank. The towing tank has thus been established as an aerodynamic test facility.

1 Introduction

For the experimental analysis of the aerodynamic behavior of vehicles wind tunnels are a well established tool. For the correct simulation of the relative motion between the vehicle and the street moving belts are used in some cases which eliminate the presence of the boundary layer on the test section ground. As moving belts are complex as well as expensive alternative solutions are some times of interest. Papenfuss [1] used the moving-model technique with a pneumatic launching system in which speeds of up to 15 m/s were realized. He pointed out that one advantage of

C.N. Nayeri (✉) · J. Glas · C.O. Paschereit
Hermann-Föttinger Institute (TU-Berlin), Müller-Breslau-Str. 8, 10623 Berlin, Germany
e-mail: christian.nayeri@tu-berlin.de

J. Glass
e-mail: jan.glas@tu-berlin.de

A. Dillmann and A. Orellano (eds.), *The Aerodynamics of Heavy Vehicles III*,
Lecture Notes in Applied and Computational Mechanics 79,
DOI 10.1007/978-3-319-20122-1_19

moving model techniques is the easier measurement of flow properties underneath the model.

Aoki et al. [2] performed towing tank experiments with a box-shaped model and a car model. They stated that besides the strong influence of the moving ground on the lift the vortices near the bottom surface of the models have a large influence on the forces. In order to remove the influence of the free surface they suggest a minimum depth d of the model $d/h > 2$ (h being the model height) leading to depth-based Froude numbers $Fd < 0.37$.

Larsson et al. [3] investigated the aerodynamic properties of a full-scale car in a water-basin where the car was rolling on the bottom. The water testing technique was found to be a valuable complement to the wind-tunnel technique.

This paper presents preliminary studies in the large towing tank facility of the TU-Berlin using a box shaped model representative of a European trailer. The goal of these studies is to evaluate the facility as a tool for aerodynamic studies on ground vehicles such as truck, automobiles and trains. At this stage no ground effect has been taken into account (Fig. 1).

2 Experimental Set-Up

The large towing tank facility of the TU-Berlin has a length of 250 m and a width of 8 m. The depth varies between 2.9 m (for the first 90 m) and 4.6 m. A 20 ton carriage (Fig. 2) equipped with eight 50kw DC-motors runs on rails and can be accelerated with 1.0 m/s^2 to a towing speed of 12.5 m/s. The facility was originaly optimized for towing of ship models. The carriage includes a working platform on which measurement equipment and up to 7 persons can be carried. The stability of the carriage is 1 mm/s.

As a bluff body model a box with the dimensions $1.36 \times 0.25 \times 0.3$ m (l $\times$ w $\times$ h) made of aluminum and corresponding to an European trailer was used. The box was equipped with an water proof internal three component force transducer to measure f_x, f_y and f_z. The transducer was attached to a strut which was attached to the carriage (shown in the white circles in Fig. 2). The strut was made from 5 mm sheet metal plates and designed to obtain a maximum hydrodynamic transparency to avoid waves on the water surface and to reduce interactions with the flow domain surrounding the model.

Force measurements where carried out at towing speeds between 0.5 and 4 m/s corresponding to a Reynolds number range of 0.6×10^6–5.2×10^6 based on the model length. The averaging time for the force measurements was 20 s.

According to Aoki et al. [2] the submergence (distance between model upper surface and water surface) of the model in the water should be at least twice the model height. In Fig. 1 the drag coefficient of the bluff body is given as a function of the Reynolds number for three different depths. It shows that for a depth of 0.6 m some deviations around $Re = 4 \times 10^6$ from a constant c_d occur. For the deeper positions

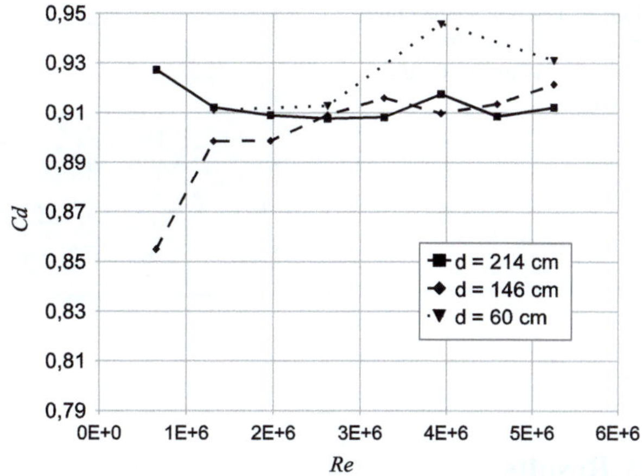

Fig. 1 The axial drag coefficient c_d as a function of *Re* and submergence *d*

Fig. 2 The towing tank facility, the carriage and the strut (*circle*)

of 1.46 and 2.14 m c_d remains constant. However, all experiments presented in this paper were carried out at a depth of 2.14 m.

As drag reducing devices short base flaps with the angles 10°, 12°, 14°, 16° were used. Their effect was investigated earlier in the wind tunnel on a tractor-trailer model also in combination with blowing as an active device by the author [4].

The towing tank results were compared with wind tunnel results obtained with the same bluff body model. The bluff body model was positioned in the center of the wind tunnel test section and was attached to an external six component balance. The maximum wind tunnel speed used was 36.2 m/s which corresponded to about 3 m/s in the towing tank for the same Reynolds number (Fig. 3).

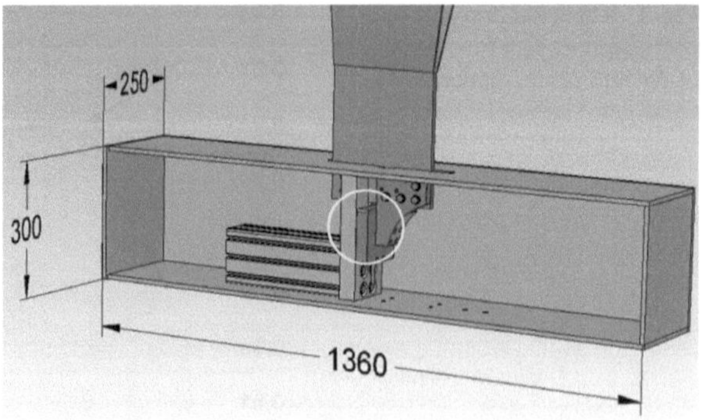

Fig. 3 The box-shaped bluff body with internal three component balance (*white circle*)

3 Results

In Fig. 4 the drag reducing effect of the base flaps is shown. The baseline case has a $c_d \approx 0,91$ which is a somewhat typical value for a box without any round corners at the leading edges. The 10° base flap reduces the drag by about $\Delta c_d = 6\,\%$, the 12° about $\Delta c_d = 7.5\,\%$ and the 16° by about $\Delta c_d = 8\,\%$. In former investigations with longer baseflaps applied to a tractor-trailer model the optimal deflection angle of the flaps was found to be 12°. This difference in the optimal base flap angle is probably attributed to the different flap length and the different boundary layer conditions (thickness, laminar/turbulent) on the models.

The curves in Fig. 4 show some fluctuation and a small tendency of increasing values with increasing Reynolds numbers. The origin of the fluctuations is not clear yet and probably represents the range of uncertainty of the set-up. The constant increase of c_d with Re is probably due to the growth of small oscillations of the model about the vertical axis of the strut due to a galloping phenomena (alternating

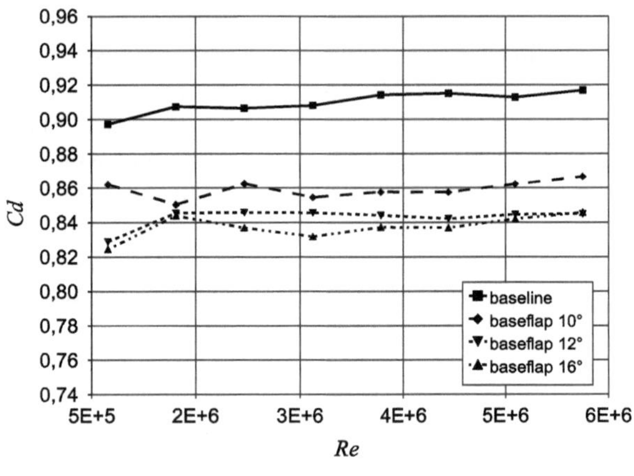

Fig. 4 Drag reduction with baseflaps as a function of the Reynolds number

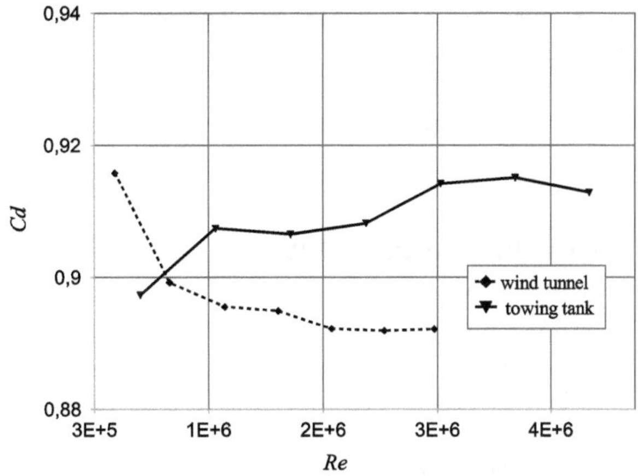

Fig. 5 Comparison of the baseline case between towing tank and wind tunnel

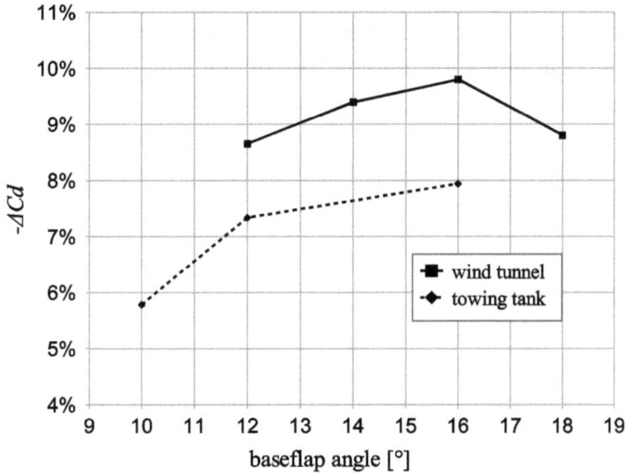

Fig. 6 Comparison of drag reducing effect of base flap with different angles between towing tank and wind tunnel. The c_d values for the towing tank were obtained through averaging the values for all experiments with $Re > 1 \times 10^6$

separation of two dimensional vortices at the vertical leading edges of the model). This phenomenon needs to be further analyzed.

The comparison of the baselines of the towing tank experiments with the wind tunnel experiments is shown in Fig. 5. There the c_d values are plotted as a function of the Reynolds number. A maximal deviation of about 2.5 % is observed at $Re > 1 \times 10^6$. However, it can be stated that for higher Reynolds numer both facilities show Reynolds number independence.

Another comparison between wind tunnel and towing tank measurements is shown in Fig. 6. The trends of drag reducing effects for different base flap deflection angles are similar. Both facilities show the best performance for a angle of 16°. However, the reductions of c_d in the towing tank are about 2 % smaller. So far, there is no concluding explanation for this discrepancy. One possible reasons is the different boundary layer conditions on the box due to different free stream turbulence levels. In the wind tunnel the turbulence level is about 0.3 to 0.5 % and thus higher than the "zero"-turbulence level of the towing tank. A turbulent boundary layer would remain longer attached to the base flaps and thus reduce the drag more. Another—and more probable—reason could be the abovementioned galloping of the towing tank model which would increase the drag.

4 Conclusions and Outlook

The large towing tank facility of the TU-Berlin was successfully introduced as a tool for aerodynamic investigations on bluff bodies. Results from drag measurements on a box-shaped model with passive drag reducing devices were compared with wind tunnel experiments. The baseline cases showed good agreement, however in the towing tank the application of the base flaps showed a less pronounced drag reduction. These discrepancies need to be resolved. A possible reason is galloping of the flow at the leading edges of the model in combination with an inadequately stiff enough strut holding the model.

For further campaigns the strut will be improved and the facility will be equipped with a plate for the simulation of the ground which is of great importance for aerodynamical studies on road or railway vehicles.

References

1. Papenfuss, H.D., Kronast, M.: Moving-model technique used in automobile aerodynamics for measurement of ground effects. Exp. Fluids **11**, 161–166 (1991)
2. Aoki, K., Miyata, H., Kanai, M., Hanaoka, Y., Zhu, M.: A water-basin test technique for the aerodynamic design of road vehicles. SAE Technical Paper (1992). doi:10.4271/92034
3. Larsson, L., Hammar, L., Nilsson, L., Berndtsson, A.: Study of ground simulation-correlation between wind-tunnel and water-basin tests of a full-scale car. SAE Technical Paper (1989). doi:10.4271/890368
4. Nayeri, C.N., Haff, J., Greenblatt, D., Paschereit, C.O., Loefdahl, L.: Drag reduction on a generic tractor-trailer using active flow control in combination with solid flaps, the aerodynamics of heavy vehicles II: trucks, buses, and trains. In: Browand, F., Ross, J., McCallen, R. (eds.) Lecture Notes in Applied and Computational Mechanics Series, vol 2, Springer, Dec 2008. ISBN: 3540850694

Part VIII
Truck Aerodynamics: Computations

What Can LES Do in Vehicle Aerodynamics?

Siniša Krajnović

Abstract The paper discusses an appropriate usage of large eddy simulation (LES) in external vehicle aerodynamics. Three different applications including wheelhouse flow, gusty flow and active flow control, are used to demonstrate how LES can be used to obtain new knowledge about vehicle flows. The three examples illustrate the information that can be extracted using LES in vehicle aerodynamics.

1 Introduction

The target of LES in vehicle aerodynamics should be based above all on the resolution requirements and the available computer resources. Before we initiate an LES investigation, we should ask the question of the purpose of a computationally demanding LES simulation. Let us start with what LES is not good for. Predictions of aerodynamic forces and moments acting on vehicles at operating speeds on the road should not be made using LES. This is based on the resolution requirements of an LES of flow around ground vehicles at operating speeds, which are beyond the capabilities of any existing computers. It is not easy to make an estimate of resolution requirements for an LES of flows around ground vehicles because of the character of these flows, i.e. flow with regions of separation. To explain this, we can begin with traditional estimates of resolution requirements found in textbooks and scientific articles that use the resolution of the near-wall streaks responsible for the maintenance of turbulence. Experimental and DNS results indicate that the sublayer and buffer region of the boundary layer consist of elongated regions of high-speed and low-speed stream-wise velocity (streaks). Analyses of DNS and experimental data indicate that the stream-wise size of these vortices is approximately 400 wall units according to DNS [1] and larger than 1000 wall units according to experiments [2] and that the span-wise spacing between them is approximately 100 [3] wall units.

S. Krajnović (✉)
Department of Applied Mechanics, Chalmers University of Technology, Gothenburg, Sweden
e-mail: sinisa@chalmers.se
url: http://www.tfd.chalmers.se/šinisa/

A. Dillmann and A. Orellano (eds.), *The Aerodynamics of Heavy Vehicles III*,
Lecture Notes in Applied and Computational Mechanics 79,
DOI 10.1007/978-3-319-20122-1_20

According to Robinson [3] "The thin, near-wall buffer region is the most important zone of the boundary layer in terms of turbulence energy production and dissipation. Buffer-region activity is characterized by a bursting process, during which low-speed fluid (provided in the form of streaks) is flung outward from the wall, generating most of the turbulence production in the boundary layer." Thus accurate representation of these vortices is of importance for a prediction of the flow.

Let us now consider the resolution requirements of these structures in LES. Pope [18] suggests that a grid that can resolve 80 % of the turbulent energy everywhere in the domain including the viscous wall region is needed for LES with wall resolution. According to Piomelli and Chasnov [17], the first grid point must be located at $y^+ < 1$ expressed in wall units $y^+ = u_\tau y/\nu$. The resolution in the stream-wise and the span-wise directions must be $\Delta x^+ \simeq 50$–150, $\Delta z^+ \simeq 15$–40 in order to accurately represent the coherent structures in the near-wall region.

To illustrate the size of the computational grid that would be needed for the resolution suggested in [17], we can take example of the computational grid for the vehicle-like body in [10]. The resolution on the body expressed in the wall units was $\langle \Delta s^+ \rangle_t = 30$–145, $\langle \Delta n^+ \rangle_t = 0.5$–0.8 and $\langle \Delta l^+ \rangle_t = 14$–35. Here $\Delta f^+ = \Delta f u_\tau/\nu$, u_τ is the friction velocity and $\langle . \rangle_t$ denotes time averaging, s is the stream-wise direction, n is the wall-normal direction and l is the span-wise direction. The resulting structured grid contains 9.8×10^6 cells, of which approximately 2×10^6 cells are located in the near-wall region ($y^+ < 20$). Note that the resolution of the boundary layers on the lateral walls and the ceiling of the wind tunnel was not considered here.

Now let us extrapolate this grid to the Reynolds number of the flow around a passenger car of approximately 5×10^6 (i.e. 24 times the Reynolds number in [10]). For this purpose we estimate the dependence of the number of the nodes with the Reynolds number. We now consider a small element ΔA of surface area L_x by L_z and let n_y be the number of y-direction grid points in the viscous sublayer. The number of grid points in the near-wall region above ΔA is then

$$\Delta N = \frac{L_x L_z n_y}{\Delta x \Delta z} = \frac{n_y}{\Delta x^+ \Delta z^+} Re_\tau^2 \tag{1}$$

Introducing the skin friction coefficient as,

$$C_f = \frac{2\tau_w}{\rho U_\infty^2} = 2\left(\frac{u_\tau}{U_\infty}\right)^2 \tag{2}$$

the simple flat plate equation from [20] $C_f = 0.0592 Re^{-1/5}$ together with Eq. 2 gives

$$Re_\tau = 0.0296^{1/2} Re^{0.9} \tag{3}$$

The substitution of the expression for Re_τ in Eq. 1 gives

$$\Delta N = 0.0296 \frac{n_y}{\Delta x^+ \Delta z^+} Re^{1.8} \quad (4)$$

Thus the resolution requirements for the representation of the near-wall region ($y^+ < 20$) increase as $\mathcal{O}\left(Re^{1.8}\right)$ and, when the Re number is increased by a factor of 24, the number of cells in the inner region will increase 305 times. The resulting resolution for only the near-wall region is thus approximately 0.61×10^9 cells. We should note that this is the lowest limit because the real car has a far more complex geometry than the simplified vehicle body considered in this paper and much more cells would be required for its representation. The analysis above considered only the influence of the Re number on the resolution in the stream-wise and span-wise directions. The number of cells in the y-direction also increases with the Re number, making the grid required for the flow in the near-wall region of a car very large. If we now add the number of cells in the rest of the domain, it is clear that the transient simulation of this problem is far from what can be computed with the largest existing computers.

The resolution requirements given above must be met in channel simulations or a flat plate boundary simulation. However, in the case of a bluff-body flow, such as the flow around a car with massive separations and regions of recirculating flow, the influence of the near-wall structures on the main flow is smaller. As follows from this observation we can relax the requirement on resolution given above; even so, however, the computational grid for the full-scale car flow is very large. This simple exercise indicates enormous resolution requirements for wall resolved LES of flows around ground vehicles. The author is of the opinion that accurate prediction of aerodynamic forces and moment requires wall resolved LES, and such simulation of the flow around ground vehicles at operating flow speeds will not be feasible in the nearest future with the present tact in the increase of computational capabilities. Thus, the authors conclusion is that LES is not a technique suitable for prediction of aerodynamic loads on vehicles.

On the other hand, Reynolds-Averaged Navier Stokes equations (RANS) commonly used in industry for prediction of aerodynamic loads are not better suited than LES for accurate predictions. The resolution requirements for RANS simulations are much lower than those for an LES, and making a good RANS simulation is certainly affordable even when low Reynolds number turbulence models are used. However, RANS simulations rely on turbulence models and, although there has been tremendous development in turbulence modeling during the past 40 years, there is no turbulence model that can deal with a complex wake flow behind for example a car or even a car wheel in a wheelhouse. It is worth mentioning here that RANS using appropriate turbulence models is very successful in predicting parts of the flow around ground vehicles where flow is attached, and its usage in studying the flow in such flow regions is appropriate. Although it is clear that RANS simulation cannot accurately predict flow around ground vehicles, these methods are almost exclusively used in the vehicle industry. Despite this fact, automotive CFD departments often say that drag coefficient is very accurately predicted using even the simplest two-equation turbulence models. If the flow cannot be predicted, then the agreement of the predicted drag coefficient with the wind tunnel data must be due to cancellations of errors mainly between the pressure predictions at the front and back ends of the vehicle. The conclusion is that even the use of RANS for prediction of aerodynamic forces is questionable, as flow cannot be accurately predicted, and that is what is required for improvement of aerodynamic performance (such as lower drag). Besides, if the only goal of a simulation is to get a drag coefficient close to the wind tunnel value and we can only prove it by comparing with the value from wind tunnel, why would one bother making a CFD simulation in the first place? After all, in such a case, we already have the accurate value from the wind tunnel. The author is of the opinion that not only LES but also CFD should be used to increase our knowledge about vehicle aerodynamic flows. The following text will illustrate the usage of LES to study flows where LES is capable of adding new insight in the flow physics of vehicle aerodynamics flows. Three applications will be illustrated: massively separated wheelhouse flow, a vehicle subjected to gusty flow and an application of active flow control in ground vehicle aerodynamics.

2 Exploration of the Wheelhouse Flow

Wheels and wheelhouses have a large influence on the aerodynamic performance of ground vehicles. Their contribution to drag is around 25–40 % [4, 22] and they are responsible for much of the mud, water and snow deposition on the vehicle. Furthermore, the design of the wheelhouse with the wheel inside has an influence on water splashes in wet road conditions on surrounding vehicles, thus affecting traffic safety. The flow around isolated wheels has received much attention by the research community [3, 14, 15, 19, 21]. Although a physical description of the instantaneous flow around isolated wheels is yet not available, the time-averaged flow is fairly well understood. The placement of wheels on vehicles has a great influence on the wheel flow for two reasons. First, the flow underneath a ground vehicle spreads outward to the sides [10]. This leads to a flow approaching a wheel at an angle of yaw. As a result of the yawed approaching flow, the flow separates on the outside of the wheels. The position of the separation depends on the value of the yawing angle (e.g. 15 % is reported in the literature [23] for passenger cars) and, with the position of separation, the size of the wheel's wake is determined, with all its influence on the vehicle, as well as on splash formation on wet roads.

Wheels are almost always placed in wheel housings (exceptions are some racing vehicles) which further complicates the flow around wheels. No complete physical model of the flow around a wheel within a wheel housing yet exists, and this is the motivation for the present work.

There is no doubt that the understanding of the complicated interaction of wheels with the wheel housing is crucial for understanding their influence on vehicles' drag and water and dirt deposition. Wheel housing often has the shape of a half-cylindrical cavity with its outer side open (but partly shielded by the wheel). The rotation of the wheel and motion of the vehicle lead the air and water swirls into the wheel housing, which are then (due to yawed flow) partly convected and partly sucked out to the outer side owing to the negative pressure in the wake on the wheel's outer side. This flow mechanism has a great influence on the drag, directional stability, mud/water deposition and water splashes. The Reynolds number of the flow around a wheel in a wheelhouse is lower than that of the flow around the vehicle itself but, even then, it is too high for the application of wall resolved LES. This chapter presents the use of LES to study a moderate Reynolds number flow of an idealized wheel in a wheel housing with the aim of increasing our knowledge of this flow. The motivation for a study of flow at a Reynolds number lower than the operational one is that the flow is governed by sharp edge separations and is therefore similar to that at operational (higher) Reynolds number.

2.1 Description of the Model

The vehicle geometry used in the present work is that used in the experimental work of Fabijanic [5]. In his work, Fabijanic [5] studied the influence of adding the wheelhouse and wheels to the basic geometry. He furthermore investigated the influence of the wheelhouse volume on the aerodynamic coefficients and pressure distribution around wheels.

The vehicle model and the computational domain are presented in Fig. 1. The basic form of the vehicle shown there is a rectangular box with a dimension of H height and W width resulting in a model aspect ratio (AR), defined as model width divided by a model height of 1.5. The model consists of rear, nose and wheel-well sections. The nose, wheelhouse and rear sections have dimensions of 95.8, 177.8 and 254 mm. Struts and holdings are used in the experiments [5] but are not included in the present simulations due to their unknown positions. Two different volumes of the wheelhouse equal to 2 (Case 1) and 2.8 (Case 2) volumes of the wheel are used in the present work. The Reynolds number based on the inlet velocity of 30 m/s and the wheel diameter was 1.6×10^5. The wheels rotated at a rotating velocity of 783.2 rad/s, which corresponds to the ground velocity equal to the inlet velocity of the air.

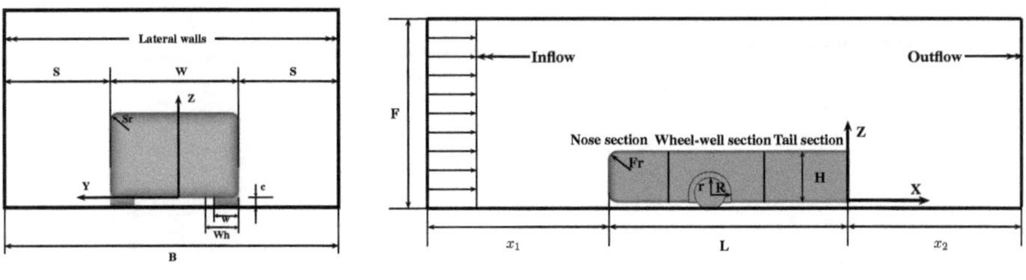

Fig. 1 Schematic representation of the computational domain with vehicle body. *Left* view side; *right* front view

2.2 Wheelhouse Flow as Seen by an LES

The advantage of flow resolved simulation such as LES it that it is capable of providing us with time dependent flow, which is the only true representation of the flow. The instantaneous flow contains all the information required to understand flow mechanisms that are responsible for the character of the flow and the resulting aerodynamic performance of the vehicle. The instantaneous flow in a wheelhouse is complex, and the purpose of the results of the author's LES presented here is to illustrate how new knowledge can be gained using LES. Figure 2 shows the instantaneous flow structures visualized using a second invariant of the velocity gradient Q. One observation of extremely complex and fine scale streaky flow structures can be made from this figure. The flow structures change over time, but here we chose to show the flow at only one time instance and two widths of the wheelhouse. One observation in Fig. 2 is the difference in the size and shape of the instantaneous structures between Cases 1 and 2. The flow structures responsible for formations of the jetting vortex on the outer side of the wheel are for example thicker in the vertical direction (z direction) and thinner in the spanwise direction (y direction), in Case 2 (Fig. 2e) than in Case 1 (Fig. 2).

One observation in Fig. 2 is the difference in the size and shape of the instantaneous structures between Cases 1 and 2. The flow structures responsible for formations of the jetting vortex on the outer side of the wheel are for example thicker in the vertical direction (z direction) and thinner in the spanwise direction (y direction), in Case 2 (Fig. 2e) than in Case 1 (Fig. 2b). The number of the flow structures on the inner side is approximately the same between the two cases (Fig. 2c, f). However, the smaller width of the wheel arch in Case 1 results in a squeezing of the flow structures in a thinner region between the wheel and the inner wall of the wheelhouse. As discussed

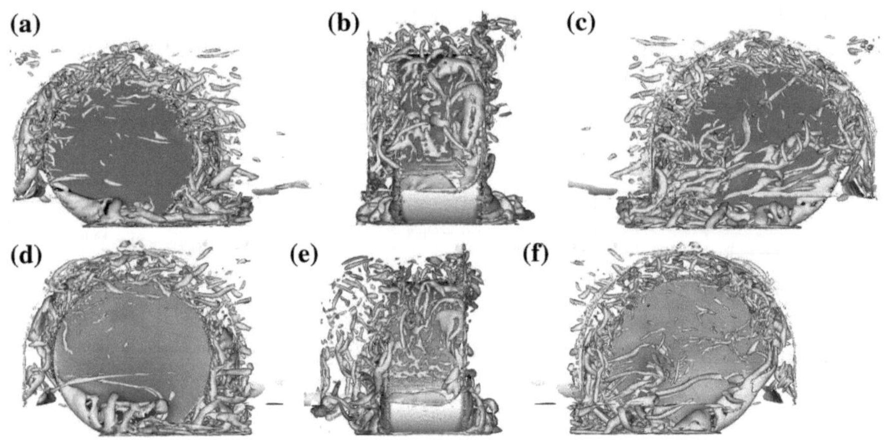

Fig. 2 Iso-surface of second invariant of the velocity gradient tensor $Q = 1 \times 10^7$. **a–c** Case 1, **d–f** Case 2. View is from outside (**a**, **d**), front (**b**, **e**) and inside (**c**, **f**) the wheelhouse, respectively

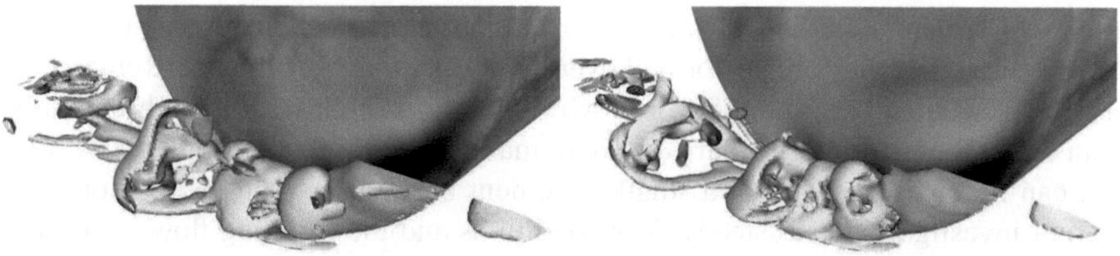

Fig. 3 Case 2. Jetting vortex on the inner side of the wheel. Iso-surface of second invariant of the velocity gradient tensor $Q = 6 \times 10^6$. The time difference between the two pictures is $t = 0.5$ or $tU_\infty/D = 0.196$, where D is diameter of the wheel

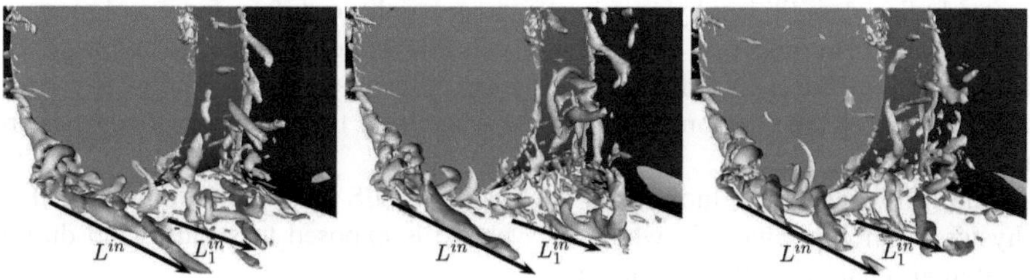

Fig. 4 Case 1. Iso-surface of second invariant of the velocity gradient tensor $Q = 2 \times 10^7$. The time difference between the two picures is $t = 0.5$ or $tU_\infty/D = 0.196$, where D is diameter of the wheel. View from outside the wheelhouse and behind the wheel

earlier, results of our LES display two pairs of trailing vortices in the mean flow. These are the result of two jetting vortices, one on each side of the wheel, that are responsible for the change in the surface pressure of the wheel through the pumping like process. This "pumping" process is illustrated in Fig. 3 and can be described as convection and breakdown of concentric vortex rings.

Figure 4 shows instantaneous flow structures L and $L1$. Two separate jetting vortices can be identified in this figure. The origin of jetting vortex L is the outer edge of the wheel near the wheels contact with the ground upstream of the wheel. The second jetting vortex has its origin just downstream of the contact of the wheel with the ground. Note that the contact of the wheel with the ground is not in only one line but on a surface similar to the real case where deflection of the tire increases the contact region of the tire with the ground.

3 Vehicles in Wind Gusts

Exploration of the flow in real gusty flow conditions is another application where LES has potential. The most common approach in studies of crosswind stability is to evaluate the aerodynamic performance using static conditions where for example the yaw angle of a vehicle or its relative position to another vehicle is changed discontinuously. The experiment or the simulation is then made for one position at a time.

Such experimental studies or numerical simulations are called "quasi-steady" (or "quasi-static") as they are performed over a range of fixed positions of vehicle, e.g. at fixed yaw angle (common for crosswind tests on vehicles). Although the whole span of positions from maximal positive to maximal negative yaw angles, for example, can be investigated with a small increment between two angle positions, this kind of investigation uses steady flow conditions and the resulting flow is different from the one in which continuous change of yaw angle is made. Several experimental and numerical investigations have recently shown that this approach of testing aerodynamic performance of vehicles results in inaccurate predictions. Tests using dynamic flow conditions have shown different forces and moment from those found in steady flow conditions; hysteresis effects and a phase shift in force and moment signals were also observed. The differences in flows between steady and transient flow conditions are expected to be due to a lag in the response of the flow to continuous changes in the position of the vehicle when truly transient flow conditions are applied.

The following text demonstrates the usage of LES for exploration of the flow physics of an experiment in which the vehicle is exposed to steady wind during continuous changes in the yaw angle.

3.1 Description of the Set-Up

The flow around the body studied in the present paper was simulated at quasi-steady conditions by [7] with a RANS approach and an explicit algebraic stress model (EASM). Guilmineau et al. [8] also presented (at quasi-steady flow conditions) DES of this flow. Krajnović and Sarmast [13] made LES of the flow around the body at quasi-steady conditions (yaw angles of 10°, 20° and 30°). The results presented in [13] showed that LES is better suited than RANS or DES for flow around the body studied here.

The model is shown in Fig. 5. It has a length of $L = 675$ mm and a width of $W = 240$ mm. The maximum height is $H = 192$ mm.

The Reynolds number based on the incoming velocity and the car length in the present paper is $Re = 9 \times 10^5$. This is the same Reynolds number as in the experimental study of [2]. The model in the experimental study was placed on a turntable that allowed the model to oscillate with the amplitude of 10° around the vertical axis. The yaw angle of the model was continuously varied between $-10°$ and $10°$ both in experiment [2] and in present LES. The frequency of the oscillation was 2 Hz corresponding to Strouhal number $St = fL/ = 0.068$ in the present LES.

Figure 7 shows a comparison between the present LES and experimental data [6] of the wall pressure at positions of pressure taps P1 and P2 (see Fig. 5). The LES results are shown for each oscillating sweep (thin gray curves) and as ensemble averaged (averaged over three oscillating sweeps) curves (thick black curves).

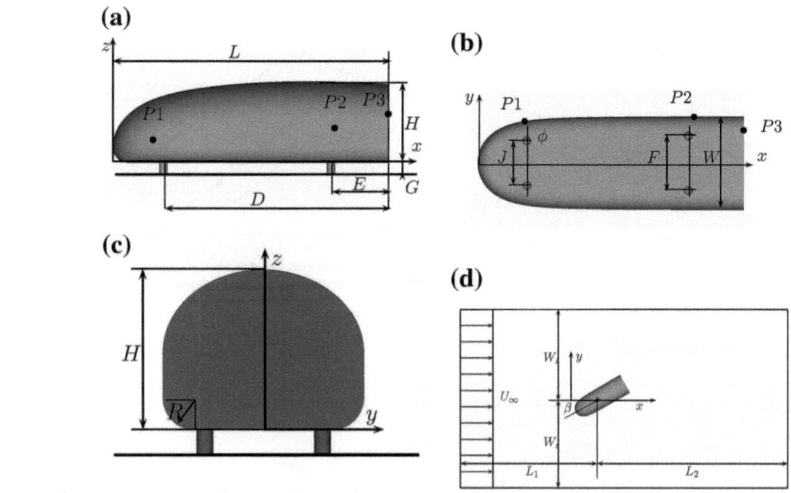

Fig. 5 **a–c** Geometry of the body. **d** Sketch of the models' set-up in the computational domain

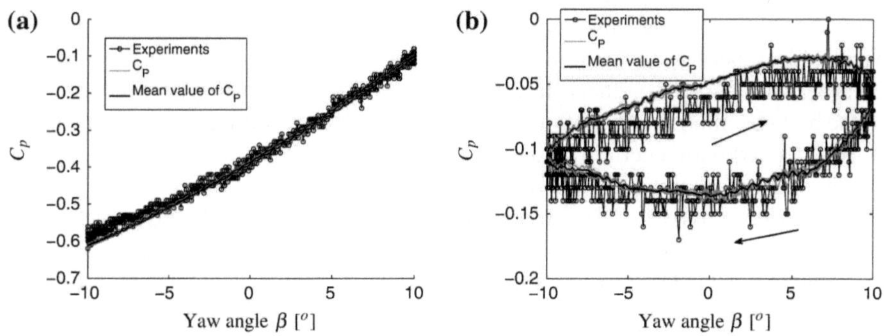

Fig. 6 Comparison between the present LES and experimental data of the C_p values at points **a** P1 and **b** P2

The first observation in Fig. 6 is that both the experimental data and the LES results oscillate during the sweep of the model. The agreement of the LES results with the experimental data is good for pressure taps P1 and P2.

Figure 7 shows the data from LES using the dynamic flow conditions and the quasi-steady experimental data for the drag and the side force coefficients. Here the LES data are presented for three sweeps between the $\beta = 10°$ and $\beta = -10°$ and back together with their ensemble averaged curves. The figure shows that the LES (using dynamic flow conditions) data and the quasi-steady experimental data are different and that the dynamic data mostly exhibit higher values, especially for small yaw angles. The second observation from Fig. 7 is that the evolution of the wall pressure coefficient and the aerodynamic forces both show the existence of the hysteresis phenomenon.

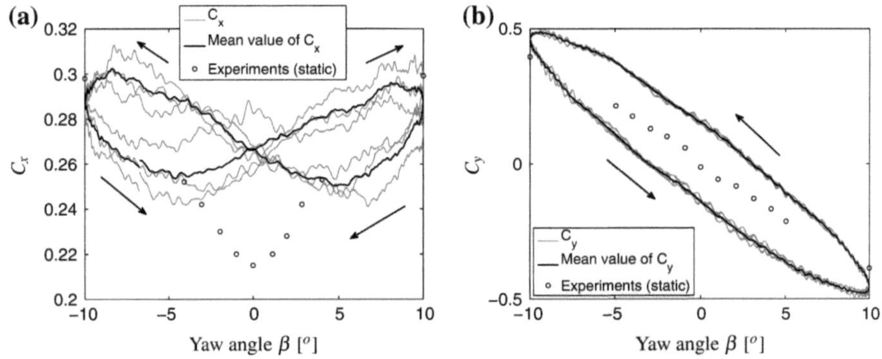

Fig. 7 Comparison of the **a** drag coefficient C_x and **b** side force coefficient C_y

All signals of the pressure and the forces show hysteresis effects in their behavior in the two parts of the sweep between the extreme yaw angles $\beta = 10°$ and $\beta = -10°$. This behavior results in different values of the surface pressure and the aerodynamic forces for the same yaw angle position but for different halves of the rotation sweep. These differences were found to be largest for small yaw angles. Furthermore, the pressure and the force signals show the existence of hysteresis, where the influence of the body oscillations on the surrounding flow and the resulting aerodynamic forces is experienced with a certain delay in time.

This results in a phase shift of the curves in Fig. 7a. Note that the phase shift is present in both halves of the sweep. A phase shift of this kind causes differences in data between the two halves of the rotation sweep and between the quasi-steady and dynamic data. The cause of the hysteresis effect can be explained by looking at the evolution of the flow during the sweep. As the body experiences a solid body rotation, the surrounding flow tries to follow up. The delay in the reaction of the flow to the body rotation is caused by the nature of the surrounding fluid, however, and is due to the inertia effects of the flow when the sign of the rotation is changed.

Figure 8 shows comparisons of the flow structures from the present LES using quasi-static and dynamic (oscillating body) flow conditions for the yaw angle of $\beta = 0°$. Owing to a lack of space, only one angle is chosen. Differences in flows between the quasi-static and dynamic flow conditions were found at all yaw angle positions, however. Figure 8 shows that, at a yaw angle of 0°, the hairpin vortices on the lee side of the body are thinner in the quasi-static case than in the oscillating case.

This difference affects the surface pressure on the body and thereby the aerodynamic forces and moments. The wake behind the base of the body is also influenced by the oscillations of the model. As seen in Fig. 8d, f, there is a curvature in the shape of the wake viewed from above when the model is oscillating. This is explained by a time lag in the response of the wake to the body oscillation. While the wake develops around the symmetry plane of the body in the LES using quasi-static flow conditions,

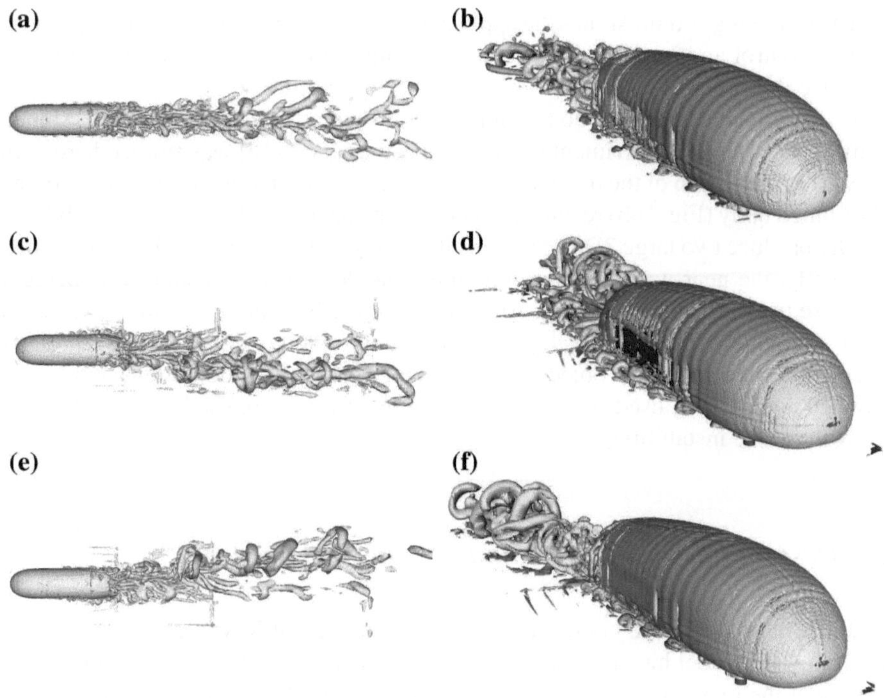

Fig. 8 Iso-surface of the second invariant of the velocity gradient tensor $Q = 2 \times 10^4$ for position $\beta = 0°$. LES of **a**–**b** quasi-steady flow; **c**–**d** dynamic flow for increasing β; **e**–**f** dynamic flow for decreasing β

the dynamic flow conditions result in a shift of the flow structures on the side of the symmetry plane of the domain in which the body is rotating.

The flow in Fig. 8e–f is a result of the oscillation of the body from $\beta = 10°$ to $\beta = 0°$. Although the flow is probably still influenced by the rotation of the body from $\beta = -10°$ to $\beta = 10°$, the "memory" of the flow from the rotation direction from $\beta = 10°$ to $\beta = 0°$ prevails, and the wake has adjusted to the direction of rotation. More recent LES insvestigation of the crosswind stability is reported in [24].

4 Development of Active Flow Control

The development of an efficient strategy for closed-loop active flow control requires a better understanding of open-loop active flow control and the flow mechanisms that are likely to produce an increase in the base pressure of ground vehicles.

The following text illustrates the applicability of the LES technique for the purpose of flow control and improving our understanding of the flow mechanisms acting in an active flow control process.

A two-dimensional bluff body with a lateral shape similar to a so called Ahmed body [1] used in the experimental study by Pastoor et al. [16] was studied here with LES. The interaction of the upper and lower shear layers after the trailing edges of the 2D Ahmed body (Fig. 10b) results in von Karman-like instabilities. Such instabilities rapidly produce two large 2D vortices in alternating order. As the vortices are formed very early, the near-wake separation bubble (the dead water) is short, producing a low base pressure and large drag. An increase of the base pressure can be achieved by an elongation of the near-wake region and suppression or delay of the shear layer interaction. To achieve this objective, the present work applies the strategy of periodic blowing and suction used in [16] to force symmetric vortex shedding and thereby delay the wake instabilities.

4.1 Description of the Model and Numerical Set-Up

Both the natural and the controlled flows were studied at Reynolds number $Re = 2 \times 10^4$. The model has the side shape of an Ahmed body [11, 12] with no slanted surface at the rear end. It extends from the one to the other lateral wall, forcing the flow above and below it. The geometry of the body is shown in Fig. 5b. It has a cross section from the side of a simplified bus with a chord length of $L = 262$ mm, height $H = 72$ mm and spanwise width $W = 550$ mm. The front of the body has a radius of 25 mm. Two trip tapes of a height of 0.8 mm and length of 5 mm were placed on the upper and the lower face of the body 30 mm downstream of the front.

The position of the body in the computational domain is shown in Fig. 9. The model was centered in the computational domain with a height of 7.7 H. The distances from the model to the inlet and the outlet were 10.25 and 20.83 H, respectively.

The boundary layers on the upper and lower surfaces separate at the rear end of the body, forming two shear layers that enclose the wake flow. These two shear layers very quickly become unstable and start to interact with each other, resulting

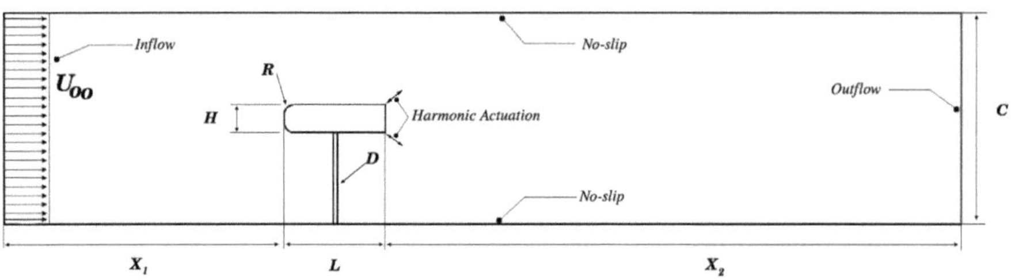

Fig. 9 Set-up of the body with active flow control placed in the wind tunnel

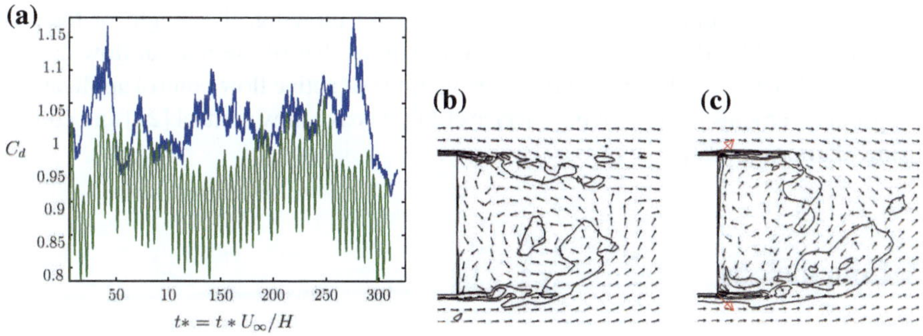

Fig. 10 **a** Time history of the drag coefficient for the controlled flow. Velocity vector field and vorticity, ω_z for **b** natural flow and **c** actuated flow

in von Karman-like vortex shedding. The shear layers are characterized by a large energy dissipation that leads to a decrease in the pressure on the rear end surface. Interaction of the shear layers will in the mean form a rather short separation bubble, producing low pressure on the rear end surface and thus high pressure drag. The idea of the flow control used in [9, 16] and in the present work is to delay the flow instabilities in the shear layers and thereby the shear layer interaction, thus making the mean separation bubble longer. This will redistribute the low pressure in the near wake, thus increasing the rear end pressure and decreasing the drag. The rear end pressure can be increased either by delaying or preventing the interaction of the two shear layers. If the latter is achieved and the vortex shedding of the upper shear layer is decoupled from the vortex shedding of the lower shear layer, less energy will be dissipated.

The alternate blowing and suction is used in the present work to delay the shear layer instabilities and thereby the interaction of the shear layers. As the actuating flow is characterized by two cycles, blowing and suction, an additional actuating parameter, the actuating frequency, must be specified together with the amplitude of the actuating flow. The actuating frequency and amplitude are characterized by Strouhal number $St_A = f_A H/U_\infty$ and momentum coefficient $C_\mu = 4su_A^2/HU_\infty^2$. Here, f_A is the frequency of the actuating flow, u_A is the actuating velocity, s is the width of one actuating slot and H is the height of the body.

Two pairs of slots extending in the spanwise direction are used for actuation of the flow. The slots, with a width of 1 mm and a spanwise length of 250 mm, were placed on the upper and the lower face (one per face) and on the rear face of the model (two slots).

Figure 10b, c shows that the actuation dominates the near-wake flow, resulting in a drag signal that has larger variation in time (green curve) than corresponding natural flow (blue curve). The LES of the natural and the controlled flows resulted in mean C_d values of 1.02 and 0.92, i.e. a decrease of about 11 % using flow actuation. A comparison of Fig. 10b, c shows that blowing postpones the rollover of the shear layers and the formation of strong 2D alternating vortices. Figure 10b, c indicate that the actuation strategy used in the present work is efficient in postponing the vortex formation. For example, it is obvious that, during blowing, the length of the free shear layer in Fig. 10c is approximately three times that of the natural flow case in Fig. 10b. LES has also been used for investigations of active flow control applications for ground vehicles using moving surfaces [25], passive flow control [26] and control using platooning [27].

5 Conclusions

The paper demonstrates the use of LES for external aerodynamics of ground vehicles. It shows that LES has great potential in producing new knowledge about vehicle flows. Similar to other tools used in vehicle aerodynamics, an application of LES must be chosen with the greatest care to avoid inappropriate usage. Flows for which it is difficult to make experimental investigation because of either physical limitations or practical problems in duplicating operational flow conditions (e.g. boundary conditions) in an experimental set-up are appropriate targets for LES in vehicle aerodynamics. The example of the wheelhouse flow is representative of flows where experimental investigations have limitations for reasons of physical constraints posed by experimental equipment and where LES is an excellent choice of numerical technique due to the separated character of the flow. The flow around the oscillating model illustrates how LES can increase our knowledge of flows where the movement of the vehicle or transient character of the upstream flow is critical for accurate representation of the flow and vehicle's performance. There is a great potential of LES for this kind of applications because of its importance for crosswind stability and vehicle comfort. The last application of the active flow control gives an idea of a future use of LES to improve the aerodynamic performance of vehicles. The usage of active flow control will gain ground in many vehicle applications, and its understanding and optimization of flow control parameters can only be achieved using experimental techniques or flow resolving numerical techniques. The latter, in form of LES, will in the future for most certainly be seen to be more cost efficient. An interesting future application of LES is in closed loop active flow control, where the whole control loop can be evaluated numerically before the physical experiment is designed. Finally, despite all the advantages of LES as compared to RANS simulations in separated flows, RANS techniques will continue to have their natural field of application in the flow regions where the flow is fully attached. The prediction of such flow regions at operational Reynolds numbers of ground vehicles will long remain out of the reach of LES due to the great resolution requirements for accurate predictions of boundary layers.

References

1. Ahmed, S.R., Ramm, G., Faltin, G.: Some salient features of the time averaged ground vehicle wake. In: SAE Paper 840300 (1984)
2. Chometon, F., Strzelecki, A., Ferrand, V., Dechipre, H., Dufour, P., Gohlke, M., Herbert, V.: Experimental study of unsteady wakes behind an oscillating car model. In: SAE Technical Paper No. 2005-01-0604 (2005)
3. Cogotti, A.: Aerodynamic characteristics of car wheels, in technological advances in vehicle design series. In: SP3 Impact of Aerodynamics on Vehicle Design, International Journal of Vehicle Design, pp. 173–196 (1983)
4. Elofsson, P., Bannister, M.: Drag reduction mechanisms due to moving ground and wheel rotation in passenger cars. In: SAE Paper 2002-01-0531 (2002)
5. Fabijanic, J.: An experimental investigation of wheel-well flows. In: SAE Paper 960901 (1996)
6. Guilmineau, E., Chometon, F.: Numerical and experimental analysis of unsteady separated flow behind an oscillating car model. In: SAE paper 2008-01-0738 (2008)
7. Guilmineau, E., Chometon, F.: Effect of side wind on a simplified car model: experimental and numerical analysis. ASME. J. Fluids Eng. 131 (2009)
8. Guilmineau, E., Chikhaoui, O., Deng, G., Visonneau, M.: Effect de vent lateral sur un modele simplifie de voiture par une methods DES. Mec. Ind. **10**, 203–209 (2009)
9. Henning, L.: Regelung abgelöster Scherschichten durch aktive Beeinflussung. Ph.D. thesis, Technischen Universität Berlin, Germany (2008)
10. Krajnović, S., Davidson, L.: Numerical study of the flow around the bus-shaped body. ASME J. Fluids Eng. **125**, 500–509 (2003)
11. Krajnović, S., Davidson, L.: Flow around a simplified car, part 1: large eddy simulation. ASME J. Fluids Eng. **127**, 907–918 (2005)
12. Krajnović, S., Davidson, L.: Flow around a simplified car, part 2: understanding the flow. ASME J. Fluids Eng. **127**, 919–928 (2005)
13. Krajnović, S., Sarmast, S.: Numerical investigation of the influence of side winds on a simplified car at various yaw angles. In: 3rd Joint US-European Fluids Engineering Summer Meeting, Symposium on Issues and Perspectives in Ground Vehicle Flows, FEDSM-ICNMM-30766, Montreal, Canada, 1–5 Aug 2010
14. McManus, J., Zhang, X.: A computational study of the flow around an isolated wheel in contact with the ground. ASME. J. Fluids Eng. **128**, 520–530 (2006)
15. Mears, A.P., Crossland, S.C., Dominy, R.G.: An investigation into the flow-field about an exposed racing wheel. In: SAE: World Congress, SAE Paper 2004-01-0446, Detroit, Michigan, USA (2004)
16. Pastoor, M., Henning, L., Noack, B.R., King, R., Tadmor, G.: Feedback shear layer control for bluff body drag reduction. J. Fluid Mech. **608**, 161–196 (2008)
17. Piomelli, U., Chasnov, J.R.: Large-eddy simulations: theory and applications. In: Henningson, D., Hallbaeck, M., Alfreddson, H., Johansson, A. (eds.) Transition and Turbulence Modelling. pp. 269–336. Kluwer Academic Publishers, Dordrecht (1996)
18. Pope, S.B.: Turbulent Flows, 1st edn. Cambridge University Press, Cambridge (2000)
19. Saddington, A.J., Knowles, R.D., Knowles, K.: Laser doppler anemometry measurements in the near-wake of an isolated formula one wheel. Exp. Fluids **42**, 671–681 (2007)
20. Schlichting, H.: Boundary-Layer Theory, 7th edn. McGraw-Hill, New York (1979)
21. Wäschle, A., Cyr, S., Timo, K., Jochen, W.: Flow around an isolated wheel—experimental and numerical comparison of two cfd codes. In: SAE: World Congress, SAE Paper 2004-01-0445, Michigan, USA, Detroit (2004)
22. Wickern, G., Zwicker, K., Pfadenhauer, M.: Rotating wheels—their impact on wind tunnel test techniques and on vehicle drag results. In: SAE paper 970733 (1997)
23. Wiedemann, J.: The influence of ground simulation and wheel rotation on aerodynamic drag optimization—potential for reducing fuel consumption. In: SAE paper 960672 (1996)
24. Krajnović S., Ringqvist P., Nakade K., Basara B.: Large eddy simulation of the flow around a simplified train moving through a crosswind flow. J. Wind Eng. Ind. Aerod. **110**, 86–99 (2012)

25. Han X., Krajnović S.: Large eddy simulation of flow control around a cube subjected to momentum injection. Flow Turbul. Combust. **92**(1–2), 527–542 (2014)
26. Krajnović, S.: LES investigation of passive flow control around an Ahmed body. J Fluids Eng. **136**(12), (2014) doi:10.1115/1.4027221
27. Uystepruyst, D., Krajnović, S.: LES of the row around several cuboids in a row, Int. J. Heat Fluid Flow **44**, 414–424 (2013)

A LES Study of a Simplified Tractor-Trailer Model

Jan Östh and Siniša Krajnović

Abstract Large-eddy simulation (LES) was used to study the flow around a simplified tractor-trailer model. The model consists of two boxes placed in tandem. The front box represents the cab of a tractor-trailer road vehicle and the rear box represents the trailer. The LES was made at the Reynolds number of 0.51×10^6 based on the height of the rear box and the inlet air velocity. Two variants of the model were studied, one where the leading edges on the front box are sharp and one where the edges are rounded. One small and one large gap width between the two boxes were studied for both variants. Two computational grids were used in the LES simulations and a comparison was made with available experimental force measurements. The results of the LES simulations were used to analyze the flow field around the cab and in the gap between the two boxes of the tractor-trailer model. Large vortical structures around the front box and in the gap were identified. The flow field analysis showed how these large vortical structures are responsible for the difference in the drag force for the model that arises when the leading edges on the front box are rounded and the gap width is varied.

1 Introduction

The aerodynamic performance of heavy vehicles is an area in which improvements can be made even though there have been great achievements in the past half century. The aerodynamic forces that act on a vehicle during driving affect the operation of the vehicle in different ways. The side and lift forces may cause instabilities and handling problems. The drag force affects the velocity and acceleration of the vehicle as well

J. Östh (✉) · S. Krajnović
Department of Applied Mechanics, Chalmers University of Technology,
SE-412 96 Göteborg, Sweden
e-mail: ojan@chalmers.se

S. Krajnović
e-mail: sinisa@chalmers.se
url: http://www.tfd.chalmers.se/~sinisa/

A. Dillmann and A. Orellano (eds.), *The Aerodynamics of Heavy Vehicles III*,
Lecture Notes in Applied and Computational Mechanics 79,
DOI 10.1007/978-3-319-20122-1_21

as the fuel consumption. For commercial vehicles overcoming the drag is a lesser part of the total fuel that goes to overcome all the driving resistances (such as drag, rolling and climbing resistance) during cruising in contrast with a passenger car, but valuable energy savings can still be made. The pressure drag, which makes up the major part of the drag on a tractor-trailer unit, essentially comes from four parts of the vehicle: the forebody of the tractor, the rear part of the trailer, the underhood and wheels and the gap between the tractor and the trailer. The latter is a large contributor to drag. The shape of the tractor cab will influence what the flow conditions will be for the trailer. The contribution to the overall drag coefficient of a tractor-trailer unit from the gap is influenced by the width of the gap between the tractor cab and trailer. The gap width will henceforth be denoted by g and expressed in non-dimensional form g/b or sometimes $g/\sqrt{A}$, where b and A are characteristic length and area, respectively, of the specific model studied. In general, the smaller the gap is, the smaller will the positive contribution to the drag coefficient of the gap be. This is of course also dependent on the relationship between the heights of the tractor and the trailer. The drag coefficient of simplified tractor-trailer units with a cab with rounded shape is smallest for zero gap widths and increases slightly with increasing gap width. Around a gap width of $g/\sqrt{A} \approx 0.5$ (A is the model cross-sectional area) a sudden increase in drag occurs [1–3]. This increase in drag is completely due to an increase in the drag of the trailer alone. The drag of the tractor cab varies very little with increasing gap widths [2, 3].

Allan [2] made aerodynamic measurements in the wind tunnel of Southampton on a simplified tractor-trailer model at Reynolds number $Re_b = 5.1 \times 10^5$ based on the height of the trailer box, b. The model he used is also the one used in the present numerical study, see Fig. 1. The model consists of one front box representing the tractor cab and a rear box representing the trailer. The front box has a height, width and depth of $0.92b$, $0.92b$ and $0.67b$, respectively. The rear box has a height, width and depth of b, b and $2.5b$, respectively. The ground clearance of the front and rear box is $0.21b$ and $0.5b$, respectively. This setup gives a realistic placement of the cab and trailer with respect to the standard shape of European tractor-trailer trucks at the

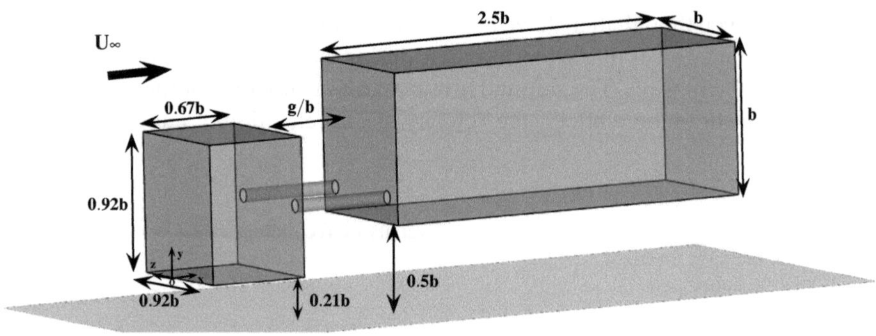

Fig. 1 Dimensions of the model used in the experiments in [2] and in the present LES simulations. Here is the variant with square leading edges on the *front box* seen

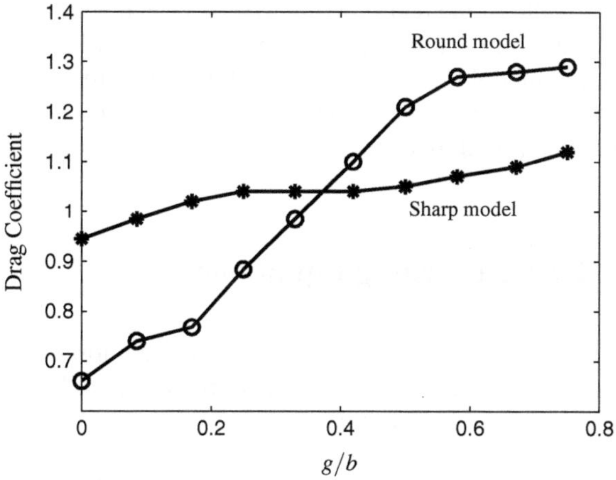

Fig. 2 Experimental results for the drag coefficient for the two models investigated, redrawn from [2]

time. Two variants of the model were investigated by Allan [2], one model with sharp leading edges on the front box and another with rounded edges. The width of the gap between the two boxes was varied and the drag was measured for 0° yaw angle. The results are shown in Fig. 2. It was found that the value of the drag coefficient of the model with sharp leading edges on the front box was more stable with respect to increased gap width. The model with rounded edges exhibited significantly smaller drag coefficient for small gap widths but the drag increased dramatically for larger gap widths and even gave a larger drag coefficient than for the sharp model. This behavior of the drag coefficient of another simplified tractor-trailer model with a rounded front was also found later in the experimental study [3].

2 Description of the Model

In the present work, LES simulations of the flow around the tractor-trailer model from [2] are conducted. At a zero yaw angle, the flow physics will be explored around the model with square leading edges on the front box and the model with rounded leading edges on the frontbox. The general dimensions of the model are shown in Fig. 1. In the round model, both the horizontal and vertical leading edges on the front box are rounded with the non-dimensional radius $0.08b$ which is the same as in the experiments [2]. This necessarily extends the length of the front box of the round model by $0.08b$. The two boxes are connected to each other with two cylinders of diameter $0.08b$, which are placed horizontally at half of the height of the front box. The complete model was suspended from the arms of an overhead balance in the experiments. No locations or details of these arms were reported, and the arms are therefore neglected in the present simulations. Two different gap widths are investigated for both models, one small gap width of $g/b = 0.17$ (b = 0.305 m is the width of the rear box) and one large gap width of $g/b = 0.67$. The experimental results from [2] indicate that the sharp model has a drag coefficient of 1.02 and the round model 0.77 for $g/b = 0.17$, based on the cross-sectional area of the rear box ($b^2 = 0.093\,\mathrm{m}^2$). For $g/b = 0.67$, the sharp model has a drag coefficient of 1.09 and the round model 1.28.

3 LES Governing Equations

The governing LES equations are the incompressible Navier-Stokes and the continuity equations filtered with the implicit spatial filter of characteristic width Δ (Δ is the grid resolution in this work):

$$\frac{\partial \bar{u}_i}{\partial t} + \frac{\partial}{\partial x_j}\left(\bar{u}_i \bar{u}_j\right) = -\frac{1}{\rho}\frac{\partial \bar{p}}{\partial x_i} + \nu \frac{\partial^2 \bar{u}_i}{\partial x_j \partial x_j} - \frac{\partial \tau_{ij}}{\partial x_j} \tag{1}$$

and

$$\frac{\partial \bar{u}_i}{\partial x_i} = 0. \tag{2}$$

Here, $\bar{u}_i$ and $\bar{p}_i$ are the resolved velocity and pressure, respectively, and the bar over the variable denotes filtering. The influence of the small scales of the turbulence on the large energy carrying scales in Eq. (1) appears in the SGS stress tensor, $\tau_{ij} = \overline{u_i u_j} - \bar{u}_i \bar{u}_j$. The algebraic eddy viscosity model originally proposed by [4] is used in the present work for its simplicity and low computational cost. The Smagorinsky model represents the anisotropic part of the SGS stress tensor, τ_{ij}, as:

$$\tau_{ij} - \frac{1}{3}\delta_{ij}\tau_{kk} = -2\nu_{sgs}\bar{S}_{ij} \tag{3}$$

where $\nu_{sgs} = (C_s \Delta)^2 |\bar{S}|$ is the SGS viscosity and

$$\bar{S}_{ij} = \frac{1}{2}\left(\frac{\partial \bar{u}_i}{\partial x_j} + \frac{\partial \bar{u}_j}{\partial x_i}\right) \tag{4}$$

is the resolved rate-of-strain tensor and $|\bar{S}| = \left(2\bar{S}_{ij}\bar{S}_{ij}\right)^{\frac{1}{2}}$. Equations (1) and (2) are discretized using a commercial finite volume solver AVL FIRE for solving the incompressible Navier-Stokes equations using a collocated grid arrangement. The convective fluxes are approximated by a blend of 95 % central differences of second-order accuracy and of 5 % upwind differences. The diffusive terms containing viscous plus sub-grid terms are approximated by a central differencing interpolation of second order accuracy. The time integration is done using the second-order accurate three-

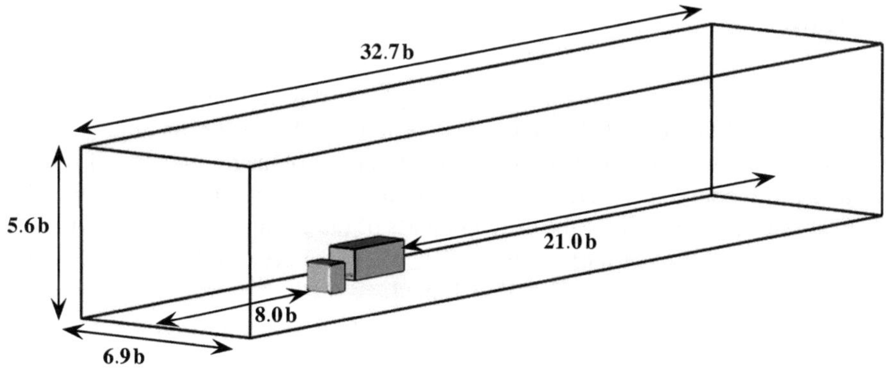

Fig. 3 The computational domain used in the present LES simulations. b = 0.305 m

time level scheme. The same numerical method was used in explorations of flows around vehicles in [5–10].

Figure 3 presents the computational domain used in the present simulations. A free stream velocity of $U_\infty = 24.4$ m/s is used at the inlet which is the same as was used in the wind tunnel tests [2]. The no-slip condition is used on the lateral walls and roof and the homogeneous Neumann condition at the outlet. On the floor the moving ground used in the experiments is simulating by setting the velocity component in the streamwise direction equal to U_∞ together with the no-slip condition. The cross-sectional area of the computational domain is the same that of the wind tunnel used in [2].

Table 1 presents the four cases (Cases 1–4) that were simulated in the present paper. For each case, two computational grids were used. The grids were made with the commercial grid generator package Ansys ICEM-CFD and consist of only hexahedral elements. For the coarse grids, the resolution in the streamwise direction is $2 < \Delta s^+ < 100$ and $2 < \Delta l^+ < 30$ in the direction parallel to the surface of the body and normal to the streamwise direction, respectively. The fine grids have $2 < \Delta s^+ < 70$ and $2 < \Delta l^+ < 20$. Here, $\Delta n^+ = n\langle u_\tau \rangle_t/\nu$, $\Delta s^+ = \Delta s\langle u_\tau \rangle_t/\nu$, $\Delta l^+ = \Delta l\langle u_\tau \rangle_t/\nu$ and $\langle u_\tau \rangle_t$ is the time-averaged friction velocity. The CFL number is below one in all the simulations.

Table 1 Definition of the parameters of the four different cases (Case 1–4) simulated in the present paper

Names	Leading edge on front box	Gap width
Case 1	Sharp	$g/b = 0.17$
Case 2	Round	$g/b = 0.17$
Case 3	Sharp	$g/b = 0.67$
Case 4	Round	$g/b = 0.67$

4 Results

The results of the LES simulations are presented in this section. Comparisons with experimental data are made in the first subsection which presents the time-averaged forces. The second subsection presents flow structures of the time-averaged velocity around the model and regards only results of the present LES simulations. The denotations that are used for the different parts of the geometry are shown in Fig. 4. The base face of the rear box, denoted RB_BF, is not seen in the figure. The four different cases, Cases 1–4, are listed in Table 1.

4.1 Time-Averaged Forces

The drag and lift force coefficients are given in Table 2. The only case for which the present LES simulations predict the value of C_D in agreement with the experimental data is Case 1, for both the coarse and fine grids. The fine grid for Case 2 yields a value fairly close to that of the experiments. For Case 3, the value is overpredicted by 6 % for both of the grids and the value is underpredicted by 12 and 10 % for the fine and coarse grid, respectively, for Case 4. No data for the lift coefficient were given in [2] and hence no comparison can be made. The results of the simulations for C_L are however presented in Table 2.

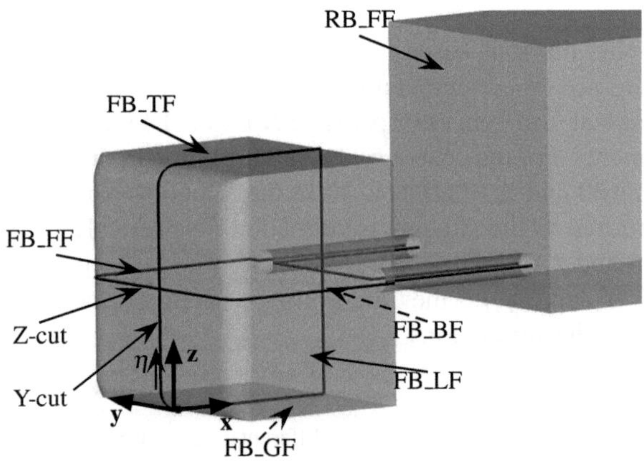

Fig. 4 Denotations of parts of the model used in the present work. Coordinate η is used to plot the pressure coefficient (C_P) and starts at origo and goes clockwise around the symmetry line on the front box. FB_FF, FB_TF, FB_BF, FB_LF and RB_FF denote the *front face* of the front box, the *top face* of the front box, the *base face* of the front box, the *lateral face* of the front box and the front face of the rear box, respectively. Y-cut cuts the front box at half of its width ($y = 0$) and is the symmetry plane. Z-cut cuts the front box at $z = 0.46b$

Table 2 Force coefficients results from the simulations together with the available experimental data from [2]

Aerodynamic coefficients	C_D	$rms(C_D)$	$\|\Delta C_D\|$	C_L	$\|\Delta C_L\|$
Case 1 exp	1.02	–	–	–	–
Case 1 fine	1.023	0.041	<1 %	−0.353	-
Case 1 coarse	1.022	0.039	<1 %	−0.365	3.5 %
Case 2 exp	0.77	–	–	–	–
Case 2 fine	0.793	0.056	3 %	−0.225	–
Case 2 coarse	0.823	0.044	7 %	−0.221	2 %
Case 3 exp	1.09	–	–	–	–
Case 3 fine	1.160	0.084	6 %	−0.447	–
Case 3 coarse	1.157	0.077	6 %	−0.490	10 %
Case 4 exp	1.28	–	–	–	–
Case 4 fine	1.124	0.041	12 %	−0.447	–
Case 4 coarse	1.146	0.048	10 %	−0.490	10 %

$|\Delta C_D|$ is the difference between the computed and the experimental value. $|\Delta C_L|$ is the difference between the coarse and fine grids

The contributions to C_D of the entire model from the four contributing faces are shown in Fig. 5. In [2] the contributions to C_D from FB_FF and RB_BF are said to be almost constant over different gap widths for both the sharp and round models. This is also true in the present simulations, which can be seen in Fig. 5d. The changes in C_D for the entire model when the gap width is varied are thus only due to changes in the contributions from the two gap faces, FB_BF and RB_FF. A great deal can be understood about the difference between the models (sharp and round leading edges on the front box) by looking at the graphs in Fig. 5. The rounding dramatically reduces C_D on FB_FF by some 55 % (see Fig. 5b) for both gap widths (as expected when rounding the front edge of a prismatic body, see [11]). C_D is however also reduced by some 50 % on the base face of the front box (FB_BF) with round front edges for both gap widths, see Fig. 5a. This was found in both the present numerical study and the experimental study [2]. The large gain in decreased drag for the entire model from the front box, with round edges is lost in an increase in the contribution to the drag from the rear box for large gap widths. Figure 5c shows the contribution to C_D from RB_FF for the sharp and round models, respectively. On the sharp model, there is actually a negative contribution to C_D from RB_FF in both the present simulations and experiments for both gap widths. For the round model with a small gap width, there is a slightly positive contribution to C_D from RB_FF. The contribution to C_D from RB_FF increases a great deal as the gap width is increased to $g/b = 0.67$ for the round model, see Fig. 5c. Thus, starting with Case 1 ($g/b = 0.17$ and a small gap width), the decrease in C_D that is gained by rounding the leading edges on the front box is lost almost entirely due to the contribution to C_D from the front face of the rear box (RB_FF) when the gap width is increased.

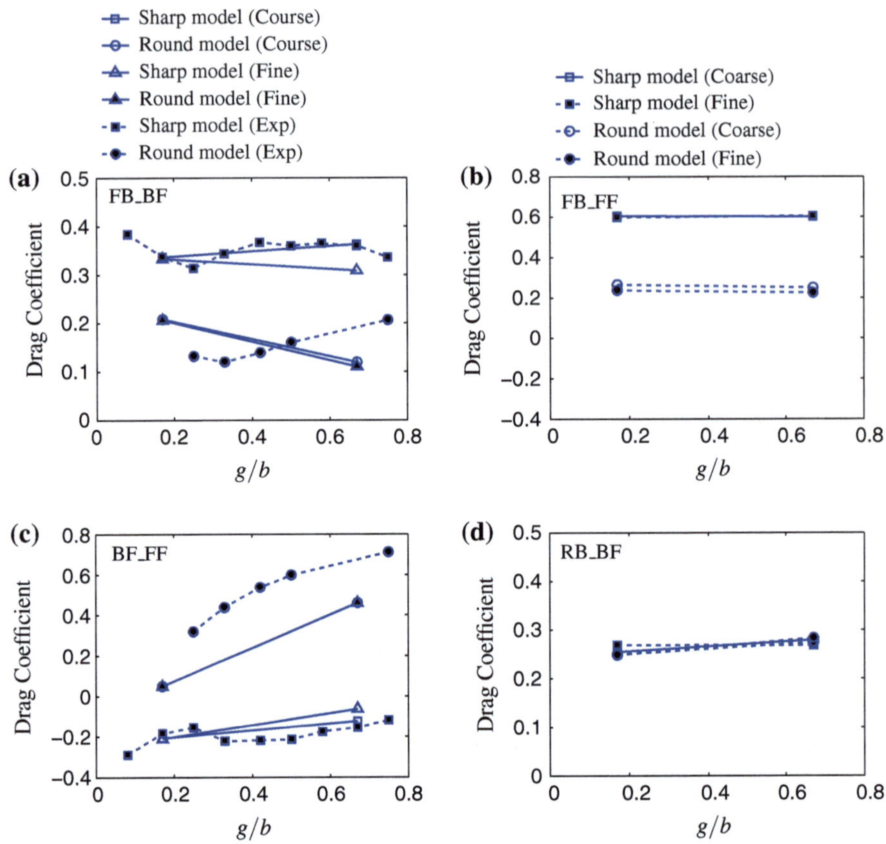

Fig. 5 Contribution to the drag coefficient of the entire model from the four different contributing faces. **a** FB_BF. **b** FB_FF. **c** RB_FF. **d** RB_BF

4.2 Flow Structures of the Time-Averaged Velocity Around the Model

The time-averaged flow field around the model is presented in this section and all the results shown come from the fine grid simulations. The four cases (Cases 1–4) will be sub-divided into the groups "small gap width" and "large gap width" to focus on the difference in the flow structures around the models with sharp and round leading edges on the front box, respectively.

4.2.1 Small Gap Width

This section presents the mean flow around the two cases (Case 1 and Case 2) with a small gap width ($g/b = 0.17$). Case 1 has a sharp leading edge on the front box and Case 2 has a rounded. Vortex cores are computed by the algorithm in the EnSight visualization package. The method for vortex core identification is based on critical-point theory (e.g. [12]) and finds eigenvalues and eigenvectors to the rate-of-deformation tensor, $\partial u_i/\partial x_j$. A full description of the algorithm is given in [13]. The algorithm can produce non-existing cores and should only be used as a tool to help localize vortices [13]. Streamlines around cores are used in the present work to assure that the cores really exist.

Time-averaged flow structures around the cab are presented for Case 1 in Fig. 6a and Case 2 in Fig. 6b. In the figures, the white curves are streamlines of the time-averaged velocity projected onto the surface of the body. The black streamlines of the time-averaged velocity are projected onto planes around the vortex cores (solid black curves). For both cases, the free stream air hits FB_FF and separates at all leading edges. On the top face of the front box (FB_TF) the flow separations form vortices V_S (see Fig. 6a) and V_R (see Fig. 6b) for Case 1 (sharp model) and Case 2

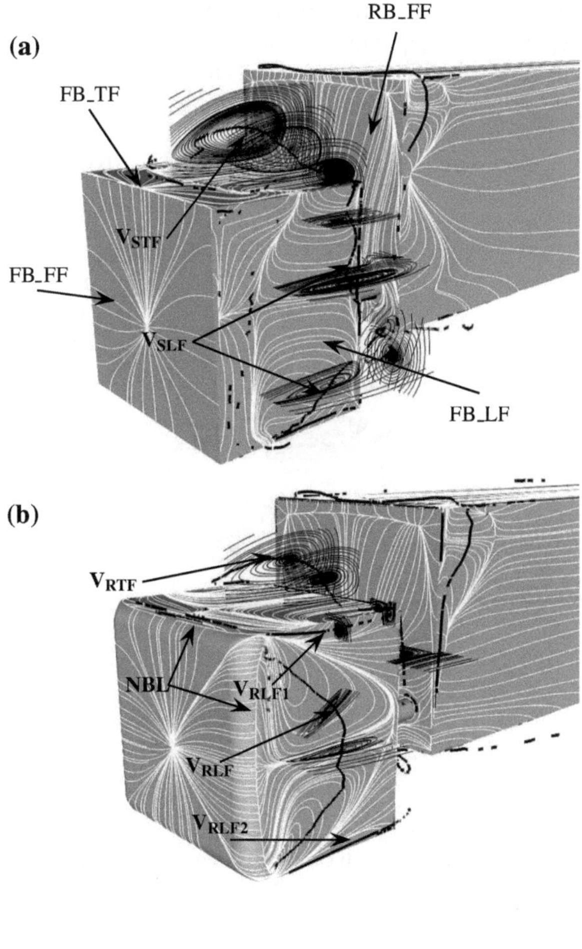

Fig. 6 Streamlines of the time-averaged velocity projected onto the surface of the body (*white curves*) together with vortex cores (*black drawn lines*) and streamlines projected onto planes around vortex cores (*black curves*). **a** Case 1 (small gap width, sharp edges). **b** Case 2 (small gap width, round edges)

(round model), respectively. For Case 1 (sharp model), vortex V_S over the top face (FB_TF) is connected with the vortex formed outside the lateral face (FB_LF) and hence is denoted the same, V_S (see Fig. 6a). For Case 2 (round model) the vortex formed outside FB_LF is called V_{LF2} (see Fig. 6b), and it is not connected with V_R (see Fig. 6b), which is formed above FB_TF. Another difference between Case 1 and Case 2 is the trailing vortices, V_{LF1} and V_{LF3} (see Fig. 6b), that extend in a streamwise direction along the upper and lower edges on FB_LF in Case 2, which were not found in Case 1.

Figure 7 shows streamlines of the time-averaged velocity projected onto plane $z = 0.46b$ for Cases 1–4. In Case 1 (sharp model) the flow does not re-attach on FB_LF after first separating on the leading edge on the front box. The large separation vortex, V_S (see Fig. 7a), extends over the whole gap and the flow finally re-attaches on the lateral face of the rear box. In Case 2 (round model) the flow does re-attach on FB_LF. This is shown in Fig. 7b and on the surface-projected streamlines on the body in Fig. 6b (white curves).

As seen in Fig. 5, which shows contributions from the faces of the model to C_D, the drag is reduced by more than some 50 % on the front face of the front box (FB_FF), regardless of gap width when the edges are rounded. Figure 8 shows streamlines of the time-averaged velocity projected onto FB_FF together with values of C_P on the

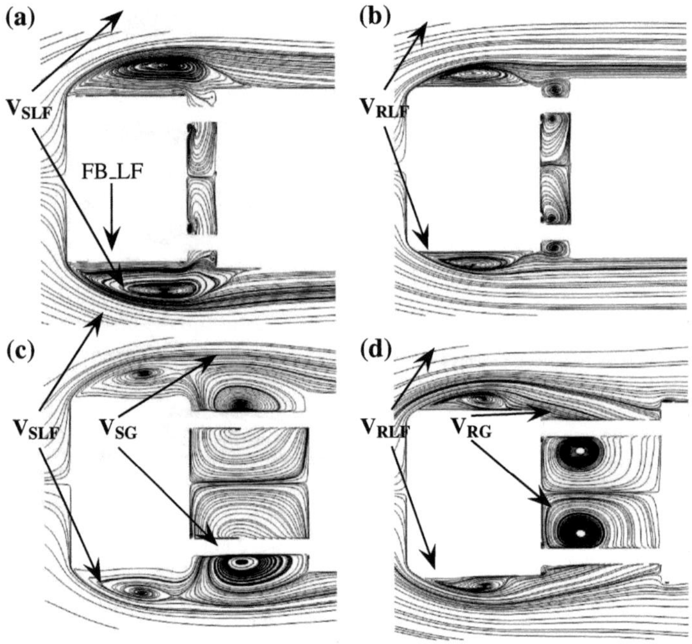

Fig. 7 Streamlines of the time-averaged velocity projected onto plane $z = 0.46b$. **a** Case 1 (small gap width, sharp edges). **b** Case 2 (small gap width, round edges). **c** Case 3 (large gap width, sharp edges). **d** Case 4 (large gap width, round edges)

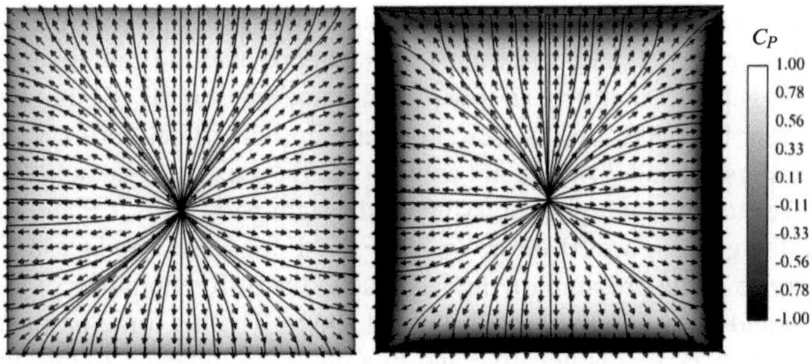

Fig. 8 Pressure coefficient on FB_FF together with streamlines of the time-averaged velocity projected onto the surface. *Left* Case 1 (small gap width, sharp edges). *Right* Case 2 (small gap width, round edges)

surface for Case 1 and Case 2. The sharp model is characterized by positive pressure everywhere on FB_FF and a value of C_P close to one (see Fig. 8), which is typical for stagnation points. On the round model, however, the flow accelerates over the curvature at the edges before separating. This separation is marked by the negative bifurcation line, NBL, in Fig. 6b. For more details on bifurcations lines and how flow topology theory are used to analyze the flow around bluff bodies we refer to [14]. The acceleration of the flow over the curvature gives rise to the area of low pressure distinguished by the black part around the edges in Fig. 8. This explains the large decrease in C_D on FB_FF (see Fig. 5b) when the leading edges are rounded.

The pressure coefficients on the base face of the front box (FB_BF) are shown in Fig. 9 for Case 1 and Case 2. As seen in Fig. 5 the contribution to C_D from FB_BF for Case 2 (round model) is some 30 % less than for Case 1 (sharp model). The primary reason for this is that the pressure along the lateral edges on the face for Case 1 is lower than in Case 2 (see Fig. 9). The reason why the pressure along the lateral edges is lower for Case 1 than Case 2 is the differences in the strength and extension of the vortices outside of FB_LF for the two cases. For Case 1, the vertical part of V_S

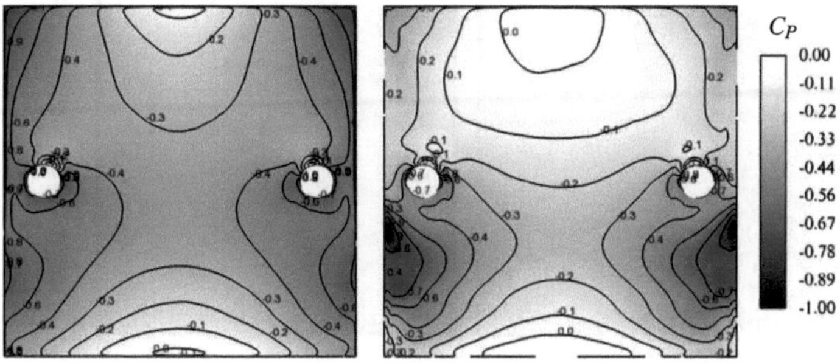

Fig. 9 Pressure coefficient on FB_BF. *Left* Case 1 (small gap width, sharp edges). *Right* Case 2 (small gap width, round edges)

(see Figs. 6a and 7a) outside FB_LF extends all over the gap. V_S creates very low pressure all over FB_LF and, since the vortex extends over the gap, the low pressure pulls into the gap and conduces to low pressure not only on FB_BF but on the front face of the rear box (RB_FF) as well. For Case 2, the vertical vortex V_{LF2} outside FB_LF (see Figs. 6b and 7b) does not extend over the gap, but re-attaches on FB_LF. This conduces to a much higher pressure on the right part of FB_LF and thus higher pressure in the gap. Figure 10 shows C_P on FB_LF for Case 1 and Case 2. Note the large difference in the pressure on the face due to the different nature of the vortices V_S (Case 1) and V_{LF2} (Case 2), as discussed above. The net force on the front face of the rear box (RB_FF) is negative and in a direction opposite to the direction of the flow in Case 1 (see Fig. 5c). It is slightly positive (see Fig. 5c) for Case 2 (round model). Figure 11 shows C_P for Case 1 and 2 on RB_FF. There is lower pressure for Case 1 for the lower half of the face compared with Case 2. This is again explained by the extension over the gap of the vertical part of vortex V_S (see Figs. 6a and 7a) compared with vortex V_{LF2} (see Figs. 6b and 7b). V_{LF2} does not extend over the

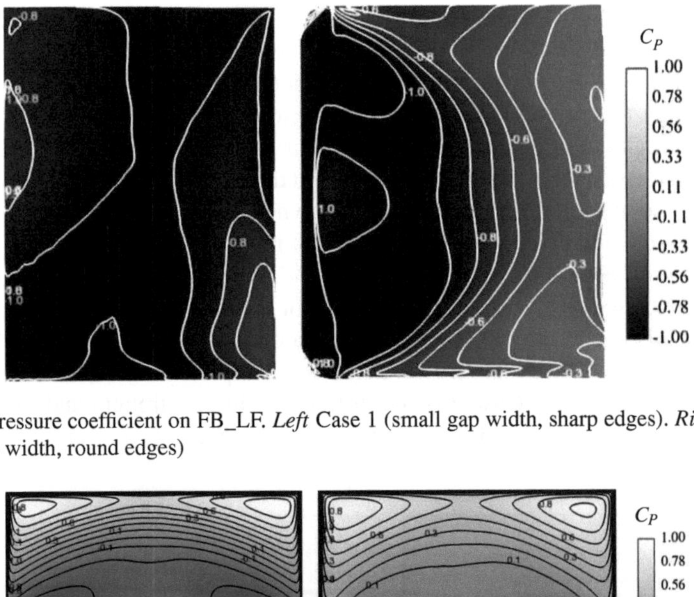

Fig. 10 Pressure coefficient on FB_LF. *Left* Case 1 (small gap width, sharp edges). *Right* Case 2 (small gap width, round edges)

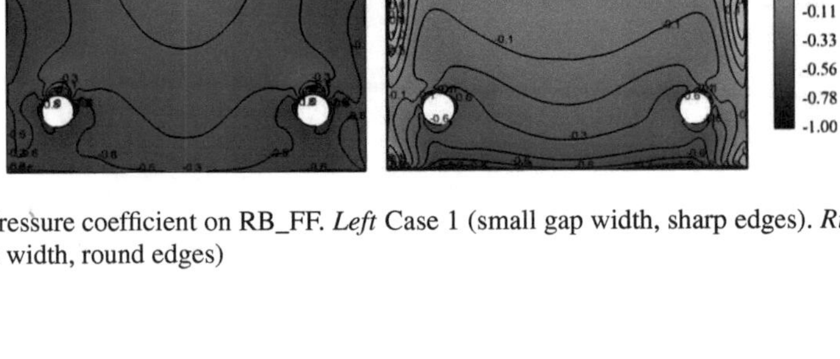

Fig. 11 Pressure coefficient on RB_FF. *Left* Case 1 (small gap width, sharp edges). *Right* Case 2 (small gap width, round edges)

gap. For Case 1, V_S shields the lateral edges of RB_FF from the airstream, and thus a low pressure is kept all the way to the outermost part of the face in a spanwise sense. For Case 2, V_{LF2} is not able to shield all of RB_FF, a part of the airstream impinges on the outermost part of the face and high pressure is formed there (see Fig. 11). The high pressure contributes to an increase in the drag of the model in Case 2. This can also be seen by comparing the streamlines in plane $z = 0.46b$ for Case 1 and Case 2 impinging on RB_FF in Fig. 7a (Case 1) and Fig. 7b (Case 2).

4.3 Large Gap Width

This section describes the mean flow fields in the two cases with a large gap width (Case 3 and Case 4). Some comparisons of differences between small and large gap widths are also made for the sharp and round models. Figure 12a shows the vortex

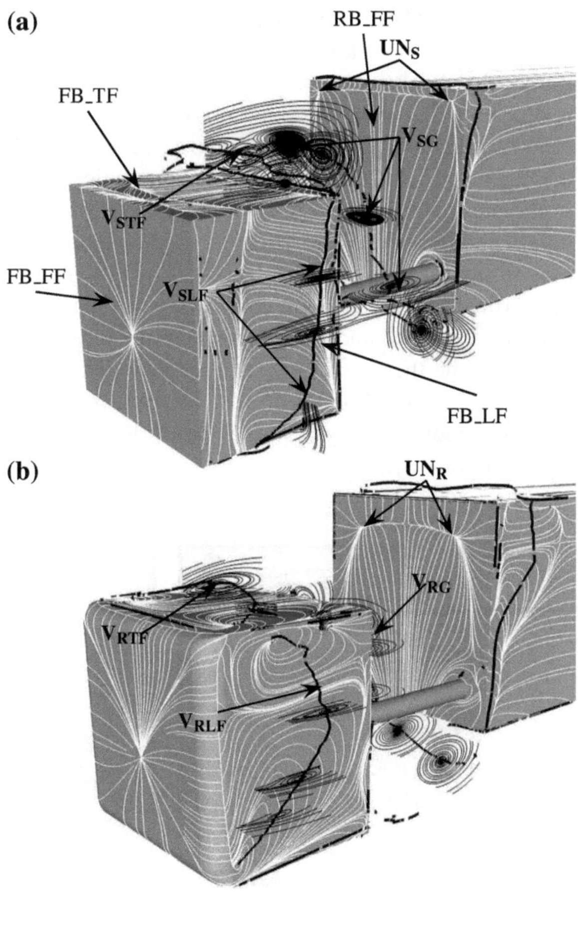

Fig. 12 Streamlines of the time-averaged velocity projected onto the surface of the body (*white curves*) together with vortex cores (*black drawn lines*) and streamlines projected onto planes around vortex cores (*black curves*). **a** Case 3 (large gap width, sharp edges). **b** Case 4 (large gap width, round edges)

system of the mean flow around Case 3 ($g/b = 0.67$, sharp edges). Above the top face of the front box (FB_TF) there's a type of vortex similar to V_S (see Fig. 6a) in Case 1 ($g/b = 0.17$, sharp edges). The vortex above FB_TF in Case 3 is called V_{STF} in Fig. 12a. The vortex outside of the lateral face of the front box (FB_LF) in Case 3 is called V_{SLF} in Fig. 12a. V_{SLF} is similar to the vertical part of V_S for Case 1 (see Fig. 6a). As seen by comparing streamlines in plane $z = 0.46b$ in Fig. 7a (Case 1) and 7c (Case 3), vortex V_{SG} is formed in the gap for Case 3 ($g/b = 0.67$, sharp edges). Vortex V_{SG} is also pointed out in Fig. 12a, and it is a large horse-shoe like vortex. It extends horizontally in the gap and then down vertically in two tails outside the two connecting cylinders between the front and rear boxes. The mean flow in the gap is completely dominated by this vortex for Case 3. V_{SG} (see Fig. 12a) is responsible for the low pressure on FB_BF (see Fig. 5a), contributing to an increase in C_D for the whole model for Case 3. However, V_{SG} is also responsible for the low pressure on the front face of the rear box (RB_FF, see Fig. 5c) which contributes to a decrease in C_D for the whole model.

Figure 12b shows the vortex system around the model for Case 4 ($g/b = 0.67$, round edges). The vortex outside of FB_LF in Case 4 is called V_{RLF} in Fig. 12b and it is very similar to vortex V_{VLF2} (see Fig. 6b) outside FB_LF for Case 2 ($g/b = 0.17$, round edges). The vortex above FB_TF for Case 4 is called V_{RTF} in Fig. 12b. V_{RTF} is much smaller than its counterpart V_R (seen in Fig. 6b) for Case 2 ($g/b = 0.17$, round edges). For Case 4 there is also a horse-shoe like vortex, V_{RG}, that forms in the gap (see Fig. 12b), as in Case 3. Topologically, there is no difference in the flow in the gap for the two cases with large gap width, Case 3 and Case 4. The gap is dominated in both cases by the horse-shoe like vortices V_{SG} and V_{RG}, respectively (see streamlines in plane $z = 0.46b$ in Fig. 7c, d). However, as seen in Fig. 7d), the vertical tails of V_{RG} are confined to the space between the connecting cylinders between the front and rear boxes (see the cylinders in Fig. 1, which describes the model) while the tails of V_{SG} are located outside the cylinders in a spanwise sense (see Figs. 7c and 12a). C_P on RB_FF for Case 3 and Case 4 is presented in Fig. 13. Note the much higher pressure in Case 4 than in Case 3, which indicates that V_{RG} is weaker than V_{SG}.

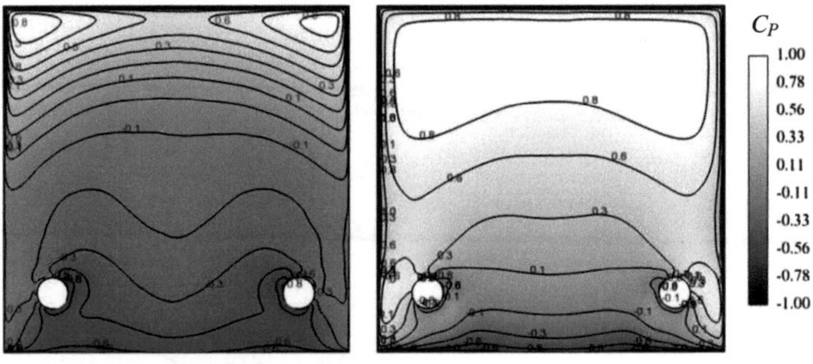

Fig. 13 Pressure coefficient on RB_FF. *Left* Case 3 (large gap width, sharp model). *Right* Case 4 (large gap width, round model)

The difference between the contributions to C_D from the front face of the rear box (RB_FF) between Case 3 and Case 4 (see Fig. 5c) is very large. Figure 5c shows that the contribution is slightly negative in Case 3 (−0.06). The contribution to C_D from RB_FF in Case 4 is 0.467. The reason why the contributions from RB_FF are so different between Case 3 and Case 4 is the difference in the vortices formed in the gap in the two cases, V_{SG} in Case 3 (see Fig. 12a) and V_{RG} in Case 4 (see Fig. 12b). Since the only geometrical difference between the models in the cases is the rounding of the leading edge on the front box, this large difference is quite remarkable. The difference is that, in Case 4, since V_{RG} is confined between the connecting cylinders, it is not able to shield RB_FF from the airstream flowing along the lateral sides. In Case 3, the gap vortex, V_{SG}, is located outside of the connecting cylinders and is much larger and stronger. This can be seen by comparing the streamlines around the vortices in plane $z = 0.46b$ (see Fig. 7c, d). Furthermore, the foci of V_{SG} are also located further away from the base part of the front box compared with the tails of V_{RG}.

The above described difference between V_{SG} and V_{RG} has the consequence that, for Case 3, the gap vortex, V_{SG}, is able to shield the exposed face, RB_FF, from the airstream impinging, and thus a low pressure is kept in the gap and the contribution to C_D from RB_FF is negative in Case 3. In Case 4, however, the gap vortex V_{RG} is not able to shield the RB_FF face and the airstream impinges on a larger area of RB_FF and thus momentum is directly transferred from the free-stream air to this exposed surface. The airstreams impinging on RB_FF in Case 3 and Case 4 can also be seen by looking at the two unstable nodes located in the upper corners of RB_FF. The unstable nodes in Case 3 are denoted UN_S (see Fig. 12a) and in Case 4 UN_R (see Fig. 12b). UN_R (Case 4) are located closer to the midpoint of the face compared with the location of UN_S (Case 3). Thus the vertical parts (the tails) of V_{SG} (Case 3) are able to shield the exposed face, RB_FF, and direct the airstream around the face so that it attaches on the lateral face of the rear box instead. There is another fundamental difference between gap vortices V_{SG} and V_{RG} as well. The first is a result of the air separating on the leading edge on the front box. This can be understood by looking at the streamlines projected in plane $z = 0$ (see Fig. 7c). V_{RG} is a result of the air separating on the base edge of the front box. The air separates on the leading edge, re-attaches on the right part of FB_LF and then separates again on the base lateral edge of the front box, and V_{RG} is formed. The strong airstream that flows outside of the FB_LF vortex V_{RLF} (see Fig. 12b) then confines V_{RG} to the area between the connecting cylinders in the gap.

5 Conclusions

The LES was successfully employed to predict the flow around a simplified tractor-trailer model. Four different variants (Cases 1–4) of the model were studied. The results of the LES agree reasonably well with experimental force measurements. The general trend in the contribution from the different faces of the model to the drag coefficient of the entire model are the same in the present LES as in the experiments. In general, the results for the two cases (Case 1 and Case 3) with a sharp leading edge on the front box are in better agreement then the two cases with a rounded leading edge (Case 2 and Case 4). The reason for this is most likely the difficulties of an exact prediction of the separation on the rounded edges for Case 2 and Case 4. In Case 1 and Case 3, the separation is exactly defined by the sharp edges of the geometry. Large vortical structures originating from the geometry were identified around the cab and in the gap. It is shown how these large structures are responsible for the difference in drag between the different variants of the model.

Acknowledgments This project is supported financially by Trafikverket (Swedish Transport Administration). Software licenses were provided by AVL List GMBH. Computations were performed at SNIC (Swedish National Infrastructure for Computing) at the Center for Scientific Computing at Chalmers (C3SE).

References

1. McCallen, R., Flowers, D., Dunn, J.O.T., Browand, F., Hammache, M., Leonard, A., Brady, M., Salari, K., Rutledge, W., Ross, J., Storms, B., Heineck, J.T., Driver, D., Bell, J., Walker, S., Zilliac, G.: Aerodynamic drag of heavy vehicles (class 7–8): simulation and benchmarking. SAE Paper 2000-01-2209 (2000)
2. Allan, J.: Aerodynamic drag and pressure measurements on a simplified tractor-trailer model. J. Wind Eng. Ind. Aerodyn. **9**, 125–136 (1981)
3. Hammache, M., Browand, F.: On the aerodynamics of tractor-trailers. In: The Aerodynamics of Heavy Vehicles: Trucks, Busses and Trains. Monterey (2002)
4. Smagorinsky, J.: General circulation experiments with the primitive equations. Mon. Weather Rev. **91**(3), 99–165 (1963)
5. Krajnović, S., Ringqvist P., Nakade K., Basara B.: Large eddy simulation of the flow around a simplified train moving through a crosswind flow. J. Wind Eng. Ind. Aerodyn **110**, 86–99 (2012)
6. Han X., KrajnovićS.: Large eddy simulation of flow control around a cube subjected to momentum injection. Flow Turbul. Combust. **92**(1-2), 527–542 (2014)
7. Krajnović, S.: LES investigation of passive flow control around an ahmed body. J. Fluids Eng. **136**(12), (2014) doi:10.1115/1.4027221
8. Uystepruyst D., Krajnović, S.: LES of the row around several cuboids in a row. Int. J. Heat Fluid Flow. **44**, 414–424 (2013)
9. Krajnović, S., Sarmast, S., Basara, B.: Numerical investigation of the flow around a simplified wheel in a wheelhouse. J. Fluids Eng. **133**(11), s. 111001. [Nr. 153736] (2011)
10. Krajnović, S., Bengtsson, A., Basara, B.: Large eddy simulation investigation of the hysteresis effects in the flow around an oscillating ground vehicle. J. Fluids Eng. **133**, s. 121103. [Nr. 153734] (2011)
11. Cooper, K.R.: The effect of front-edge rounding and rear edge shaping on the aerodynamic drag of bluff vehicles in ground proximity. SAE Paper No. 850288 (1985)
12. Perry, A.E, Chong, M.S.: A description of eddying motions and flow patterns using critical-point concepts. Ann. Rev. Fluid Mech. **19**, 125–155 (1987)
13. Sujudi, D., Haimes, R.: Identification of swirling flow in 3-D vector fields. AIAA Paper 95-1715 (1995)
14. Krajnović, S., Davidson, L.: Flow around a simplified car, part 1: large eddy simulation. ASME J. Fluids Eng. **127**, 907–918 (2005)

Accurate Drag Prediction Using Transient Aerodynamics Simulations for a Heavy Truck in Yaw Flow

Per Elofsson, Guillaume Mercier, Bradley D. Duncan and Samuel Boissinot

Abstract Aerodynamic development of a full-scale truck presents a challenge for experimental testing due to the scale of the vehicle relative to most wind-tunnel test facilities. Numerical simulation is becoming more prevalent for assessing design changes and improving vehicle aerodynamic drag. In this process, the cumulative effects of small design changes are needed. Furthermore, the drag must be considered both at zero crosswind and with five degrees crosswind yaw angle in order to properly represent typical driving conditions. It is well-known that the aerodynamics of heavy trucks are complicated by a very transient wake flow that causes large fluctuations in base pressure, and therefore in the drag coefficient. This effect is often even more prevalent at non-zero yaw angles. The transient wake flow presents a challenge for effectively using simulation tools to predict the drag effects of small design changes, which may have some influence on the wake flow and base pressure.

In this study, long transient simulations are used to assess the variation in wake flow leading the drag fluctuations. This information is used to estimate the confidence interval for the drag coefficient prediction as a function of time-averaging period for 5° yaw. Analysis of the force and pressure histories, as well as transient flow visualization and spectral analysis are used to identify the source of unsteadiness in the wake flow regions, and the impact of the unsteadiness on the surface pressure distribution. This study provides useful information on the range of certainty of drag values for future simulations, and facilitates using simulation to isolate the effects of small design changes from the inherent variability in the base pressure drag. Finally,

P. Elofsson (✉) · G. Mercier
Scania, Truck Chassis Development, SE-15187 Södertälje, Sweden
e-mail: per.elofsson@scania.com

G. Mercier
e-mail: guillaume.mercier@scania.com

B.D. Duncan · S. Boissinot
Exa Corporation, 55 Network Drive, Burlington, MA 01803, USA
e-mail: brad@exa.com

S. Boissinot
e-mail: sboissinot@exa.com

A. Dillmann and A. Orellano (eds.), *The Aerodynamics of Heavy Vehicles III*,
Lecture Notes in Applied and Computational Mechanics 79,
DOI 10.1007/978-3-319-20122-1_22

improved understanding of the transient wake effects on the drag distribution leads to a new approach for using reduced domain and reduced time simulations for rapid optimization of the truck cab and chassis.

1 Background

1.1 Aerodynamics Using Simulation

For heavy trucks, using simulation to incorporate aerodynamics data earlier in design process provides benefits for fuel-economy improvements, product development cost savings, and reduction of parts costs. Innovative solutions can be found by closely coupling the vehicle design and engineering with aerodynamics analysis at each stage of development.

Using simulation to complement physical testing processes later in the design process potentially saves costs for testing of production prototypes, removes uncertainty from physical test procedures, and keeps pace quickly with changes in the design. Since full-scale testing of heavy trucks at moving ground conditions is currently not possible, aerodynamics simulation of detailed truck models in on-road conditions including cross-wind is vital for optimization of the cab, roof fairing, engine bay cooling flow, wheel housings and fairings and chassis. This requires high fidelity of complex, turbulent, transient flow effects for robust evaluation of small geometry changes as the vehicle model is refined during the design process, some changes being very small relative to the overall vehicle.

Including cross-wind effect is key to estimating true on-road performance using estimation of wind-averaged aerodynamic drag. This is usually performed using yaw angles less than 10° to match typical driving conditions. Figure 1 shows a simulation of a full-scale truck with 5° cross-wind (meaning that flow approaches from driver's side). Cross-wind sensitivity is also important at higher yaw angles for analysis of vehicle stability.

1.2 Lattice-Boltzmann Method

The Lattice Boltzmann Method (LBM) has been used extensively in ground transportation for predictive simulation of aerodynamics, thermal cooling flow, and aeroacoustics. LBM has been widely validated across many cases and applications. References describing LBM are found in [1–5]. References describing application of LBM to aerodynamics for cars and trucks are found in [6–10]. LBM uses a fundamentally transient flow solver, and the primary benefit is the low-dissipation time-accuracy for flow structures and surface pressures. The method is able to predict a large range of time scales on a full-scale, detailed truck model, from small wind-noise fluctuations

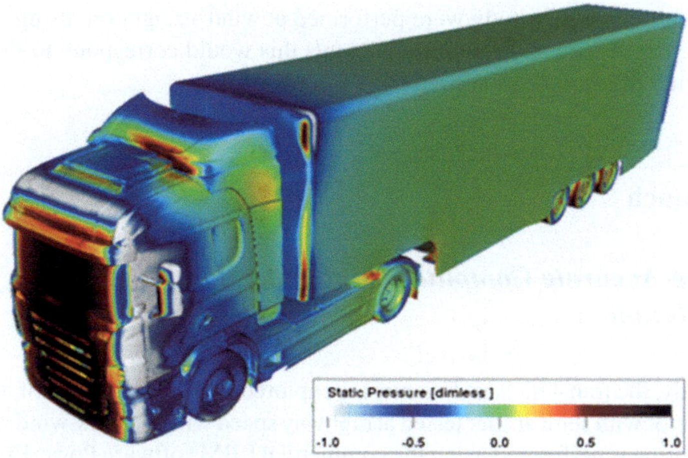

Fig. 1 Non-dimensional static pressure (Cp) for the Scania truck simulation at 5° yaw

at 5 kHz to the dominant truck wake oscillations at <1 Hz. Consequently, LBM can robustly handle diverse flow regimes and distinguish aerodynamic effects of small design changes.

1.3 Transient Wake Flows for Heavy Trucks

For heavy trucks and other vehicle shapes with a vertical rear face (SUVs, buses, some simplified body shapes), the vehicle produces a long and highly unstable wake which causes variable pressure on the back face. The wake interacts with the ground layer and produces a variety of complex flow structures such as reversed flow, trailing vortices behind the wake, and shedding vortices around the perimeter of the wake. The complex small-scale turbulence that occurs on a real truck due to surface details of the cab, mirrors, engine bay flow, wheels and underbody also contribute to the wake instability. In cross-wind, the side of the vehicle also represents an obstacle to the flow and induces roll-up into large-scale vortices which further contribute to wake instability.

The time scales and spatial scales of the unstable wake of a full-scale truck (with or without cross-wind) are not well understood. This presents a challenge for simulation methods in that force oscillations occurring with a long time-scale on the order of several seconds, can create uncertainty in the time-averaged results when insufficient time-averaging is used. In order to get an idea of the time scale it is relevant to look into previous studies performed on other bluff bodies, see references [11–13]. For example, an experimental study in the near wake of a generic ground vehicle body by Duell and George [11] shows fluctuation of the wake closure point corresponding to a pumping effect between the upper and lower trapped vortices at Strouhal number

0.069. Even though this study were performed at wind straight on, using the square root of the trailer area and free stream 25 m/s this would correspond to fluctuations with a period of a couple of seconds.

2 Approach

2.1 *Time-Accurate Computation of Truck Flow and Wake Evolution*

In this study, the transient aerodynamics is explored using simulations of a full scale Scania tractor with semi-trailer tested at highway speed with 5° cross-wind yaw angle. The simulation is performed using the commercial LBM software PowerFLOW from Exa Corporation. The spatial grid arrangement for the simulations is setup using a common practice for all Scania trucks, which results in consistent comparisons across models, and design variants. For cross-wind setup, the common practice uses a larger region of high resolution in the down-wind regions in order to follow the angle of the separating flow.

The simulations use a starting flow condition obtained from a coarser-resolution, and then run in time-accurate mode. A short initial transient is required as the flow evolves from the coarse solution, and then fully-developed flow is simulated for the required time to produce smooth time-averaged flow results and high-confidence averaged force values. The flow variables on the surface of the truck and in the flow region around the truck are sampled using spatial and time-averaging according to user inputs to reduce the data to a tractable size compared to the complete simulation. The sampled data is averaged in time and used for post-processing. The un-averaged data can also be used for transient flow visualizations and further analysis of the fluctuations in the flow variables. Finally, the overall force values are also sampled during the simulation at high frequency and can be averaged or the time-accurate fluctuations in forces can be analyzed.

2.2 *Determination of Relevant Time Scales*

For practical design studies it is important to be able to a priori establish the desired confidence interval of the averaged force values needed for each simulation. Typically, a confidence interval of body-aligned drag coefficient (C_D) of 0.005, 0.002 or 0.001 may be needed for evaluation of iterations in the vehicle shape. Since some significant wake oscillations for the truck have some time scales longer than 1 s, the primary consideration in setting the simulation length is to utilize sufficient time-averaging to narrow the range for the predicted averaged drag coefficient to the needed level. In this study, this issue is explored using a long-time simulation to study the long time scales, as well as high-frequency dataset to create more detailed visualizations of the transient wake flow structure.

(1) Long-time simulation: A long-time simulation was performed for 60 s of real time. The flow results and in particular the drag time history was used to thoroughly evaluate the oscillations in drag that occur over long periods of time. The target time scale of interest was between 1 and 5 s, based on previous observations of the strongest fluctuations in the wake. The 60-s run time provided ample data for evaluating the target time range using multiple averaging windows. The flow data was also sampled at 3 Hz for the simulation to allow visualization of the low-frequency flow structures in the target time scale range. The instantaneous force history was also recorded for the 60-s period of the simulation.
(2) High-frequency measurement: In addition, a high-frequency measurement was sampled for 10 s at 50 Hz. This provided flow structure data for visualization and spectral analysis to understand the fundamental flow structures in the target time scale range in more detail.

The data captured from the simulation was used to quantify and identify the features of the fluctuating flow field. Moving averages of the force history were calculated using different averaging window width and evaluated to measure the maximum and minimum averaged values of drag coefficient with each averaging window width. Next, the spectrum of the force history was used to determine the contributions of each time-scale to the overall fluctuation level. The fluctuation in force levels were related to the pressures at different regions of the vehicle using a spectral filter to show the amplitude of pressure fluctuations at different frequencies. Correlation analysis was also used to find the spatial correlation between the fluctuating pressure field at the base of the vehicle and upstream regions near the vehicle or further downstream within the wake.

Finally, the high-frequency transient data was used to find the source of transient flow structure in the wake and other regions around the truck. Visualizations of vortex structures and of separated wake structures were used. Spectral analysis was then utilized on the high-frequency transient data in order to isolate the important fluctuations in the wake in the time scale of interest (1–5 s) from the higher-frequency turbulent vortex and wake structures.

2.3 Methodology for Rapid Design Iterations

The investigation of time-averaging requirements for simulations merits a parallel investigation into methods to reduce the simulation time for typical use for design of the front of the truck. Often features of the design are intended to have a local effect on wind-resistance of components, reduce flow impingements, manage cooling flow, redirect part of flow approaching the tires, and to control the flow between the cab and the trailer. Though some unintended adverse effects on the wake flow and the drag at the rear face of the truck may be caused by these changes, improving the local drag effect on the cab usually produces a net drag savings. Therefore the typical design process for the truck design does not require that the impact on the transient wake be considered for every design iteration. The understanding of the transient flow structure can be used to knowledgeably guide a rapid design iteration process in which any changes to the wake are neglected while small changes are made on the front portion of the truck. Then, only the best designs are tested using full truck simulations.

In this study, alternative approaches using front half drag in combination with reduced time, and a new sub-domain method are investigated. The feasibility of a rapid design process is demonstrated in which a sub-domain of the flow region around the truck is simulated. The flow domain is truncated to simulate a region of interest, and the inflow/outflow boundaries of this fluid domain are replaced by steady boundary conditions with prescribed velocity and pressure, except on the outlet face which is replaced by a pressure boundary condition only. The boundary conditions are prescribed using time-averaged data from a previous simulation. This method was used with an outlet face prescribed at a location at the middle of the trailer and not close to any of the wheels. Since the sub-domain simulations do not include any of the wake effect, the required simulation time is shorter, allowing many more design iterations to be investigated with the same turn-around time. The results are shown below showing the feasibility of this method.

3 Results for 5° Yaw

3.1 Effect of Averaging Time Scale on Forces Statistics

The time history of the drag coefficient for the 60-s simulation is shown in Fig. 2. Some time-averaging is applied to the raw data to improve the readability of the graph (0.12 s). Moving averages are also shown with time averaging windows ranging from 1 to 10 s. With 1-s averaging the mean value varies about 0.010 from minimum to maximum value. This is considered as a confidence range of $C_D = \pm 0.005$ counts of the total drag. While this is sufficient for some comparisons in which large drag differences are considered between different vehicles or large geometry changes, often a finer confidence range is needed. Doubling the averaging time to 2 s narrows the range to $C_D = \pm 0.003$. Doubling again to 4 s narrows the range to $C_D = \pm 0.002$ of the total drag. Finally, increasing the averaging window to 10 s produces approximately $C_D = \pm 0.001$ of the total drag. Certainly this tight of a confidence range is not always required, but is used for distinguishing the effects of small design changes with high precision.

The time history and moving average curves show a large range of time scales at work in the drag force. The most dominant fluctuations are removed by averaging more than a few seconds, indicating that most fluctuations have a time scale less than 5 s. The time-scales are calculated using the force spectra for drag (C_D), side force

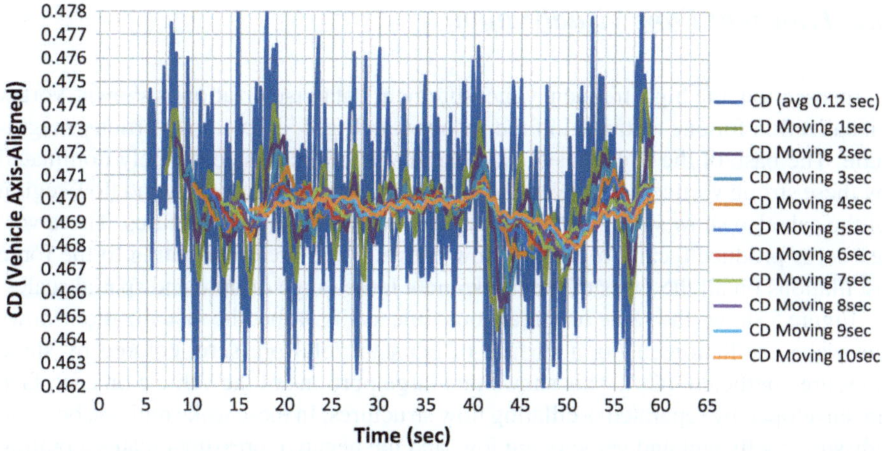

Fig. 2 Time history and moving averages of drag coefficient (C_D) with different averaging widths, for the long-time run at 5° yaw

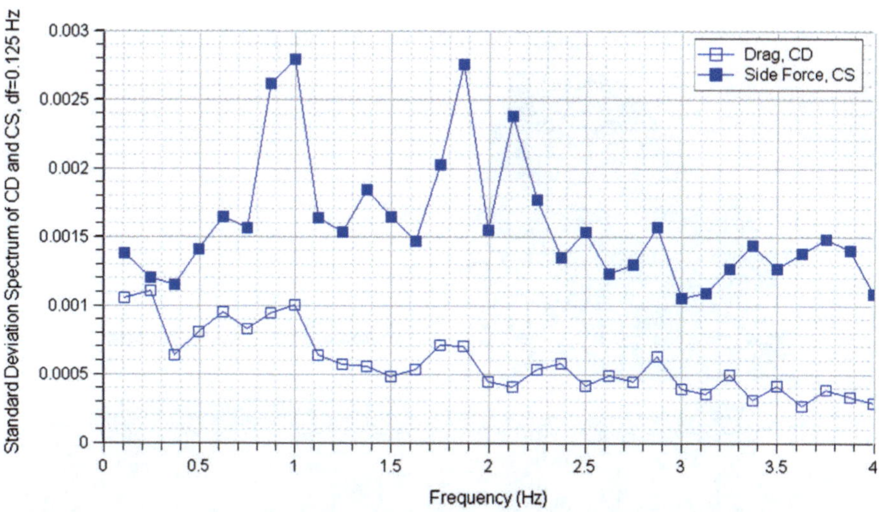

Fig. 3 Force spectra for drag coefficient (C_D) and side force coefficient (C_S) for the long-time run a spectrum bandwidth of 0.125 Hz. The spectra are scaled into units of standard deviation of force coefficient

(C_S) force coefficients, as shown in Fig. 3. The side force fluctuations are strongest at frequencies near 1.0 and 2.0 Hz. These fluctuations are dominated by the separating flow on the downwind side of the cab and trailer. The drag spectrum shows the strongest oscillations at a low frequency of about 0.25 Hz, indicating a dominant time scale of 4 s. This is consistent with the findings of the moving averaging results. At least 4 s of averaging is needed to include one cycle of oscillation, and produce a reliable average of the drag coefficient (within $C_D = \pm 0.002$).

3.2 Transient Flow Visualization

In this section, the high-frequency measurement data visualization is shown. Vortex visualization for one instance in time is shown in Fig. 4, colored by vorticity magnitude. The base of the truck has the strongest vortices and is completely dominated by these strong vortices. However, the underbody region and down-wind flow region of the cab also show strong turbulence and small scale vortex structures. These vortex cores produce locally regions of low pressure and are contributing to the force oscillations on all the faces of the truck and a large range of time and spatial scales.

Figure 5 shows the same instantaneous time as Fig. 4, but the visualization shows the wake oscillations using a surface of constant total pressure, colored by static pressure coefficient (Cp). This shows that large portions of the vehicle on each face are enveloped by separated oscillating flow structures. In these wake regions, both the pressure coefficient and velocity are low, and the negative pressure creates a pulling force on each surface. This force fluctuates due to the motion of the boundary of the wake, caused by absolute instability of the separated flow structures.

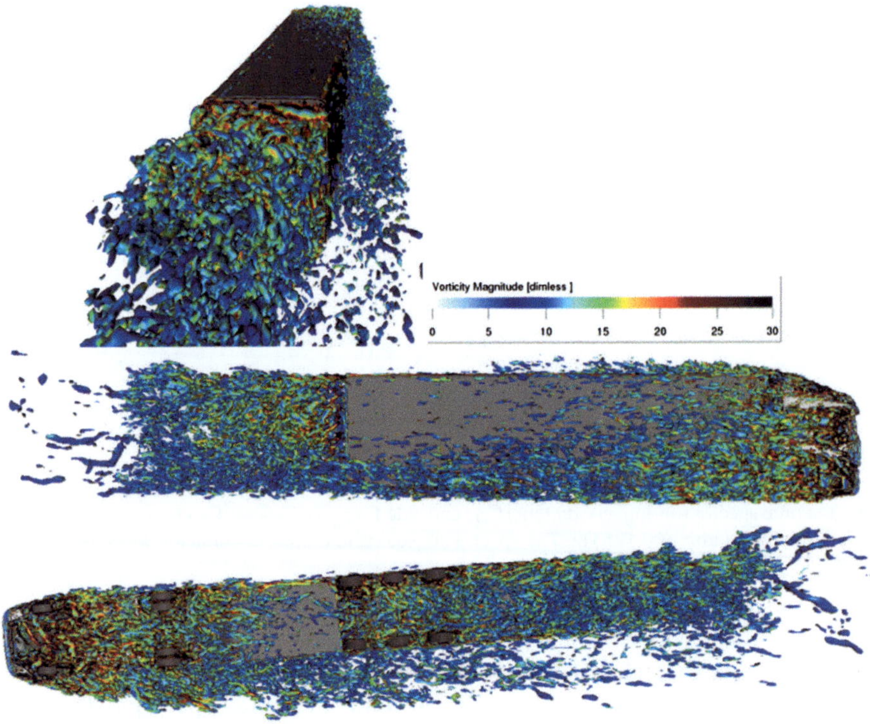

Fig. 4 Vortex visualization using isosurface of non-dimensional Lambda 2 = −20. Isosurface is colored by non-dimensional vorticity magnitude. Data was sampled at higher frequency (50 Hz) for the transient visualization

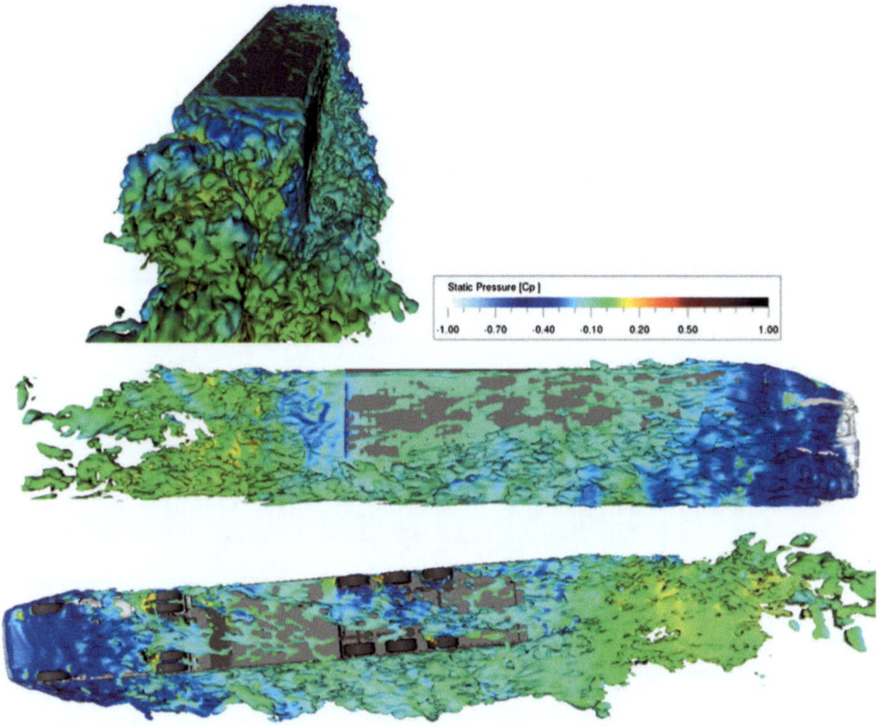

Fig. 5 Transient wake visualization using isosurface of non-dimensional total pressure = 0.25. Isosurface is colored by non-dimensional static pressure. Data was sampled at higher frequency (50 Hz) for the transient visualization

3.3 Spectral Analysis of Pressure Fluctuations and Wake Oscillations

In this section, spectral and statistical analysis of the flow data around the vehicle is used to more fully explain the low-frequency oscillations in the forces. In Fig. 6, the time histories of pressures at 3 points are shown using a spectral filter. The graphs show clearly the amplitude and time-scale of different dominant pressure fluctuations in each region. The back face pressure point shows the strongest pressure fluctuations, with the fluctuations filtered to a range of 0.01–0.5 Hz showing a fluctuation amplitude of $Cp = \pm 0.010$ to 0.015 with a time scale of 2–6 s. If this oscillating pressure amplitude applies on the entire back-face it can contribute a drag effect of comparable amplitude (since $C_D = \Delta Cp^{*}Area$). In contrast, the pressure fluctuations on the sides of the trailer are much smaller.

Figure 7 shows pressure correlations. The correlation coefficient is simply the non-dimensional correlation between the pressure at a reference point near the trailer back face and the pressures at other points on the same plane (horizontal or vertical plane).

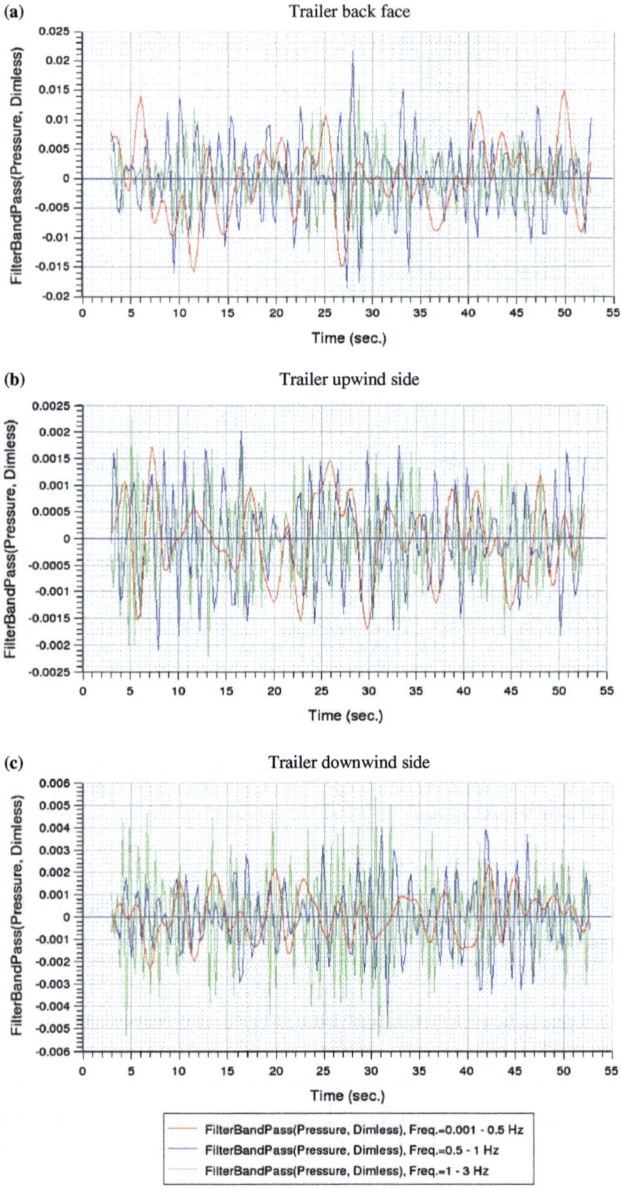

Fig. 6 In these graphs, pressure time histories from the long transient simulation are filtered using a spectral filter to show the contribution of each frequency range to overall pressure fluctuations. Note the large difference in scale for each graph. Pressures were extracted from a *horizontal plane* (located at the *top* of the side mirror), at points near the trailer backface, upwind side of the trailer, and downwind side of the trailer. **a** Trailer back face. **b** Trailer upwind side. **c** Trailer downwind side

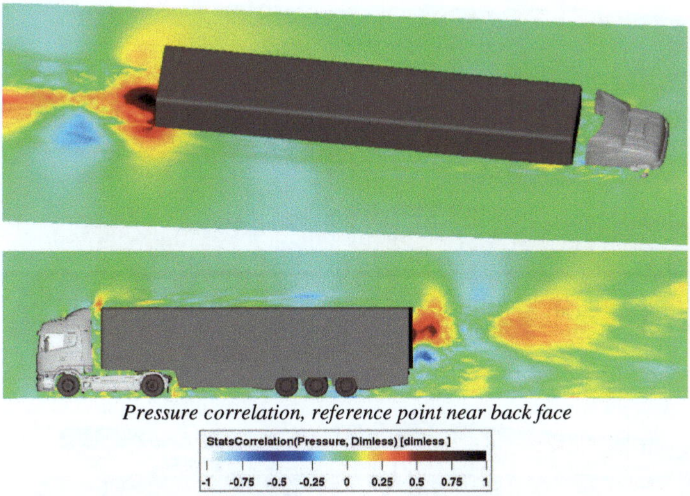

Fig. 7 Analysis of correlation between the history of static pressure (Cp) near the back face, and Cp at other points on a horizontal plane or *vertical* plane. Calculation was performed using the long transient simulation; the fluid data was sampled at averaged at 0.3 s intervals and represents only low-frequency fluctuations (<1.5 Hz)

The correlation coefficient has a value of 1.0 at the reference point. The images help to show that the wake pressure oscillations are not strongly influenced by upstream flow. The horizontal and vertical plane pressures do not show any regions of strong correlation with the back face region. However, the wake region itself contains structure that is shown in the correlation. The pressure near the back face is strongly correlated to the wake flow several meters behind it. This means that the large-scale oscillations in the wake contribute significantly to the fluctuations in the base pressure drag.

Next, Figs. 8 and 9 show results from spectral analysis of the flow variables around the truck. The low-frequency, long-time flow structures are evaluated by using either a band-pass spectral filter, with frequency range 0.01–1.0 Hz, or a low-pass filter which also includes the mean components and has a frequency range of 0.0–1.0 Hz. The low-pass filter looks more like a smoothed transient flow, whereas the band-pass filter shows only fluctuations which are mostly obscured by the mean flow during unfiltered animations. Figure 8 shows the band-pass filter applied to the wake visualization, producing surface of constant total pressure fluctuation. It is colored by static pressure fluctuations which can represent the transient forces applied onto the vehicle surface. The figure is remarkable in that it shows instability of the entire separated flow with very long vortex patterns the length of the whole truck. These streaky vortices travel back into the wake region perimeter and in particular in the region to the down-wind side of the wake. These oscillations produce the strong oscillations in the side force as well as drag oscillations on the front half of the truck. However, the amplitude of static pressure fluctuations is much higher in the back

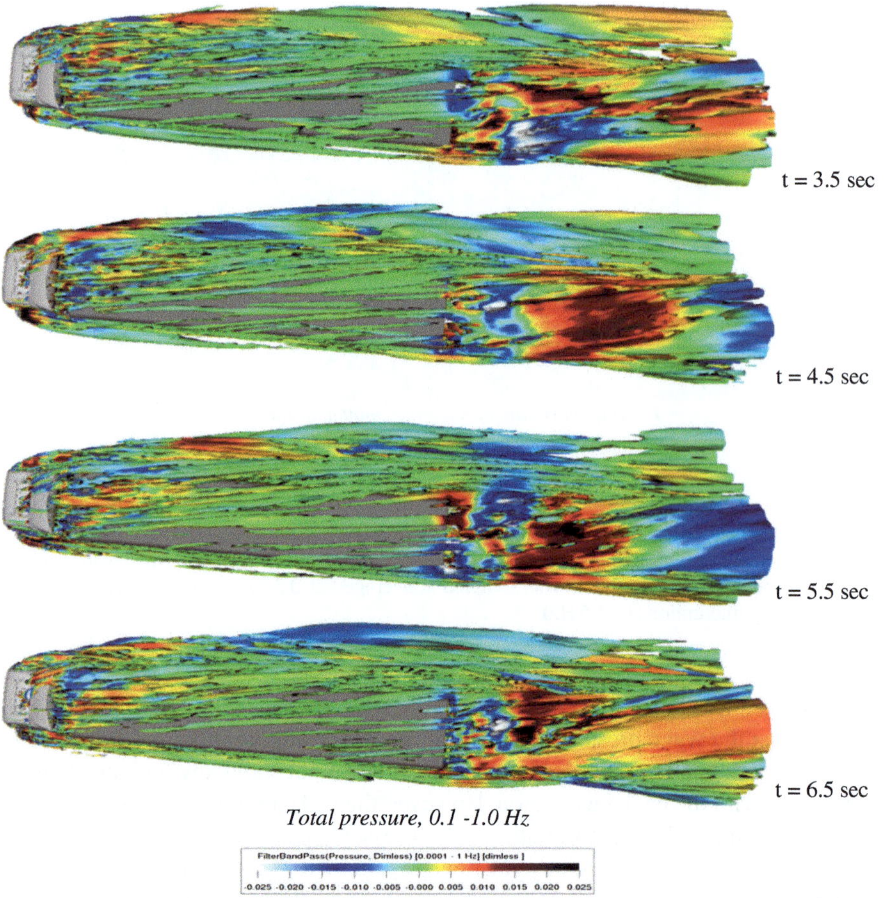

Fig. 8 Large-scale wake oscillations for 4 instances in time. Results are obtained using spectral band-pass filter from 0.1 to 1.0 Hz, calculated using the high-frequency dataset sampled at 50 Hz for 10 s. Isosurfaces of total pressure fluctuations are shown, colored by pressure fluctuation

face region, with amplitude of approximately $\Delta Cp = \pm 0.020$ and the trailing vortex structures are much larger and more organized. This indicates that the trailing vortex structure inside the base region wake produces the dominant force oscillations on the truck (consistent with previous findings).

Figure 9 shows the long-time oscillations in the wake using a low-pass filter, on a horizontal plane. The oscillations are subtle when shown along with the mean flow, but clearly the wake tapers to a closure point that is not fixed in space but oscillates left and right.

Finally, Fig. 10 shows the amplitude of pressure oscillations for different frequency ranges using the standard deviation of Cp computed from the pressure spectrum at each point. In spite of the significant effect on the drag, these images show

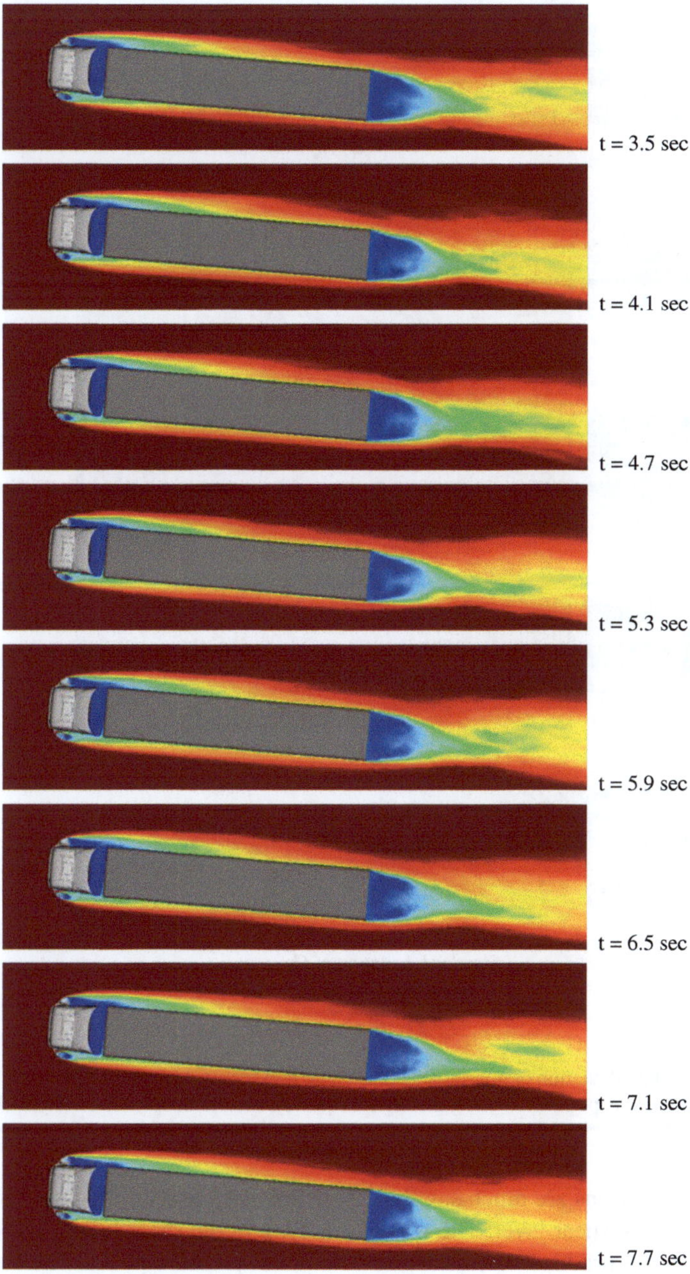

Total pressure, Low-pass filter 0.0 to 1.0 Hz

Fig. 9 *Horizontal* plane (at mid-height of side mirror), showing total pressure using low-pass filter for instances in time over a span of 4.2 s. The filter removes oscillations above 1 Hz and provide smoothed visualizations of the wake oscillation pattern. The calculation was performed using the high-frequency dataset sampled at 50 Hz for 10 s

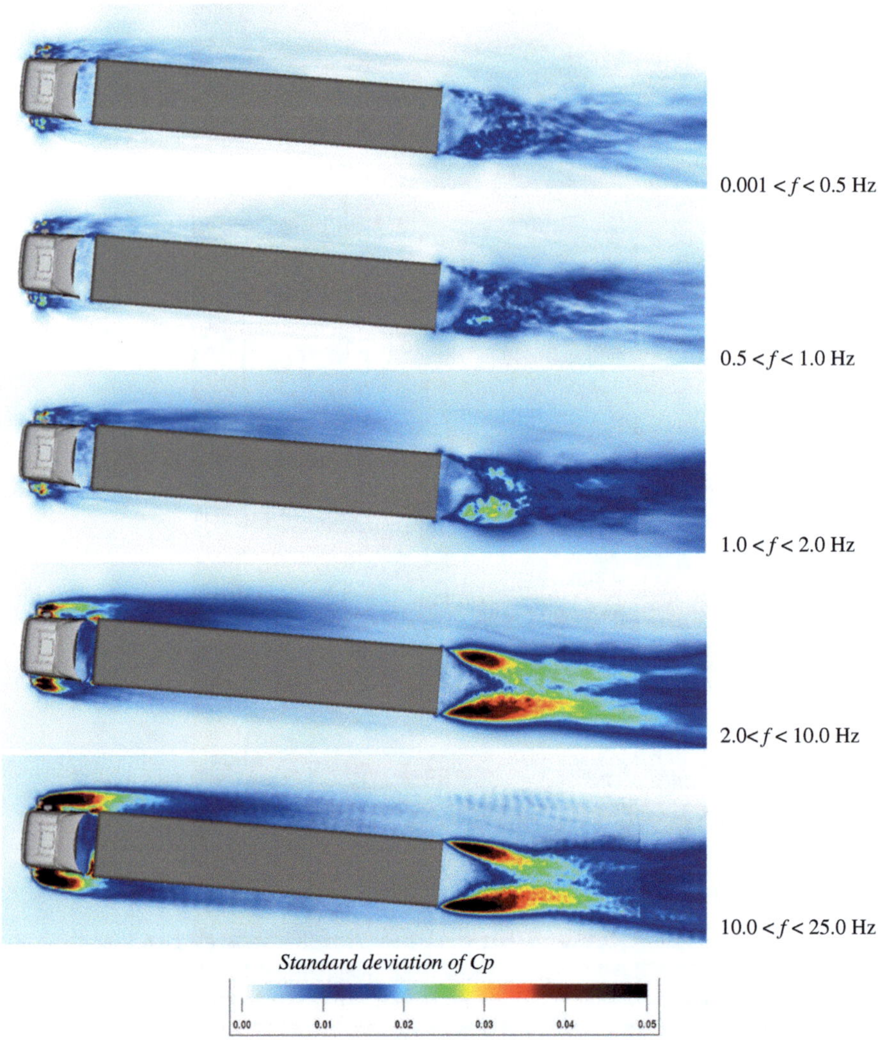

Fig. 10 Standard deviation of non-dimensional static pressure (Cp) for different frequency ranges. The calculation was performed using the high-frequency dataset sampled at 50 Hz for 10 s

that the fluctuations are much smaller inside the wake and on the back face than in the surrounding shear layer. The vortex shedding in the shear layers in the perimeter of the wake produce very strong low and high pressure peaks. The small shifts in pressure near the base of the vehicle are actually a side effect of a much strong dynamic system surrounding this region and extending rearward around the vehicle wake.

3.4 Other Geometry and Yaw Angle

To verify if the results are dependent on the Scania tractor shape and the yaw angle, a generic basic tractor combined with the same semi-trailer was simulated at 0° yaw angle like shown in Fig. 11 (left hand side). The drag spectrum shows a frequency shift of high fluctuation peaks and that the amplitude of the drag coefficients fluctuation is kept in the same order of magnitude than the simulation with the Scania tractor at 5° yaw angle.

3.5 Rapid Design Iterations

As described in *Approach*, the implications of the long transient behavior of the wake on simulation time and turn-around time can be significantly mitigated for design iterations on the front of the vehicle.

The first direct approach is to adapt simulation time according to the variability studied here and the expected drag delta between two geometry configurations. By defining an accuracy deviation due to the simulation time one is able for a given expected drag delta to define the minimum simulation time to be sure to capture the effects in terms of drag. This means that for an expected change of $C_D = 0.010$ one could reduce the physical simulated by half compared to changes with expectation in the $C_D = 0.002$.

Another approach is simply to ignore the drag on the rear half of the vehicle. Figure 12 shows the effect of drag fluctuations on the front half of the truck. A special

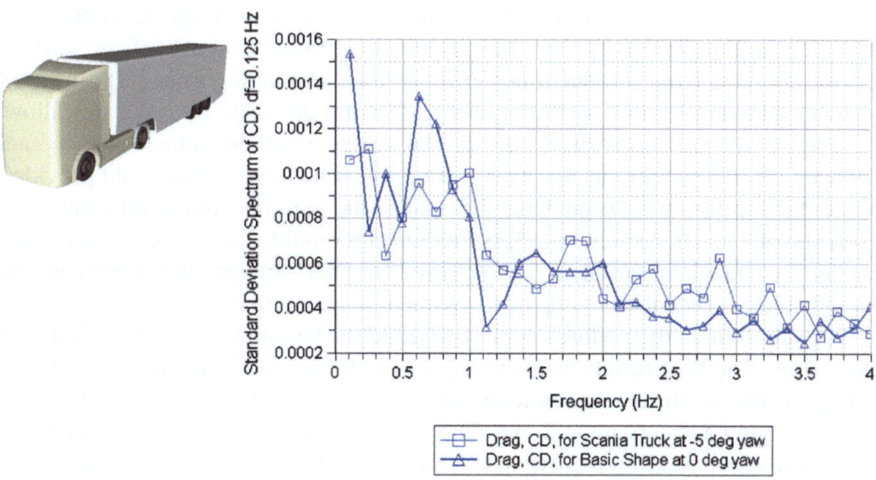

Fig. 11 Geometry of the generic basic tractor and comparison of drag spectra for current results on detailed vehicle at 5° yaw angle to a simulation on a simplified truck shape at 0° yaw angle

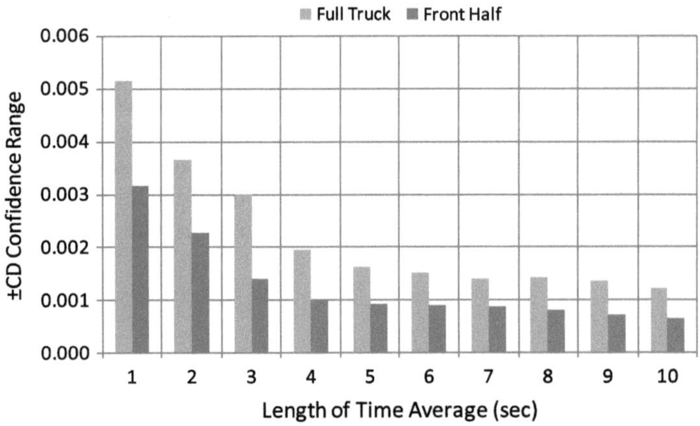

Fig. 12 Comparison of deviation in moving average values for the front half of the truck compared to the full truck. For the front half, the 4-s moving average has a maximum deviation of $C_D = \pm 0.001$, whereas for the full truck the confidence range with 4-s averaging is $C_D = \pm 0.002$, and with 10-s averaging is approximately $C_D = \pm 0.001$. The full truck data is also shown in Fig. 2.

cropped region was used to sample the instantaneous forces up to the truck midpoint, on the trailer behind the chassis wheels, in order to remove the drag effect of the unsteady base drag. The analysis is the same as in Fig. 2. It shows that significant oscillations remain on the front portions of the truck. Figure 12 shows that a narrow drag confidence range of $C_D = \pm 0.001$ of total drag can be obtained with 4 s of averaging on the front half of the truck.

Another approach is similar but further reduces the simulation effort by removing the rear portion of the flow domain. This sub-domain method is based on limiting the simulated volume around the vehicle and in this case by imposing an outlet boundary condition before the end of the trailer. Depending on the grid cells distribution, it can further reduce the computational effort by a factor ranging from 1.1 to 1.3. The retained spatial reduction exempting the model from the unsteady wake area allows the usage of the previously discussed approach on runtime reduction to convergence allowing an additional reduction factor of up to 3 which combined could provide a cost reduction factor between 3 and 4,5 against the high resolution full simulation. More specific location studies (such as roof fairings) could also reduce the simulation domain by a much larger factor (such as used for DOE studies), hence reducing the overall study costs but reducing the assessment potential to that localized area.

The sub-domain method was applied to simulate the same geometry as the previous results. The flow results were averaged from time 1 to 5 s and are shown below in Fig. 13. The results are similar to the full run. These results show that both shorter time-averaging and sub-domain simulations can be used for rapid design iterations, with significant speed-up relative to full simulation runs. In order to determine the final drag effect of optimization changes, the full simulation should be run to properly include any effects of the geometry changes on the wake.

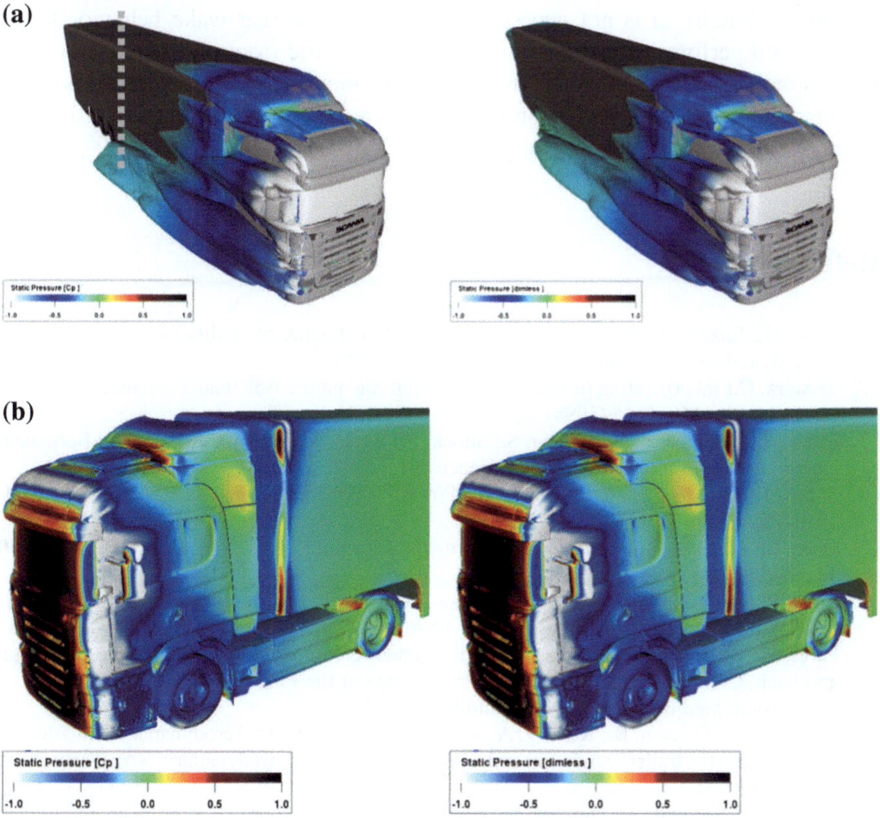

Fig. 13 Mean-flow visualization using fluid sub-domain simulation (*left* hand side), compared to full simulation. The sub-domain simulation is averaged for 4 s, while the full-simulation is averaged for 10 s. **a** Isosurface of non-dimensional total pressure = 0.0, colored by Surface Cp. **b** Surface Cp

4 Conclusions

The study reported here provides valuable insight and information into the use of LBM aerodynamics simulation for drag prediction of a full-scale fully-detailed truck including effect of cross-wind. The results show that large-scale oscillations in the flow structure occur predominantly in with a time scale range of 1–5 s and lead to corresponding averaging requirements for precise prediction of mean force values. Furthermore, the oscillations were identified to be due primarily to instability of the wake closure point using the correlation analysis and band-pass filter. However, fluctuations on the sides and top of the trailer are dominated by an instability consisting of longitudinal vortices in the perimeter of the separated flow regions. These produce large oscillations in pressure affecting the lift and side force, and have a secondary effect on the drag by contributing to additional oscillations of the perimeter of the rear wake. Finally, it is not necessary to compute the rear wake behavior for all cases, when performing iterations on the design of the front portions of the truck. Using sub-domain simulations or front-half drag, many more design iterations can be explored. This enables greater innovation and in the end better solutions to reduce the drag and improve fuel economy.

References

1. Chen, H., Teixeira, C., Molvig, K.: Realization of fluid boundary conditions via discrete Boltzmann dynamics. Int. J. Mod. Phys. C **9**(8), 1281 (1998)
2. Teixeira, C.: Incorporating turbulence models into the lattice-Boltzmann method. Int. J. Mod. Phys. C **9**(8), 1159–1175 (1998)
3. Chen, H., Kandasamy, S., Orszag, S., Shock, R., Succi, S., Yakhot, V.: Extended Boltzmann kinetic equation for turbulent flows. Science **301**, 633–636 (2003)
4. Shock, R., Qian, Y., Chen, H., Zhang, R.: PowerFLOW Simulations for 2D Cavity and Backward Step Flows (2002)
5. Chen, S., Doolen, G.: Lattice Boltzmann method for fluid flows. Ann. Rev. Fluid Mech. **30**, 329–364 (1998)
6. Lietz, R., Mallick, S., Kandasamy, S., Chen, H.: Exterior airflow simulations using a lattice Boltzmann approach. SAE Paper 2002-01-0596
7. Fares, E.: Unsteady flow simulation of the Ahmed reference body using a lattice Boltzmann approach. Comput. Fluids **35**(8–9). In: Proceedings of the First International Conference for Mesoscopic Methods in Engineering and Science (2006)
8. Horrigan, K., Duncan, B., Keating, A., Gupta, A., Gargoloff, J.: Aerodynamic simulations of a generic tractor-trailer, validation and analysis of unsteady aerodynamics. SAE Paper 2008-01-2612 (2008)
9. Fischer, O., Kuthada, T., Mercker, E., Wiedemann, J.: CFD approach to evaluate wind-tunnel and model setup effects on aerodynamic drag and lift for detailed vehicles. SAE Paper 2010-01-0760 (2010)
10. Duncan, B.D., Golsch, K.: Characterization of separated turbulent flow regions in CFD results for a pontiac NASCAR race car. SAE Paper 2004-01-3556 (2004)
11. Duell, E.G., George, A.R.: Experimental study of a ground vehicle body unsteady near wake. SAE Technical Paper Series No. 1999-01-0812, Detroit (1999)
12. Al-Garni, A.M.: Experimental investigation of the flow around a generic SUV. SAE Technical paper Series No. 2004-01-0228, Detroit (2004)
13. Heinecke, M., Beedy, J., Horrigan, K., Sengupta, R.: Aerodynamics study of a production tractor trailer combination using simulation and wind tunnel methods. SAE Paper 2010-01-2040 (2010)

Part IX
Truck Aerodynamics: Active Flow Control

Analysis of the Active and Passive Drag Reduction Strategies Behind a Square Back Ground Vehicle

Charles-Henri Bruneau, Emmanuel Creusé, Delphine Depeyras, Patrick Gilliéron and Iraj Mortazavi

Abstract A square back Ahmed body is used to mimic a heavy vehicle, bus or truck, in order to derive control processes that yield a significant drag reduction. The first step is to analyse carefully the structures in the flow that have a strong impact on the drag forces. Then, active and passive controls using blowing jets or porous medium layers are presented and discussed. The results are analyzed showing the direct impact of the control on the flow behaviour.

1 Introduction

The research developed today in bus and truck aerodynamics is carried out from the point of view of the durable development. Some bus and truck companies have the objective to develop control solutions able to reduce at least 20 % of the aerodynamic drag of the vehicles without constraints on the comfort, the storage or the safety. Thus, it is necessary to modify locally the flow, to remove or delay the separation position

C.-H. Bruneau · D. Depeyras · I. Mortazavi (✉)
IMB, UMR 5251, Univ. Bordeaux, F-33400 Talence, France
e-mail: mortaz@math.u-bordeaux1.fr

C.-H. Bruneau
e-mail: bruneau@math.u-bordeaux1.fr

C.-H. Bruneau · D. Depeyras · I. Mortazavi
CNRS, IMB, UMR 5251. INRIA, Team MC2, Talence, France

E. Creusé
LPP (UMR CNRS 8524), Université Lille 1, Villeneuve-d'Ascq, France
e-mail: creuse@math.univ-lille1.fr

E. Creusé
Team SIMPAF INRIA Lille Nord Europe, Villeneuve-d'Ascq, France

P. Gilliéron
Renault, Direction de La Recherche, 1, Avenue du Golf, F-78288 Guyancourt, France
e-mail: patrick.gillieron@renault.com

A. Dillmann and A. Orellano (eds.), *The Aerodynamics of Heavy Vehicles III*,
Lecture Notes in Applied and Computational Mechanics 79,
DOI 10.1007/978-3-319-20122-1_23

or to reduce the development of the recirculation zone at the back and of the separated swirling structures [12].

A large part of the drag coefficient around a bluff body is due to the pressure forces on the front and back walls. In particular the vortices generated on the sharp corners of the geometry are moved in the near wake, inducing strong pressure forces at the back. Consequently, each time a big vortex stands close to the wall the drag coefficient increases. A way to reduce this drag coefficient is to control the flow so that the vortices are either smaller or convected away faster. To reduce the size of the vortices and to convey them faster, one way is to use porous slices to create a low speed flow parallel to the main flow [6, 8]. Then Kelvin-Helmholtz instabilities generate small eddies that are less strong and the pressure forces are reduced. This approach is well adapted to heavy vehicles like trucks or buses.

To convey faster the vortices, active control using blowing jets can also be used, but it is maybe more difficult to introduce in a truck than a passive device with no energy supply [10, 17]. In this work, the bluff body is a simplified truck (or bus) called the square-back Ahmed body [1]. The first step is to analyse carefully the structures in the flow that have a strong impact on the drag forces. Then, active and passive controls using blowing jets or porous medium layers are presented and discussed. Moreover, a theoretical study shows the strong relationship between the distance of an inviscid vortex and the pressure forces on the back wall. The results given by the active and passive control processes are analyzed to prove that this relationship is true in a real flow. Indeed, the variations of the drag coefficient are directly linked to the distance of the strong shedding vortices to the back wall.

2 Modeling and Numerical Simulation

In this section, the method used to simulate the flow past Ahmed body on top of a road using Cartesian grids is presented. To compute the flow around solid bodies an immersed boundary model is used, penalizing Navier-Stokes equations [2] for the velocity and pressure (U, p) as unknowns that read on the non dimensional form based on the density ρ, the velocity at the entrance section U_∞ and the height H of the Ahmed body [7]:

$$\partial_t U + (U \cdot \nabla)U - \frac{1}{Re}\Delta U + \frac{U}{K} + \nabla p = 0 \ \ in\ \Omega_T = \Omega \times (0, T) \tag{1}$$

$$\nabla \cdot U = 0 \ \ in\ \Omega_T \tag{2}$$

where Re is the non dimensional Reynolds number based on the length L of the body as usually done in the car industry, K is the non dimensional coefficient of permeability of the medium and Ω is the full domain including the porous layer and the solid body. In the fluid the permeability coefficient goes to infinity, the penalisation term vanishes and we solve the non dimensional Navier-Stokes equations. In the solid body the permeability coefficient goes to zero and it has been shown in [2] that solving these equations corresponds to solve Darcy's law in the solid and that the velocity is proportional to K. For numerical simulations we set $K = 10^{16}$ in the fluid and $K = 10^{-8}$ in the solid body. For values bigger than 10^{-5} the body moves towards a porous medium.

The Eqs. (1), (2) above are associated to an initial datum ($X = (x, y, z)$): $U(X, 0) = U_0(X)$ *in* Ω and the following boundary conditions: $U = U_\infty = (u_\infty, 0, 0) = (1, 0, 0)$ on the entrance section and on the road; $\sigma(U, p)n + \frac{1}{2}(U \cdot n)^-(U - U_{ref}) = \sigma(U_{ref}, p_{ref})n$ on the open boundaries to convey properly the vortices through the artificial frontiers [3], where $\sigma(U, p) = 1/Re(\nabla U + \nabla U^t) - pI$ is the stress tensor, n is the unit normal pointing outside of the domain and for a real number a the notation $a = a^+ - a^-$ is used.

Then a numerical simulation is performed solving directly the Navier-Stokes equations and using a second-order Gear scheme in time with explicit treatment of the convection term. All the linear terms are treated implicitly and discretized via a second-order centered finite differences scheme. The CFL condition related to the convection term requires a time step of the order of magnitude of the space step as U is of order one. A third-order finite differences upwind scheme is used for the space discretization of the convection terms. The efficiency of the resolution is obtained by a multigrid procedure using a cell-by-cell relaxation smoother [9].

All the simulations are performed on the domain $\Omega = (0, 15) \times (0, 5)$ in two dimensions (see the Fig. 1) with a 1920×640 cells uniform mesh. The grid convergence for the same geometry has been studied in [8] and the grid 960×320 corresponds already to the finest grid required. The velocity vector is $U = (u, w)$ in two dimensions. The Reynolds number based on the length $L = 3.625H$ of the body is 30,000 that corresponds to $Re = 8{,}275$. The choice of this medium Reynolds number ensures the grid convergence in 2D on a fine grid.

For the active control, we add an horizontal jet $U_j = (u_j, 0, 0)$ at the back of the body. The horizontal jet has a non dimensional thickness $h_j = 0.024$ and an amplitude A ($u_j = Au_\infty$) which corresponds to the forcing intensity

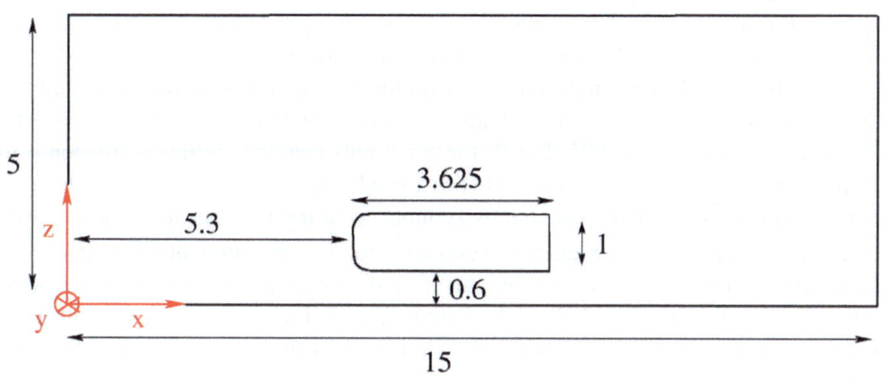

Fig. 1 Computational domain

$C_\mu = \frac{h_j}{H}\left(\frac{u_j}{u_\infty}\right)^2$. As the jet is coming out of the body, taken into account by the penalization method, the divergence-free condition (2) is locally modified as

$$\nabla \cdot U = \frac{u_j}{\delta x}. \tag{3}$$

with δx the horizontal space step in order to avoid the flow come back into the body. This modification is coupled to the penalization method in the solid and does affect neither the mass conservation on the other cells nor the flow computation.

For the passive control, porous layers of thickness $h = 0.1H$ are inserted on some parts of the Ahmed body [6]. Using the above mentioned meshes, a significant number of points are included in the porous layer. With the penalization method, solving the flow inside a porous medium of high porosity is reduced to set an intermediate value of K in the range 10^{-3} to 1 at the velocity points inside the porous layers. Previous works have shown that a good value for the control is $K = 10^{-1}$ [6] related to a high intrinsic permeability medium. The addition of such layers is equivalent to solve the Navier-Stokes equations in the fluid with a Fourier-like boundary condition instead of the no-slip boundary condition [11]. Thus the shear effects in the boundary layer are reduced. Moreover a low speed laminar flow is created inside the layer. The conjunction of these two phenomena decreases significantly the drag force.

3 Behaviour of the Flow

For a square back Ahmed body, the flow separates at the back and is similar to a two-dimensional base flow, dominated by counter-rotating structures. Experimental and numerical studies confirm the two-dimensional behaviour of the detached near-wall flow [1, 10, 12, 13]. Here, a two-dimensional behaviour means that the flow is fully detached at the back without longitudinal structures. These characteristics have been verified, computing the flow around the three-dimensional square back Ahmed body on top of a road at $Re = 30000$ in [4]. Therefore, the control strategies will be implemented on the two-dimensional configuration on a fine 1920×640 grid.

To analyse the flow we have chosen two monitoring points in the vicinity of the lower and the upper corners of the back to detect their individual frequencies. The dominant frequency $f = 0.25$ (Fig. 2) for two points near the corners corresponds to the mechanism of eddy formation and vortex shedding. These two points are located in the shedding layers that create the two counter-rotating vortices at the back. Using the corresponding eddy size and the averaged velocity, a Strouhal number $St = \frac{f\mathscr{L}}{\mathscr{V}}$ (based on the mean recirculation height $\mathscr{L}$ and on the mean velocity at the point (9.0,1.6) $\mathscr{V}$) of value $St = 0.22$ is found. This value agrees with the Strouhal number observed in other experimental and numerical studies concerning flows past bluff-bodies [14].

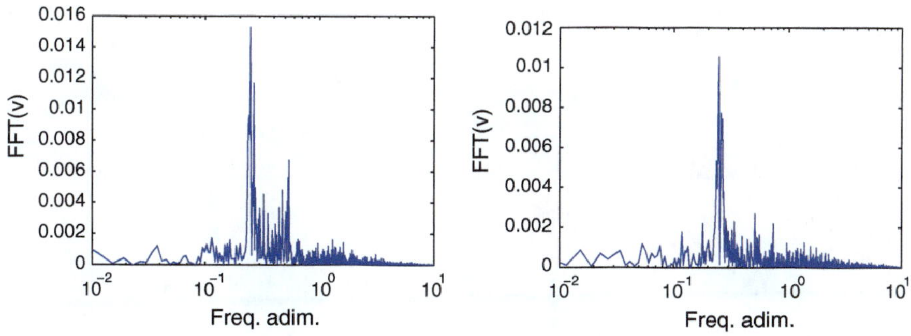

Fig. 2 Spectrum of the vertical velocity component for two lower (*left*) and upper (*right*) corner points of the uncontrolled flow

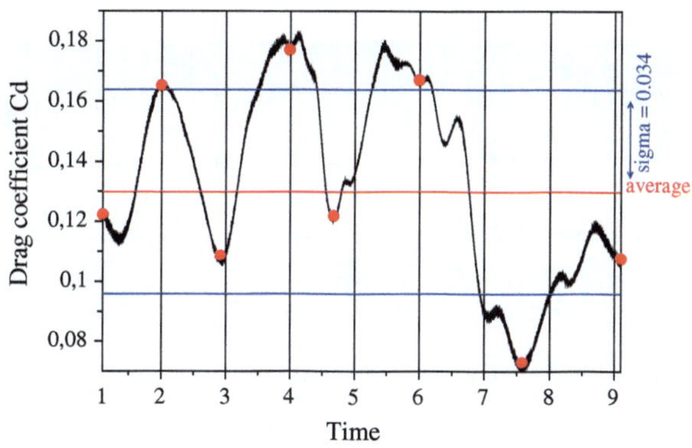

Fig. 3 C_d history during two shedding cycles for the uncontrolled flow

An efficient way to understand the impact of the flow behaviour on the body forces is to plot the C_d versus the time for two consecutive shedding cycles in the near wake (Fig. 3) and to explore the correlations between the dynamical evolution of the flow and the variations of the drag coefficient. As about two thirds of the drag forces are on the back wall, the discussion focus on this part of the flow. The pressure and the vorticity fields are plotted at different times corresponding approximately to the mean value or to local extrema of the drag coefficient representing the main evolution steps in the wake dynamics: vortex creation, development, emission and convection. As the Figs. 4 and 5 show counter-rotating vortices are alternatively generated at the lower and upper corners of the body as observed in the experiments reported in [16]. The lowest C_d value corresponds to a shedding moment when the core of the vortical structures is farther to the obstacle (t = 7.6) and the new structures are not yet formed. Oppositely, the highest C_d values occur when the shedded vortices strongly interact with the body (t = 2, 4 and 6). This behaviour of the flow confirms a strong correlation between the vortex core distance from the back wall and the drag forces generated around the body. It is therefore interesting to obtain a deeper understanding of the vortex removal procedure from the wall and the force distribution related to it.

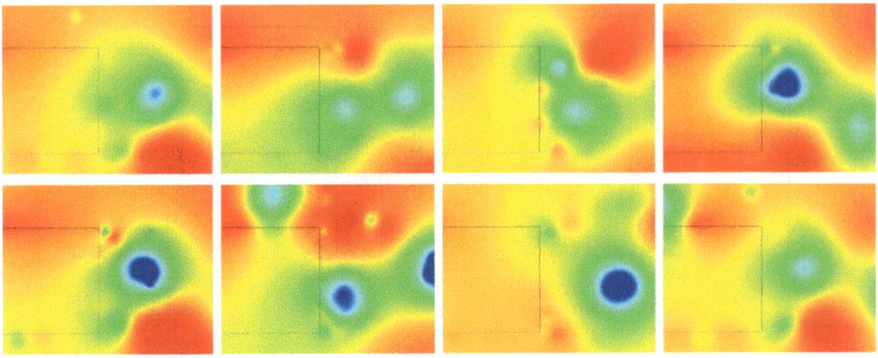

Fig. 4 Evolution of the pressure field during two shedding cycles for the uncontrolled flow. Times t = 1.1, 2, 2.9, 4, 4.7, 6, 7.6 and 9.1 are plotted from *top left* to *bottom right*

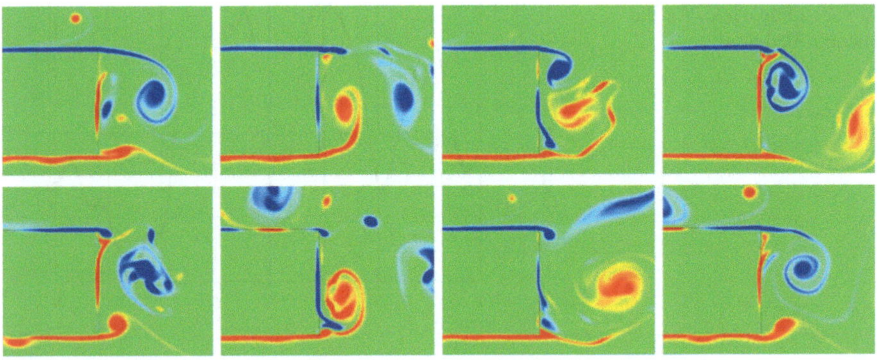

Fig. 5 Evolution of the vorticity field during two shedding cycles for the uncontrolled flow. Times t = 1.1, 2, 2.9, 4, 4.7, 6, 7.6 and 9.1 are plotted from *top left* to *bottom right*

4 Active Control Versus Vortex Motion

The real wake behind a bluff body is composed of vortical structures with different circulations and periods of shedding depending on the geometrical constraints and the Reynolds number. They have a mutual interaction with the near wake flow feeding the vortex street and being themselves conditioned by the former generated vortices [15]. However, even if this high order real motion topology is very complex, its main characteristics is the alternate vortices behind the body which induce a strong pressure well that is responsible of a significant part of the drag forces [18]. A direct benefit of the control is to push away this pressure well to decrease the drag coefficient. In the following the effect of an active control jet on the vortex removal from the back wall is studied.

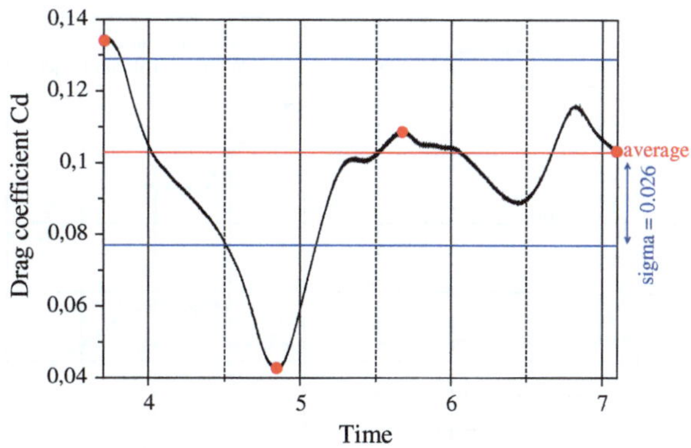

Fig. 6 C_d history during two shedding cycles for the flow with the active control

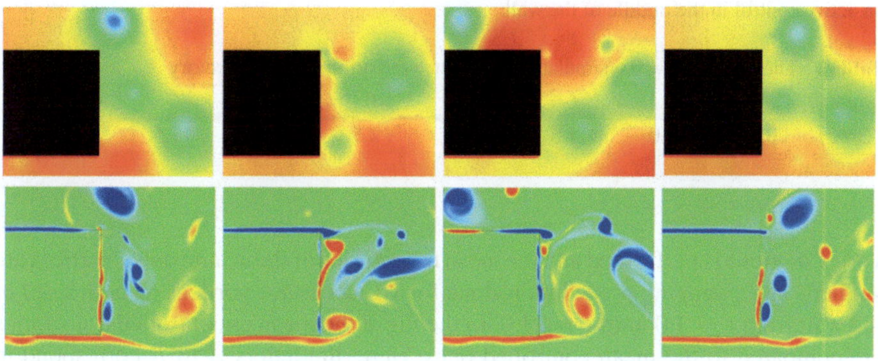

Fig. 7 Evolution of the pressure (*top*) and vorticity (*bottom*) fields during two shedding cycles for the flow with the active control. Times t = 3.7, 4.8, 5.7 and 7.1 are plotted from *left* to *right*

Starting from the same initial solution in both cases, the flow without control is compared to the controlled flow with a closed-loop actuator in the middle of the back wall. This position is chosen in order to take into account the shedding of vortices on both sides of the back wall. Sensors are placed in the bottom and the top of the back wall. Like for the uncontrolled flow, we explore the link between the flow evolution and the drag coefficient. The history of the drag coefficient during a shedding cycle is plotted in Fig. 6. Four times corresponding to the extrema and the mean value are selected but this mean value is close to 1.3 when the mean value of the uncontrolled case is about 1.7. The plots in Fig. 7 show clearly that the alternation of big vortical structures (Figs. 4 and 5) is replaced by a more chaotic shedding of smaller vortices, especially on the top where the flow is weaker than at the bottom due to the presence of the road. There is no more a deep well of pressure in the near wake. Additionally, to get a general view of the relationship between vortex kinematics (for more details see [5]) and the drag reduction due to the control, an averaged estimation of several

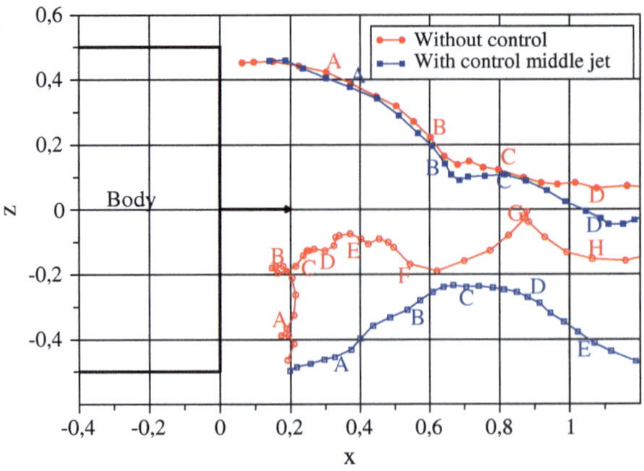

Fig. 8 Comparison of averaged trajectories of the vortices for the cases without control and with an active control in the middle of the wall

Table 1 The letters show the simulation time since the origin of the trajectory

Points	*A*	*B*	*C*	*D*	*E*	*F*	*G*	*H*
Time *t*	0.5	1.0	1.5	2.0	2.5	3.0	3.5	4.0

vortex motions is performed and the mean trajectories of the up and down vortices are studied. The averaging procedure is done for 10 successive vortices on both sides of the wall from $t = 3.0$ until $t = 23.0$. In the Fig. 8 the mean uncontrolled trajectories are compared to the mean controlled ones. Here, the letters A to F correspond to definite times on the trajectories of the vortex for both flows (Table 1). As the figure shows, the trajectories are about the same for the upper shedding processes. But, the lower trajectories are drastically modified as the vortices are expelled from the body much more quickly with the control. In the Fig. 9, for the upper vortex, the uncontrolled and controlled cases removal speeds are very similar and show almost a linear behaviour (a). While, for the lower shedding the uncontrolled case is removed from the wall with a polynomial law close to t^2 whereas for the controlled case, the removal procedure is much more accelerated and is defined by a quasi-linear law with even a smaller slope compared to the linear motion t (b). Let us note that the active control induces a strong pressure forces reduction at the back and consequently a 20 % reduction of the whole drag coefficient C_d. This result shows the efficiency of the active control on the lower shedding process. The difference between the upper and lower sheddings is due to the presence of the road as the jet flow under the body enhances the control jet effect. It is thus possible to connect the control efficiency to the removal speed of the vortices from the wall.

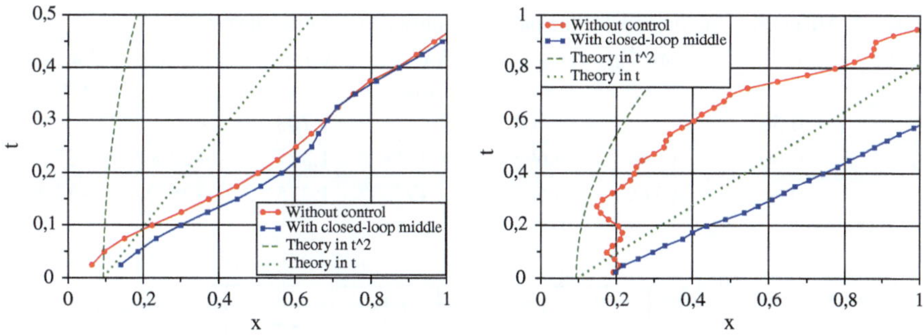

Fig. 9 Comparison of averaged removals of the vortices from the wall for the cases without control and with an active control in the middle of the wall

5 Passive Control with Porous Layers

Here, the idea is to implement porous layers in some locations of the body in order to reduce the drag forces. This approach seems quite convenient for bus or truck geometries because the aesthetic criteria are less important in that cases and introducing porous materials in some locations seems more feasible than on car geometries. Moreover, energy generation for active devices in buses and trucks is a quite complicated and expensive procedure. Therefore passive strategies are more applicable for this kind of vehicles.

The study is focused on two efficient geometries (Fig. 10) corresponding to the cases giving the best drag reductions in [8]. The main difference with the previous study is the distance between the body and the road that is equal to H in [8] and to $0.6H$ here as in the original benchmark [1]. It should be noted that the distance to the road modifies the efficiency of the passive control. According to the Table 2, the drag reduction is altered when the body is closer to the road as the velocity of the jet flow under the body is conversely proportional to this distance. Therefore it is more difficult to control the bottom shedding. Nevertheless, the decrease of the drag coefficient (up to 25 %) is still high.

To better understand the passive control effect, the drag history for two shedding cycles and corresponding instantaneous pressure and vorticity fields are studied in Figs. 11 and 12 for the case 2. As the figures show the porous layers modify com-

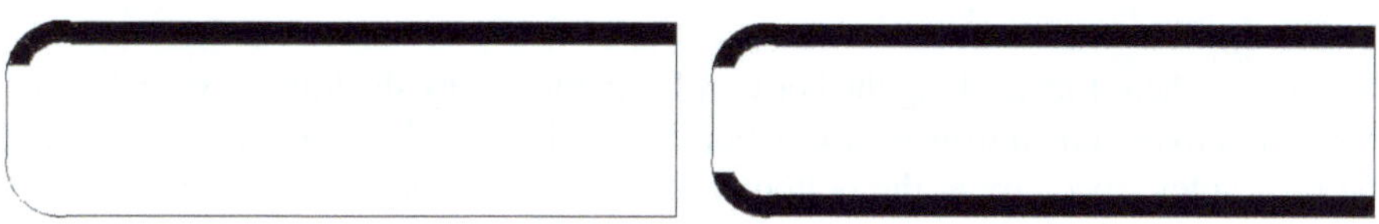

Fig. 10 Porous layers configurations for the passive control: case 1 (*left*) and case 2 (*right*)

Table 2 Mean drag coefficients for the passive control using porous layers

Case	$C_{d_{up}}$	Var. $C_{d_{up}}$	$C_{d_{down}}$	Var. $C_{d_{down}}$	C_d	Var. C_d	Var. C_d from [8]
0	0.565	–	1.104	–	1.708	–	–
1	0.604	+7 %	0.696	−37 %	1.327	−22 %	−37 %
2	0.683	+21 %	0.565	−49 %	1.275	−25 %	−31 %

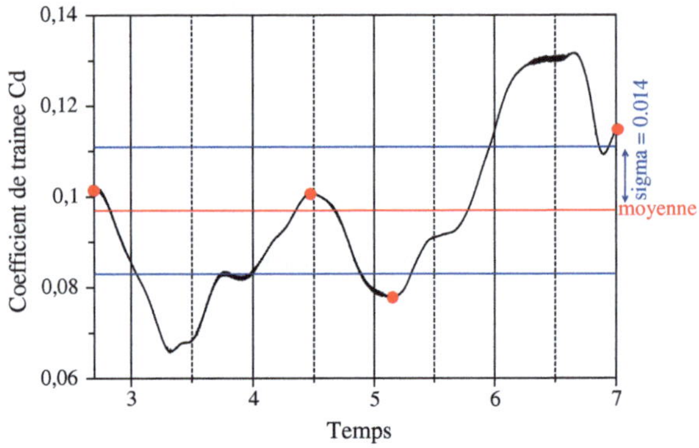

Fig. 11 C_d history during two shedding cycles for the flow with passive control (case 2)

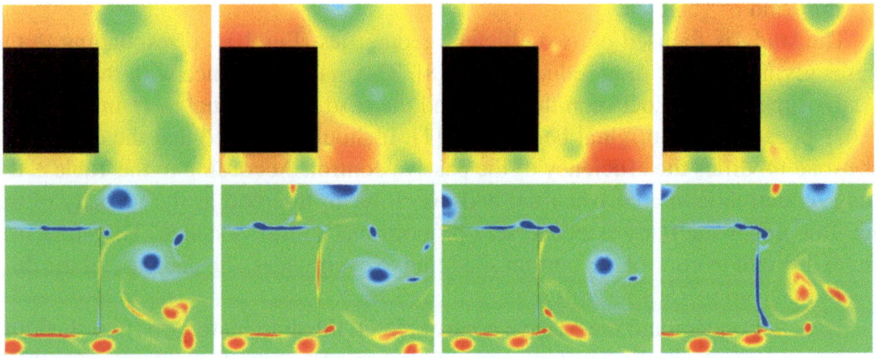

Fig. 12 Evolution of the pressure (*top*) and vorticity (*bottom*) fields during two shedding cycles for the flow with passive control (case 2). Times t = 2.7, 4.5, 5.2 and 7 are plotted from *left* to *right*

pletely the shear forces along the body and then the vortex shedding is very different compared to the uncontrolled case 0. For the passive case 2, there are small vortices on both sides, specially at the bottom wall as small vortices are reinforced by the jet under the body. Indeed, introducing porous layers induces a low velocity flow in the porous medium parallel to the main flow in the fluid. At the interface between these two flows there is a Kelvin-Helmholtz instability that yields small vortices that are clockwise on top of the body (blue in the figures) and counter clockwise below. These small vortices, that replace the uncontrolled large structures, are convected to the back and change drastically the flow behaviour, participating to a very strong decrease of the down drag coefficient (Table 2). The pressure fields (Figs. 4 and 12) reveal that the pressure is much higher than for the uncontrolled case. The wells centers are further to the body and have very few influence on the body wall itself. The plot (d) of Figs. 4 and 12 that correspond to the higher values of the drag coefficient (Fig. 11) show a similar well close to the body. For case 2, it corresponds to the merging of three small eddies.

6 Coupled Control

To achieve a more efficient control, in this section, the coupling of passive and active control techniques is proposed. The choice is to keep a porous layer on the roof and to add an action at the bottom back to improve the efficiency. Keeping the porous layer yields a beneficial change of the shear forces along the roof and split the upper strong shedding vortex into smaller eddies. This is possible for the square back Ahmed body as there is no rear window. We have seen in the previous section that the big bottom shedding vortex remains with a porous layer on the roof. It is thus necessary to make an action on it. Using a porous layer at the bottom is not sufficiently efficient for the bottom shedding, so the idea is to use a closed-loop blowing jet to push it down in the wake. The position of this jet, that is introduced into the square back body with the case 1 passive control, is at $H/3$ from the bottom corner that corresponds to the main position of the vortex center and the sensor is located at the down position. The results of this coupled active/passive control are astonishing as the drag coefficient reduction reaches -31 % that is better than the sum of the two gains and corresponds to our hopes (see Table 3). Indeed, adding the actuator to the configuration with the porous layer on the roof, reduces the bottom shedding vortices slipping motion on the wall, pushing away generated vortices far from the back wall. The averaged drag coefficient is very low around 1.18 (Fig. 13). This is due to the fact that the vortex structures at the back are much smaller, on top due to the porous layer and at the bottom because the jet split the big vortex into smaller parts (see Fig. 14).

Table 3 Mean drag coefficients for the passive control (case 1), the closed-loop active control at the down point and the coupled passive-active control

Case	$C_{d_{up}}$	Var. $C_{d_{up}}$	$C_{d_{down}}$	Var. $C_{d_{down}}$	C_d	Var. C_d
0	0.565	–	1.104	–	1.708	–
1	0.604	+7 %	0.696	−37 %	1.327	−22 %
J_{down}	0.539	−5 %	1.064	−4 %	1.629	−5 %
Coupled	0.552	−2 %	0.591	−46 %	1.183	−31 %

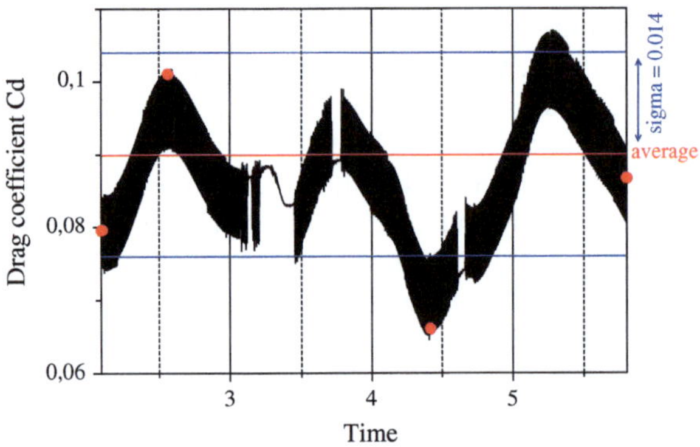

Fig. 13 C_d history during two shedding cycles for the flow with the coupled active/passive control

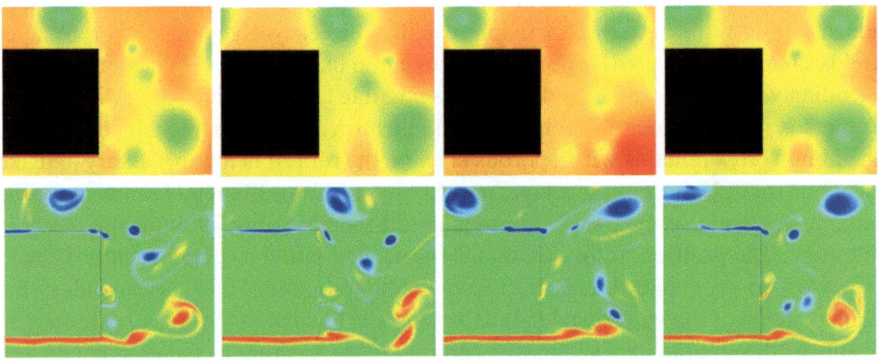

Fig. 14 Evolution of the pressure (*top*) and vorticity (*bottom*) fields during two shedding cycles for the flow with the coupled active/passive control

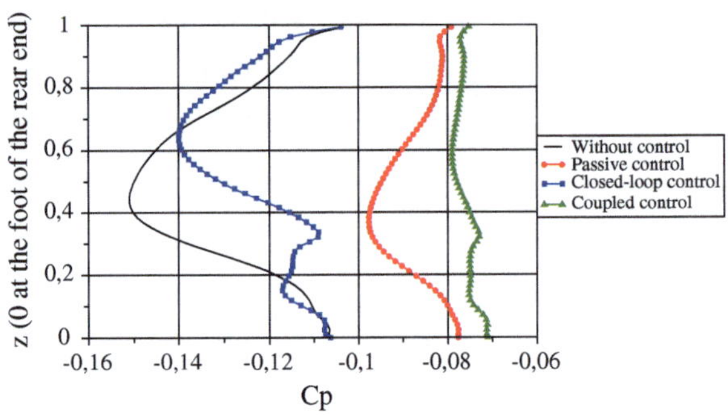

Fig. 15 C_p profile at the rear end of the body for the uncontrolled case, the passive control (case 1), the closed-loop active control at the down point and the coupled passive-active control

Finally, in Fig. 15 are plotted the C_p profiles at the rear end of the body for the uncontrolled case and the three above test cases. The results show clearly the improvement achieved by the coupling as the profile is almost flat and very high, that means that the large recirculation zone in the vicinity of the back wall has dramatically decreased.

7 Conclusions

In this paper, active and passive techniques to control the flow around the square back Ahmed body, that can be assimilated to a simplified bus or truck, on top of a road are analyzed and compared to the uncontrolled case. The results concerning the active control with a closed-loop blowing jet at the middle of the back wall give a good drag decrease, as it synchronizes itself to the flow but its implementation in a bus or truck back wall seems quite difficult.

The porous layers passive control is more robust, easy to handle and efficient and the obtained drag reduction results are promising. To improve its results a possibility is to couple it to a closed-loop control with a good position of the actuator and sensor. The results obtained in this last case are very promising as a 31 % reduction of the drag coefficient is reached. These procedures should give similar results in 3D around a square back Ahmed body.

Acknowledgments The simulations were run on PLAFRIM platform supported by IMB University of Bordeaux and INRIA Bordeaux - Sud Ouest.

References

1. Ahmed, S.R., Ramm, G., Faltin, G.: Some salient features of the time-averaged ground vehicle wake. SAE-Paper 840300 (1984)
2. Angot, Ph., Bruneau, Ch.-H., Fabrie, P.: A penalization method to take into account obstacles in incompressible viscous flows. Numer. Math. **81** (1999)
3. Bruneau, Ch.-H.: Boundary conditions on artificial frontiers for incompressible and compressible Navier-Stokes equations. Math. Model. Num. Anal. **34**(2) (2000)
4. Bruneau, Ch.-H., Creuse, E., Depeyras, D., Gilliéron, P., Mortazavi, I.: Coupling active and passive techniques to control the flow past the square back Ahmed body. Comput. Fluids **39** (2010)
5. Bruneau, Ch.-H., Creuse, E., Depeyras, D., Gilliéron, P., Mortazavi, I.: Analysis of the drag reduction using vorex kinematics behind a bluff-body. ASME Paper FEDSM-ICNMM2010-30716 (2010)
6. Bruneau, Ch.-H., Mortazavi, I.: Passive control of bluff body flows using porous media. Int. J. Num. Methods Fluids **46** (2004)
7. Bruneau, Ch.-H., Mortazavi, I.: Numerical modelling and passive flow control using porous media. Comput. Fluids **37**, n^0 5 (2008)
8. Bruneau, Ch.-H., Mortazavi, I., Gilliéron, P.: Passive control around the two-dimensional square back Ahmed body using porous devices. J. Fluids Eng. **130** (2008)

9. Bruneau, Ch.-H., Saad M.: The 2D lid-driven cavity problem revisited. Comput. Fluids **35**, n^0 3 (2006)
10. Brunn, A., Wassen, E., Sperber, D., Nitsche, W., Thiele, F.: Active Drag Control for a Generic Car Model. In: King (ed.) Active Flow Control, Notes on Numerical Fluid Mechanics and Multidisciplinary Design, vol. 95, pp. 247–259 (2007)
11. Carbou, G.: Brinkmann model and double penalization method for the flow around a porous thin layer. J. Math. Fluid Mech. **8** (2006)
12. Gilliéron, P., Chometon, P.: Modelling of stationary three-dimensional separated air flows around an Ahmed reference model. ESAIM **7** (1999)
13. Krajnović, S., Davidson, L.: Numerical study of the flow around the bus-shaped body. ASME J. Fluids Eng. **125** (2003)
14. Martins, L.F., Ghoniem, A.F.: Simulation of the non-reacting flow in a bluff-body burner: effect of the diameter ratio. J. Fluids Eng. **115** (1993)
15. Mortazavi, I., Giovannini, A.: The simulation of vortex dynamics downstream of a plate separator using a vortex-finite element method. Int. J. Fluid Dyn. **5** (2001)
16. Pastoor, M., Henning, L., Noack, B.R., King, R., Tadmor, G.: Feedback shear layer control for bluff body drag reduction. J. Fluid Mech. **608** (1984)
17. Rouméas, M.: Contribution à l'analyse et au contrôle des sillages de corps épais par aspiration ou soufflage continu, Ph.D. thesis Toulouse, France (2006)
18. Sipp, D., Coppens, F., Jacquin, L.: Theoretical and numerical analysis of wake vortices. ESAIM Proc. **7** (1999)

Heavy Trucks Fuel Savings Using the SaOB Actuator

A. Seifert, I. Dayan, C. Horrell, J. Grossmann and A. Smith

Abstract This paper describes a development program taking small scale Aerodynamic laboratory experimental technology to full-scale road tests. The fuel saving concept is based on attaching a 135 mm radius, quarter circle cross-section device, to the rear-side of truck-trailers. A full-scale conceptual prototype was designed and characterized by TAU and adopted as a full-scale adjustable and cost-effective prototype by ATDynamics. Bench-top tests at TAU validated the performance of the prototype as sufficient to warrant full-scale test success. Based on the bench-top tests it was decided that full scale inlet pressures of 3–6 psi at flow rates of 1–1.5 L/s per actuator are required. The full-scale prototype device comprised of some 100 suction and oscillatory blowing (SaOB) actuators' array with a common compressed air supply. A positive displacement pump operated by a gasoline engine supplied the compressed air. As part of an ongoing ATD research project, a series of road tests were performed at the Goodyear Proving Ground, San Angelo, TX. Two identical trucks were tested. One truck-trailer was standard, while the other was equipped with the TAU-ATD device. Gauges located just downstream of the pump and at 5 locations along the supply ducts measured the supply pressures. Portable sensors measured the device suction pressure and pulsed blowing frequency. It was found that the pressure drop in the supply ducts was 10–15 %. However, additional 35 % pressure drop existed in the flexible tubes between the ducts and SaOB actuators. Out of the 81 possible configurations, determined by a 3 by 3 parameter space, 5 configurations were actually tested with valid results. One configuration, measured twice at a driving speed of 65 MPH, provided 5 % increase in fuel economy (not counting the input pump energy). This translates to a 1.75 L/100 km savings or 1 L/100 km taking into account the flow power invested. This improvement was obtained with inlet pressure lower than 4 psi, marginal according to all previous tunnel and bench-top tests. Furthermore, it is still open how close to optimal is this device configuration.

A. Seifert (✉) · I. Dayan
The Meadow Aerodynamics Laboratory, School of Mechanical Engineering, Tel Aviv University (TAU), 69121 Tel Aviv, Israel
e-mail: seifert@eng.tau.ac.il

C. Horrell · J. Grossmann · A. Smith
ATDynamics (ATD), Inc., South San Francisco, CA, USA

A. Dillmann and A. Orellano (eds.), *The Aerodynamics of Heavy Vehicles III*,
Lecture Notes in Applied and Computational Mechanics 79,
DOI 10.1007/978-3-319-20122-1_24

With significantly reduced pressure losses, resulting in 5–6 psi inlet pressure at 15 % the current required input energy it is expected that 6–9 % net fuel saving would be obtainable in future road tests, potentially leading to the most compact commercial product to date.

1 Scientific Background

The use of Active flow control (AFC) for environmental protection is novel in many ways. Commonly, the capability to alter flow states was considered relevant to the aerospace industry. However, the growing need to reduce fossil fuel burn and emissions called for unconventional look at the way we approach large vehicles fuel efficiency [10]. Flow control was simultaneously introduced with the boundary layer concept by Prandtl [12] at the turn of the 20th century. In the period leading to and during WW II, as well as in the Cold War era, flow control was extensively studied and applied mainly to military fluid related systems. Though the fluid mechanics aspect can be robust, steady-state flow control methods were proven to be of inherently marginal power efficiency, and therefore limited the implementation of the resulting systems. Unsteady flow control using periodic excitation utilizing flow instability phenomena [17] has the potential of overcoming the efficiency barrier. Separation control using periodic excitation at a Strouhal number of order unity, can save 90–99 % of the momentum required for separation control using steady tangential blowing [17]. The feasibility of increasing the efficiency and simplifying fluid related systems (e.g., for high-lift, [7, 11, 13]) is very appealing. This becomes even more appealing if one considers that *1 % saving* in the fuel consumption of the US fleet of large trucks (about 8 million trucks in 2004, [18]) is worth more than *a Billion$ per year*, with huge environmental and political benefits. The progress in miniaturization, actuators, sensors, simulation techniques and system integration enables the utilization of wide bandwidth unsteady flow control methods in a closed-loop AFC (CLAFC) manner. See [3, 4] for reviews of the subject. Experimental demonstrations are required to close the gap between the current theoretical understanding, the computational capabilities and real-world problems. The described study brings together AFC expertise, specifically actuator development and implementation for separation control, to real-life road demonstration.

The essential ingredients of a full-scale active separation control system, packaged as an "add-on" device that will be attached to the rear-end of large truck trailers, in order to significantly reduce the aerodynamic drag were recently completed and will be presented in this paper. At highway speeds, the aerodynamic drag is responsible for roughly 65 % of the fuel consumption, making the potential of 12–20 % aerodynamic drag reduction, translating to 8–12 % fuel saving [15, 16]. There has been considerable effort in the US to reduce the fuel consumption of trucks using shape changes, simple add-on devices and steady state AFC methods [8], but those have inherent limitations and can lead to only half the expected benefit of our suggested method (considering rear-end devices alone). Hsu et al. [8] cite several research efforts focused on truck-trailer drag reduction. Adjustable inclination flat plates could be attached at the truck lee-side. These could reduce drag but are rather large, heavy and expensive. Their size raises functionality and compatibility issues. When the length is limited to a practical one, about 6.5 % fuel saving can be obtained at 65 mph [9]. Periodic excitation was also mentioned in [8] and is also presently considered in [5]. Steady blowing [6] was applied to a modified aft-region of a truck-trailer and resulted in significant aerodynamic drag reduction, but at a marginal to zero or mostly negative energy efficiency due to reasons identified already by our research group [17]. It is expected, and substantiated ample experimental evidence [15, 16], that the combination of steady suction and pulsed blowing could overcome the above identified efficiency barrier.

The current study is aimed at transitioning laboratory developed small scale active flow control (AFC) technology as an "add-on" device attached to the aft-face of tractor-trailer and track testing the fuel economy of the full-scale device. The TAU-developed Suction and Oscillatory Blowing (SaOB) actuators are used for drag reduction of heavy ground transportation systems. The above fluidic device is a combination of an ejector (jet-pump) and a bi-stable fluidic amplifier that were thoroughly studied [1, 2]. After completing the actuator development and adaptation to the speed range relevant to trucks, three stages of experiments were performed on a circular cylinder, the generic bluff-body [15, 16]. These studies resulted in a significant reduction of drag due to delay of boundary layer separation. A wide range of boundary conditions were tested and only the common results to all conditions were considered as valid. Successful wind tunnel demonstration on a two-dimensional (2D) equivalent of a blunt truck-trailer model was subsequently performed and reported [15, 16] This stage also included a complete, small-scale, truck model with drag reducing devices. The current paper describes the development of a fullscale prototype of the "add-on" device for road testing, its track tests and additional bench-top experiments conducted to better understand the results of the road tests and troubleshoot possible shortcomings of the current design. The technology could lead to a revolutionary aerodynamic drag reduction of heavy road vehicles with an order of magnitude smaller device dimensions than the current art. It could lead to new efficiency levels of aeronautical systems as well. The current research is especially appealing since it provides a valuable contribution to a slower production of greenhouse gases.

2 The Conceptual Device Prototype

A conceptual design of a full-scale prototype device was developed by the TAU team. The subsonic SaOB actuator developed by Arwatz et al. [1, 2, 14] was scaled up by a factor of about 50 % such that it fitted inside a 1/4 circle with R = 135 mm cross-section device. Figure 1a shows a cross-section of the device attached to the rear-end of a conceptual trailer. To maintain maximum flexibility, 4 rows of staggered suction holes 4 mm diameter each were drilled in the arc shaped- cover over the range 110–125° (where 90° is the vertical—plane of trailerend). The suction locations were determined from the 2D truck experiments where the device radius was 37.5 mm and holes' diameters' was 2 mm [15, 16]. Based on the same set of experiments it was decided to place the pulsed blowing slots 10–20 deg downstream of the mean suction holes' locations. Three alternatives for the pulsed blowing slot locations were designed: 130, 135 and 140°. The minimum arc-distance between the holes and slots was 10°. It was decided to use a slot width of 2.5 mm, with a constant cross-section nozzle transitioning the SaOB exits to the two exit slots. The width of each unit to be controlled by one actuator was determined to be 75 mm, a conservative size that will be tested, and hopefully reduced, in the future.

Figure 1b shows an isometric plot of a small section of the prototype device, built by rapid prototyping technique and bench-top tested at TAU. Figure 1c shows the layout of the full scale SaOB actuator. A critical factor in the design and operation of the system is the narrowest diameter of the ejector jet, currently 2.8 mm. This is a decisive factor in the creation of pressure losses of the current AFC system, and therefore the energy efficiency of the entire actuation system.

3 Description of the TAU Bench-Top Experiments

The device shown in Fig. 1b was covered from both sides by plates and tested in the actuators' lab of TAU. The compressed air was introduced from the sides of the supply channel (see also in Fig. 1a). The entire volume of the prototype was sealed to assure fluid inhaling only through the suction holes. The inlet pressure, the suction pressure, the oscillation frequencies and exit velocities were all measured. During the experiments, a series of increasing inlet pressures, in the range of 0 to about 0.5 bar were applied. The number of open suction holes, as shown in Fig. 1b, were altered to be one, two or three (with four holes in each row) by covering those desired "close" by an adhesive tape attached to the cover over the holes.

Figure 2a shows the variation of the suction pressure in the device as a function of the inlet pressure and number of open suction holes. A desired trend of an almost linear variation of the suction pressure with inlet pressure can be observed. Clearly, the increased number of open suction holes reduces the suction pressure as the flow rate increases. The suction velocities into the suction holes were also measured, by placing a calibrated hot wire in the 4 mm diameter suction hole inlet plane. Though this is not an accurate measurement due to the large sensitivity to position in this sink flow, the measured values of 5–18 m/s for the entire range of parameters assured us that we shall be able to achieve sufficient suction magnitudes, corresponding to the wind tunnel conditions also in the road test campaign.

The oscillation frequency is shown in Fig. 2b. Again a clear trend of increased frequency with the increase of the inlet pressure can be observed. This range of frequencies generates a Strouhal number between 0.3 and 0.6 based on the device radius and a freestream velocity (driving speed) of about 30 m/s (100 km), suitable for separation control [17]. Note also that the oscillation frequency is robust, such that it does not depend on the suction pressure and to a lesser extend on the suction flow rate.

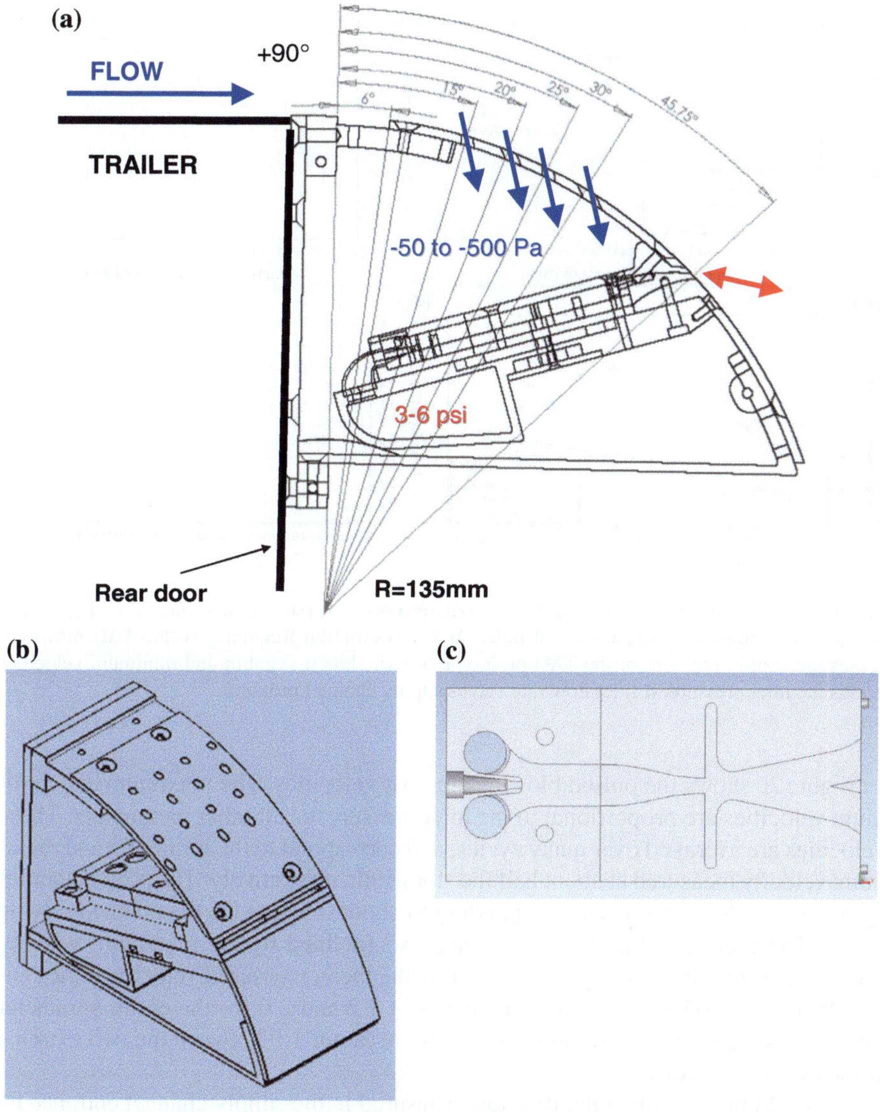

Fig. 1 **a** A cross section of the conceptual "add-on" device, attached to the rear-end of the trailer. Four rows of staggered suction hole and one pair of pulsed blowing slots shown. Holes and slot locations are altered by altering the location of the screws attaching the arc to the frame. The actuators and supply channel "block" are attached only to the arc. **b** An isometric view of the TAU conceptual prototype device. **c** A layout of the SaOB actuator. Length is 60 mm. Flow inlet is from *left*

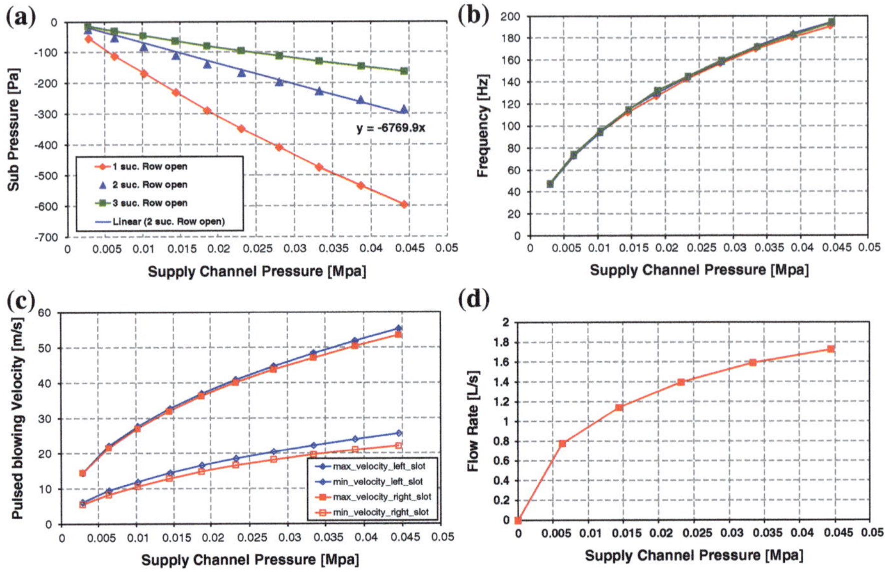

Fig. 2 **a** Suction pressure generated by the TAU prototype versus the inlet pressure. The legend indicated the number of open rows of holes. **b** The oscillation frequency of the TAU prototype versus the supply pressure of the TAU prototype. **c** Exit slots maximum and minimum velocities versus the inlet pressure. **d** Inlet flow rate versus supply channel pressure

Figure 2c shows the pulsed blowing slot exit velocities. The maximum and minimum velocities are proportional to the inlet pressure, as all other parameters. These velocities are averaged over many cycles, and correspond to the minimum and maximum velocity measured at about half the slot width, respectively. The peak velocities are between 15 and 55 m/s, corresponding to about 0.5 to twice the planned driving speed. The corresponding "switching quality" (defined by the maximum velocity minus the minimum velocity normalized by the mean measured velocity at the center of each slot exit) was found to be between 2.2 and 2.6 for the entire parameter range. This leads to ejected momentum ratio of about 1:6 between the two extreme velocities in the cycle.

Figure 2d presents the inlet flow rate measured in the supply channel entrance by an orifice flow meter versus the inlet pressure. Having defined both the inlet flow rate and inlet pressure required to achieve the above performance goals from the device prototype enabled us to proceed to the design of the road test AFC system.

4 Design of the Road Test Flow Control System

The design of the road tests AFC system had to take several considerations into account. The main ones are:

- The circumference of the trailer is about 7.5 m long.
- The supply of the inlet flow has to be from the lower "legs" of the device.
- The sides and top edges of the trailer should be controlled (bottom part will not be controlled based on the 1:12 scale model tests).
- Flexibility of suction holes and pulsed blowing slot locations has to be maintained from the lab to prototype.
- Low fabrication cost was a prime drive of the design.
- Sensors had to be placed to assure performance, at least at the start and end of each track test ("point").

The device basic "cell" width of about 75 mm indicated the need for about 100 SaOB actuators for the entire full-scale device. Therefore the max required flow rate was about 150 L/s at about 0.5 Bar. A suitable pump (positive displacement, 7.4 psi max, 300 ACFM at 3600 RPM, with a corresponding $\Delta T = 45°C$) and a 16HP gasoline engine to drive it were identified and placed under the trailerend in a special cage. Figure 3 shows the drawings of the flow supply system. The selection of such a pressure supply system imposed a low efficiency (lower than 25 % indicated also by the very significant temperature rise of the pumped air) penalty on the AFC system, significantly reducing the potential for net increase in fuel efficiency. But this aspect deemed secondary as compared to the pioneering aerodynamic nature of road test demonstration of this innovative AFC system. The weight of the pressure

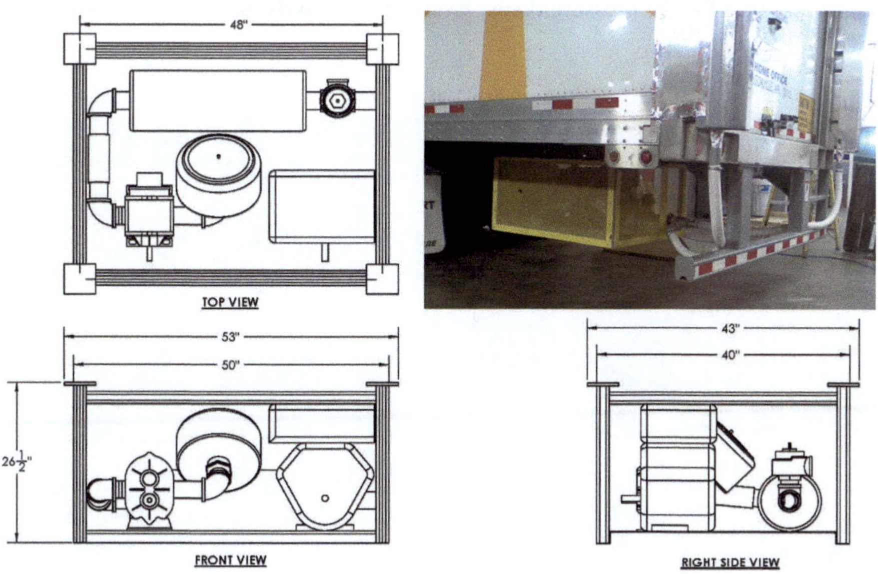

Fig. 3 The pressure supply system, showing the inlet filter, the pump, the Gasoline engine, the silencer and the relief valve. Tube exit diameter was 3". Dimensions of the drawings are in Inches. Exit pipe was 2.5" and 2" flexible tubes routed the flow to the two inlets of the device

supply system was about 400 kg. No provision was made to increase the weight of the "reference" truck by the weight of the AFC system.

Observing the pressure supply system, housed under the trailer in the road tests, as shown in Fig. 3, one notes that a 2" flex hoses route the flow into the device. It was decided that a single duct would be routed around the inverted U shaped device. The inner dimensions of the supply duct cross section were 2" by 1". The TAU prototype concept was altered by ATDynamics as shown in the crosssections of Fig. 4.

The concept chosen for the air supply was to use a rectangular cross section duct, stationary to the inner of the device and connecting the vertical legs with the horizontal section of the supply channel by Aluminum welding. The movable central body of the device is firmly connected only at its exits to the Stainless external cover. The larger fraction of each curved cover plate contained the suction holes. Three interchangeable sets of cover plates allowed a large combination of suction holes and pulsed blowing slot locations. Figure 5 shows the three alternative pulsed blowing locations. About 400 screws connected the interchangeable curved plates to the static L shaped part of the device and to the main body of the actuator.

A special transition section, with two mirror image duplications were designed and fabricated to connect the vertical parts with the horizontal part (see Fig. 6a, b). While the suction chamber was made continuous, no valves were placed very close to the edges. The weight of the AFC drag reducing device was about 150 kg.

Due to the complex, restricted and thin plates made device, special care had to be taken during the assembly process to good seal between all the device parts. This is especially true for the pressure ducts. Every small leak from the high pressure duct

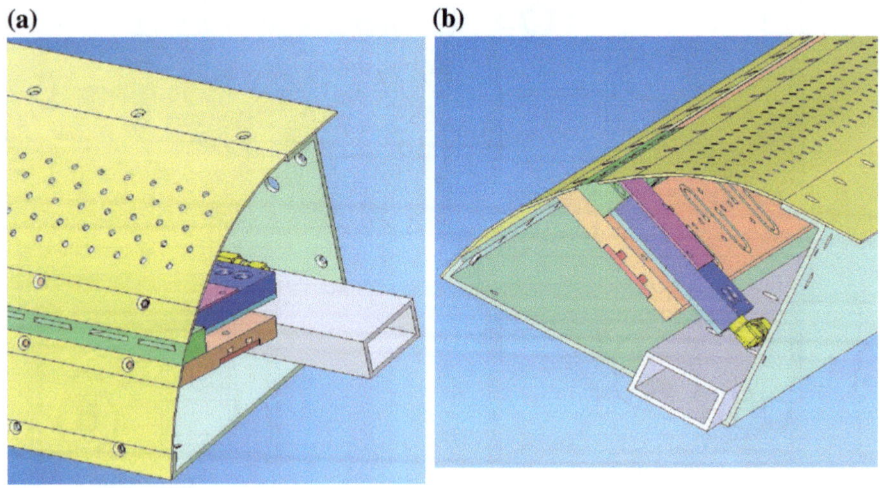

Fig. 4 **a** A cross-section of the lower edge of the road test device, showing the inlet duct and cover details. **b** A cross-section of the road test device, showing only one actuator. The actuators are connected to the duct with pneumatic connectors and flexible 1/8" I.D. 6" long tubes. The pattern of the feedback tubes is also visible

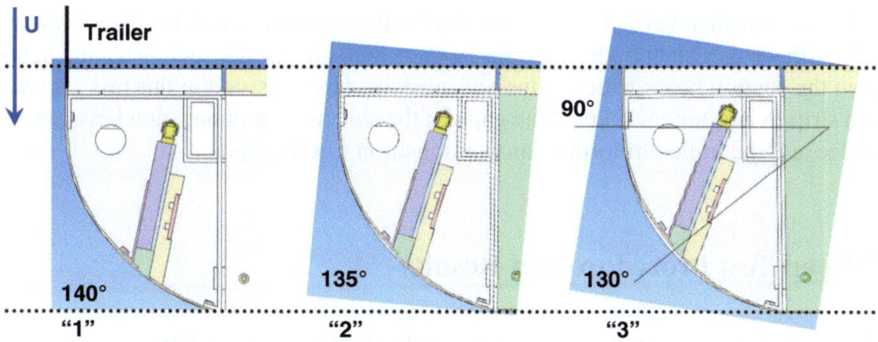

Fig. 5 A top-view cross-section of the 3 device configurations

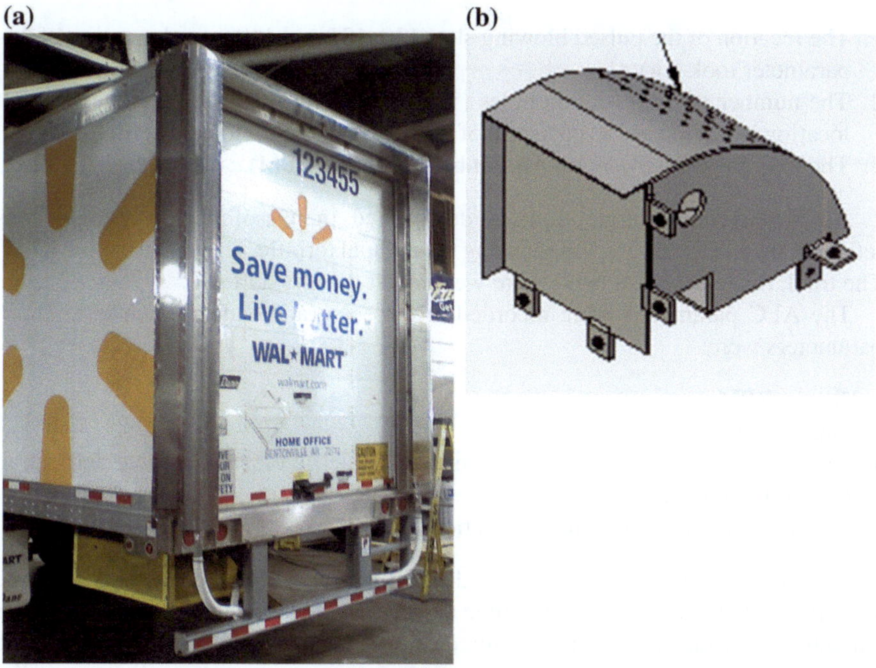

Fig. 6 **a** Showing an isometric rear-view of the AFC device installed on the trailer. Note that there is no obstruction to the roll-up cargo door. **b** The design of the *upper-right* corner piece. Note the rectangular openings for the pressure duct and the circular openings for the continuity of the suction chamber

to the suction chamber will be detrimental for the performance of the device. Same, only with weaker magnitude, are leaks from the outside into the suction chamber, not from the suction holes. Silicon glue and adhesive tapes assisted in this task. A wide thin strip of Aluminum bridged the gaps on the side and top connection between the device and the trailer, to form a smooth transition (see Fig. 6a).

5 Road Test Procedure and Results

The road tests were conducted according to the following procedure:
The device configuration had to be defined, according to the Taguchi L9 method. Three parameters, with three levels each, were chosen. These parameters were:

1. The location of the pulsed blowing slot (130, 135 and 140°, an alternation of this parameter took 4 h).
2. The number of open suction holes (1, 2 or 3 open rows, The "average" suction location was fixed at 15° upstream of pulsed blowing slot location, 1/2 h change).
3. The inlet pressure (4, 5 or 6 psi, set by the engine RPM).

After the device configuration was determined, the Gasoline blower engine was refueled, the inlet pressure was set using the manual throttle, setting the engine RPM. The truck removable fuel tanks were weighed at the test start.

The AFC parameters were recorded at the start (and end) of each test. Those parameters were:

- Blower RPM
- Tank pressure
- Device pressures (5 points) along the supply duct, 2 on each vertical and one in the middle of the top section.
- Device suction pressures and oscillation frequency.

The latter measurements were achieved using an Endevco 1 Psi unsteady pressure sensor, coupled with a 1.6 I.D mm tube at its positive side. The output of the sensor was gained by a factor of 50 and measured using a hand held multi-meter and portable scope.

For measuring the frequency, the edge of the tube was positioned at the pulsed blowing slot exit and for measuring the suction pressure inside the device chamber, the tube was inserted some 2 cm into one of the suction holes.

Thereafter, the trucks drove for 41 miles around the track at a constant speed of 65 MPH. A careful procedure for exit and return of the trucks from the track was kept and repeated.

At the end of each test run, all the parameters were measured again and the fuel tanks were disconnected from the trucks and weighted again.

If the averaged wind during a run exceeded 5 MPH, usually the test was scraped. Any other peculiarities were record. Before decision on the next test condition, the fuel consumption data was processed.

Overall, 22 runs were performed in 3 days. One day was very windy resulting in mostly "bad" points. Seven out of the 22 runs were determined "good", where these included runs with at least 2 repeats.

Two repeat runs produced about 5 % fuel savings. These can be translated to about 1.75 L/100 km saved. Considering the net power invested calculated by the inlet pressure time the flow rate, assuming a pumping efficiency of 66 %, more than 1 L/100 km were still saved. The pumping power was estimated by (ACFM*ΔP/0.66). The actual fuel burned by the Gasoline engine coupled with the positive displacement pump was significantly larger, since the overall AFC system efficiency is estimated at 15 % or less.

The device configuration and AFC system conditions for this case were:

- Pulsed blowing slot at 135°,
- Two open suction rows of holes (115°, 120°).

The inlet pressure was about 6.5 psi (0.045 MPa) and the flow rate was estimated from the measured pump RPM. The measured suction pressure was 150 Pa and the oscillation frequency was about 110 Hz. See Fig. 7a, b for the successful configuration.

It was determined, based on the field measurements, that both the suction pressure and oscillation frequency were lower by about 30 %, compared to the expected results for such an inlet pressure (measured around the supply duct), based on the bench-top tests. It was suspected that the 3 mm I.D., 150 mm long tubes at the inlet to the valves were responsible to this significant pressure loss, not present in the TAU pre-test bench-top tests.

Following the test, while the device was being dismounted from the trailer, we checked the interior lower part of the device for debris and other fluids, solids and articles that might have been sucked in and could clog the device. After 3 days of tests in the dry track in San Angelo TX, nothing significant was found in the device (see Fig. 7c), negating fears regarding the vulnerability of the suction system to debris.

6 TAU Post-test Bench-Top Experiment

Following the discrepancy identified between the pre-test bench-top data and the measured frequency and suction pressure during the road tests it was decided to build a mock-up of the road test device (see Fig. 8a) and bench-top test it at TAU. Those measurements used exactly the same equipment and procedures as the TAU prototype device characterization. It was found that indeed the small diameter tube used in the road tests is responsible to almost 35 % pressure drop. When replaced by a shorter, 6 mm I.D. tube, the pressure loss was almost completely eliminated (Fig. 8b).

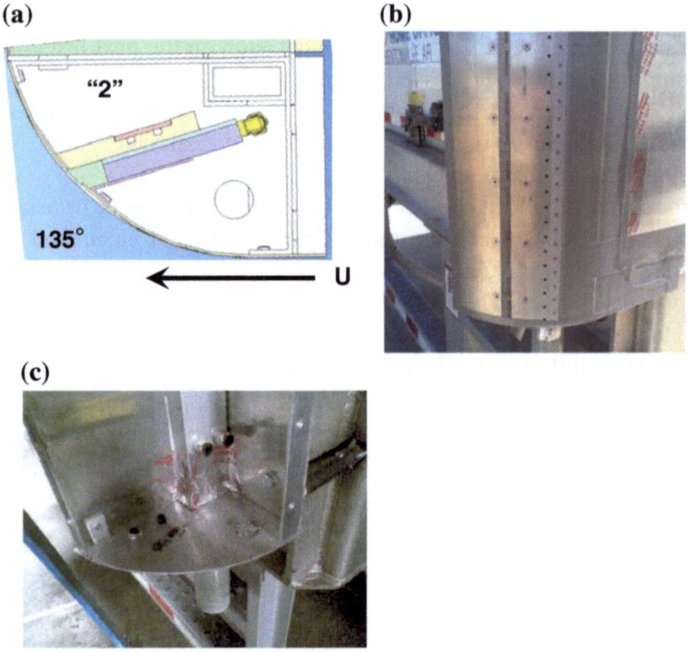

Fig. 7 **a** The best configuration identified during the test campaign. *Internal view*. **b** An external image of the device at the best test condition discussed above. Two open row of holes (currently transparent tape is applied on one slot, it was not present during the successful test) Flow from *right to left*. The interior of the AFC device (with actuator block and covers removed)

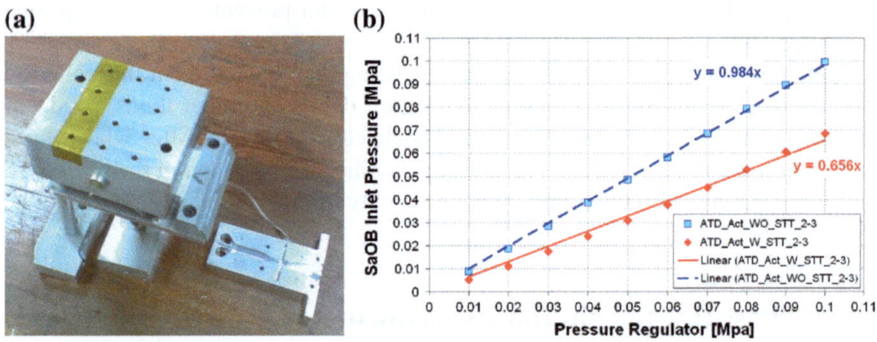

Fig. 8 **a** The post-test device mock-up, showing one actuator and its exit slots. A simulated suction pressure with 3 rows of holes rides on top of the actuator. Another actuator is open on the table. **b** The inlet pressure versus regulator (supply pressure) with the original 3 mm tube as tests and with a 6 mm ID tube

This set-up was also used to test certain modifications and upgrades to the device as will be tested in following road tests. The measurements made by the portable equipment for identifying the frequency and suction pressure were validated and found to be within 5 % of the lab equipment measurements.

7 Conclusions and Recommendations

An aggressive and challenging study took a laboratory technology to road tests with a leap frog jump from 1:6 scale 2D and 1:12 3D scaled experiments directly to full scale tests. The technology is aimed at increasing the trailer's rear-end static pressure using an active flow control device housed in a 1/4 circle with a radius of 135 mm. This is by far the shortest penetrating device either to the back or to the center of the rear-end plane. The AFC system contained the no-moving-parts TAU developed Suction and Oscillatory blowing (SaOB) actuator and a dedicated pressure supply system. A conceptual design phase scaling up the actuator and pressure supply system in order to minimize pressure losses and maintain performance, primarily of the boundary layer suction mechanism, was conducted by the TAU team. A comprehensive bench-top characterization study, performed on a single AFC system block—containing one actuator made by rapid prototyping technique, assured suitable performance over the entire parameter space. This design was then adapted by the ATD team for ease of assembly, cost savings and functionality in the road tests, at the cost of increased pressure drop. Due to the huge parameter space, at least 81 data points with repeat runs were needed to cover the test matrix. The Tagucci L9 matrix was attempted, but even that required some 11 points with at least 2 "good" repeats each. Eventually 7 "good runs" were performed. Out of these, one condition reduced the fuel consumption by about 5 %, or 1.75 L/100 km. When accounting for the flow power and pump efficiency this is reduced to 1 L/100 km. However, due to poor overall system efficiency, net energy saving was not achieved yet. All performance parameters were measured during the road tests. This enabled us, using post-test bench-top experiments to validate the assumption of 33 % pressure drop due to small diameter flexible tubes at the actuator inlet. This could be easily replaced with double diameter tube, reducing the pressure loss on that tube to a mere 1.6 %.

Taking into account that we are far from exploring the parameter space and the potential for increasing the pumping system efficiency, we are confident that 7–10 % net fuel savings using this innovative, yet complex, technology is within reach. Furtheremore, there is no competitive technology so close to market for roll-up door trailers.

Acknowledgments The authors would like to thank O. Stalnov, I. Fono, V. Palei, S. Pastuer, A. Blas, A. Kronish, S. Balvis, S. Moshel and M. Vaserman for the great scientific and technical support. The study was supported by AFC Technologies, Israel in collaboration with Ephicas, Delft, NL and managed by RAMOT of Tel Aviv University. Four related patents (one approved and three pending) exist. The road tests were funded in part by grants from New York and California states.

References

1. Arwatz, G., Fono, I., Seifert, A.: Suction and oscillatory blowing actuator. In: Morrison, J.E., Birch, D.M., Lavioe, P. (eds.) Proceedings of IUTAM Symposium on Flow Control and MEMS, Sept 2006, pp. 33–44. Springer, London (2008)
2. Arwatz, G., Fono, I., Seifert, A.: Suction and oscillatory blowing actuator. AIAA J. **40**(5), 1007–1017 (2008)
3. Cattafesta, L.N., Sheplak, M.: Actuators for active flow control. Ann. Rev. Fluid. Mech. **43**, (2011)
4. Collis, S.S., Joslin, R.D., Seifert, A., Theofilis, V.: Issues in active flow control: theory, simulation and experiment. Prog. Aero Sci. **V40**, N4–5 (2004) (previously AIAA paper 2002–3277)
5. El-Alti, M., Chernoray, V., Kjellgren, P., Hjelm, L., Davidson L.: Computations and full-scale tests of active flow control applied on a Volvo truck-trailer. In: Dillmann, A. (ed.) Proceedings of the 3rd Heavy Vehicle Drag Reduction Conference (2010)
6. Englar, R.J.: Advanced aerodynamic device to improve the performance, economics, handling and safety of heavy vehicles. SAE paper 2001-01-2072 (2001)
7. Heinz, N., King, R., Gölling, B.: Robust closed-loop lift control on an industry-relevant civil aircraft half model. In: King, R. (ed.) Active Flow Control II 2010. NNFM, vol. 108, pp. 125–139. Springer, Heidelberg (2010)
8. Hsu, T.-Y., Hammache, M., Browand, F.: Base flaps and oscillatory perturbations to decrease base drag. In: Proceedings of the UEF Conference on the Aerodynamics of Heavy Vehicles:Trucks, Buses and Trains, Lecture Notes in Applied and Computational Mechanics, Springer, Heidelberg, Sept 2004
9. Kehs, J.P., Visser, K.D., Grossman, J., Horell, C., Smith, A.: Experimental and full scale investigation of base cavity drag reduction devices for use on ground transport vehicles. In: Dillmann, A. (ed.) Proceedings of the 3rd Heavy Vehicle Drag Reduction Conference (2010)
10. McCallen, R.C., Salari, K., Ortega, J., DeChant, L., Hassan, B., Roy, C., Pointer, W.D., Browand, F., Hammache, M., Hsu, T.Y., Leonard, A., Rubel, M., Chatalain, P., Englar, R., Ross, J., Satran, D., Heineck, J.T., Walker, S., Yaste, D., Storms, B.: Doe's effort to reduce truck aerodynamic drag-joint experiments and computations lead to smart design. AIAA paper, June 2004
11. Pack Melton, L.G., Yao, C.S., Seifert, A.: Active control of flow separation from the flap of a supercritical airfoil. AIAA J. **44**(1), 34–41 (2006) (previously AIAA Paper 2003–4005)
12. Prandtl, L.: Motion of fluids with very little viscosity. In: Third International Congress of Mathematicians at Heidelberg, 1904, from Vier Abhandlungen zur Hydro-dynamik und Aerodynamik, pp. 1–8, Gottingen, 1927, NACA TM-452, March 1928
13. Seifert, A., Pack, L.G.: Active control of separated flow on a wall-mounted "hump" at high reynolds numbers. AIAA J. **V.40**(7), 1363–1372 (2002) (Part of AIAA paper 99–3430)
14. Seifert, A., Pastuer, S.: Method and mechanism for producing suction and periodic flow. US Patent 2006–0048829-A1. Granted 2005
15. Seifert, A., Stalnov, O., Sperber, D., Arwatz, A., Palei, V., David, S., Dayan, I., and Fono, I.: Large trucks drag reduction using active flow control. In: McCallen, B. (ed.) Proceedings of the 2nd heavy vehicle drag reduction conference, pp. 115–134. ISBN 978-3-540-85069-4 (2007)
16. Seifert, A., Stalnov, O., Sperber, D., Arwatz, A., Palei, V., David, S., Dayan, I., Fono, I.: Heavy trucks drag reduction by active flow control. AIAA paper 2008–0743, Jan 2008
17. Seifert, A., Darabi, A., Wygnanski, I.: Delay of airfoil stall by periodic excitation. J. Aircraft. **33**(4), 691–699 (1996)
18. Traffic safety facts: large trucks. DOT HS 810 619 (2005)

Part X
Bluff Body, Wake and Jet Flows

Comparison of Experimental and Numerically Obtained Flow Properties of a Bluff Body

Gandert M.R. van Raemdonck, Peter van Leeuwen and Michel J.L. van Tooren

Abstract The expected road transport demand in the next twenty years and the increasing environmental constraints together with the rising fuel prices has renewed the interest in truck design; any reduction in truck fuel consumption can be associated with large annual fuel cost reduction and considerable emission savings. Within the development of aerodynamic solutions numerical analysis tools, based on RANS equations, are often used to indicate flow phenomena and characteristics to design low drag bluff bodies. The presented work will discuss the similarities, but mainly the differences between wind tunnel experiments and the time-averaged numerical analysis. Rear pressure distributions are completely different when the numerical outcome is compared with the wind tunnel experiments. The CFD analysis of the boundary layer thickness is within acceptable resemblance with the wind tunnel measurements and the analytical power law model results. Stereoscopic PIV results show different wake structures.

1 Introduction

The expected road transport demand in the next twenty years and the increasing environmental constraints together with the rising fuel prices has renewed the interest in aerodynamic truck and trailer design in the last decade; any reduction in truck fuel consumption can be associated with large annual fuel cost reductions and considerable emission savings for the transport sector. The environmental concerns and the

G.M.R. van Raemdonck (✉) · M.J.L. van Tooren
Faculty of Aerospace Engineering (TU Delft), Kluyverweg 1, 2629 HS Delft, The Netherlands
e-mail: g.m.r.vanraemdonck@tudelft.nl

M.J.L. van Tooren
e-mail: m.j.l.vantooren@tudelft.nl

P. van Leeuwen
Faculty of Mechanical, Martime and Materials Engineering (TU Delft), Mekelweg 2, 2628 CD Delft, The Netherlands
e-mail: p.m.vanleeuwen@tudelft.nl

A. Dillmann and A. Orellano (eds.), *The Aerodynamics of Heavy Vehicles III*,
Lecture Notes in Applied and Computational Mechanics 79,
DOI 10.1007/978-3-319-20122-1_25

harsh competition force the transport companies to reduce fuel consumption in an economically and environmentally sustainable way.

Generally there are two ways to reduce the fuel consumption of a vehicle. One can improve the efficiency of the engine delivering power: improvements on the side of the available power. Or one can lower the different forces acting on a truck travelling over the road: the required power side. The latter can be achieved by reducing the weight of the vehicle, reducing its aerodynamic drag and by reducing the friction resistance of the tires.

Improving the fuel economy of trucks by aerodynamic means has become an accepted practice in the last decades. Truck manufacturers improved the aerodynamic performance of their tractor by applying roof deflectors, side fenders and corner vanes. Many aerodynamic solutions have been developed for the front and top of the tractor and for the gap between the tractor and trailer, [1, 6, 19].

Within the development of these aerodynamic solutions numerical simulations, based on steady Reynolds Average Navier-Stokes equations (RANS), are often used to indicate flow phenomena and characteristics. In order design low drag bluff bodies more insight in the flow behaviour is required as well as in the prediction capability of the used numerical tools.

In 2005 a research program was initiated at the faculty of Aerospace Engineering to reduce the fuel consumption of heavy duty vehicles by lowering the aerodynamic drag of articulated trucks. These types of vehicle are mostly used for long-haul road transport and have a high aerodynamic drag contribution due the bluff shape of the trailer and its high average travelling velocity. The presented research discusses the flow characteristics of a standard model, called Generic European Transport System, by comparing numerical simulations and wind tunnel experiments. The obtained (dis)similarities of several design examples will be discussed.

2 Set-Up

2.1 Truck Model: Generic European Transport System

A new wind tunnel model, in analogy with the GTS model analysed by Gutierrez [7] and Storms [15], is designed that meets the different requirements in order to perform the desired research. This new generic model is based on a European tractor semi-trailer combination used for international road transport and is referred to as GETS, Generic European Transport System, see Fig. 1. The guide lines of the Society of Automotive Engineers concerning wind tunnel testing of bluff bodies SAE J1252 [14] are considered with respect to Reynolds number and blockage effects. The dimensions of the new model are defined and displayed in Fig. 2

The Reynolds number for a European full-scaled truck, based on the square root of the frontal area of $A = 10.34$ m^2, a driving velocity of 25 m/s, air density of 1.225 kg/m^3 and an air viscosity of 1.7894×10^{-5} kg/ms, becomes 5.5×10^6. The

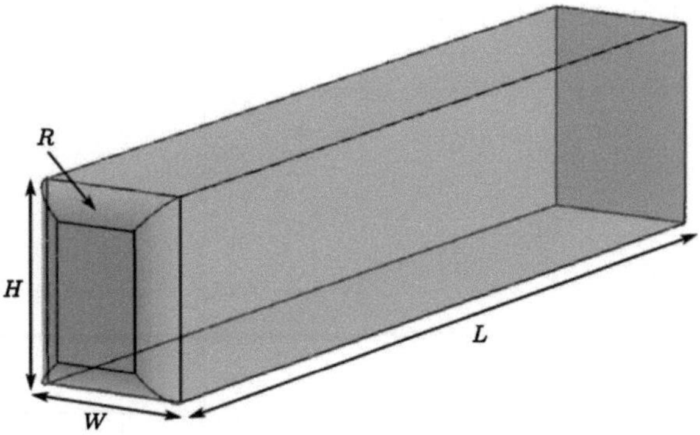

Fig. 1 Generic European transport system model

Fig. 2 Wind tunnel model dimensions

length L [mm]	1100
width W [mm]	173
height H [mm]	234
ground clearance [mm]	33
front radius R [mm]	36
scale	1:15
ratio frontal area model-section	0.02
ratio width model-section	0.22
ratio height model-section	0.24
test velocity V_t [m/s]	60
Reynolds number, $\sqrt{A}$	0.8264×10^6

Reynolds number in the wind tunnel, with a testing velocity of 60 m/s is 0.8264×10^6 and meets the requirements set by SAE J1252 [14] for bluff bodies.

Front edge separation is undesirable within this research, therefor several radius ratios based on the width of the body and the frontal radius are recommended. Within the experiments of Cooper [3] and Henneman [8] a study is executed on front edge separation of bluff bodies within a certain range of radius ratios. Their experiments showed that a certain transcritical Reynolds number $(Re_r)_t$ based on the radius of the front edge has a constant value of 1.24×10^5 for the turbulence intensity present in the used wind tunnel test section. This transcritical Reynolds number can be defined as

$$(Re_r)_t = \frac{\rho r V_t}{\mu} \tag{1}$$

where ρ is the air density, μ the air viscosity, V_t the test velocity and r resembles the radius of the front edge. In order to calculate the radius of the frontal edges with the above Eq. 1, V_t is set on 50 m/s to increase the velocity range where the boundary layer stays attached. This gives a radius of 36 mm for all the frontal edges.

With a microphone is indicated that an attached and turbulent boundary layer is present at side of the model. Oil visualization identified a small separation bubble just behind the front radius. The separation bubble disappeared after adding zig-zag tape (0.65 mm thickness) in front of the curvature, van Raemdonck [20].

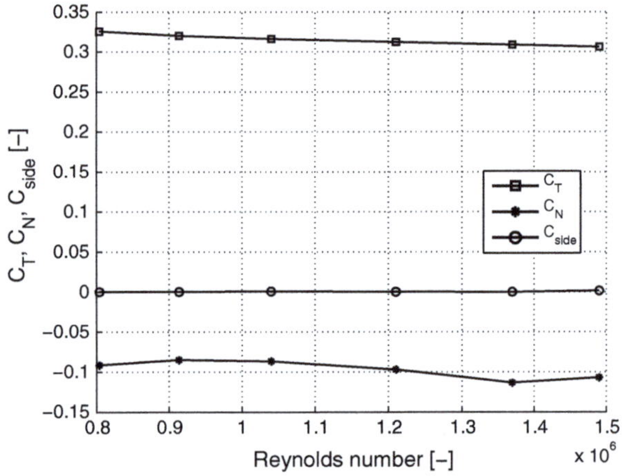

Fig. 3 Reynolds effects

The GETS model is tested within a velocity range of 50–100 m/s in order to detect possible Reynolds effects. As Fig. 3 illustrates no large Reynolds effects are detected within the velocity range.

2.2 Evaluation of Numerical Tools

A surface model of the GETS model is generated based on the dimensions summarized in Fig. 2. Discretization of the volume around the model is done by introducing blocks. A structured surface mesh is extruded normally to create a structured boundary layer mesh. The rest of the volume around the model (length = 25 width, half width = 8 width; height = 11 width) is filled with tetrahedrals. A half model, except for the yaw angle variation, is used for the simulations in order to reduce the amount of cells (total number of cells: 6 million) and the corresponding processing time (Fig. 4).

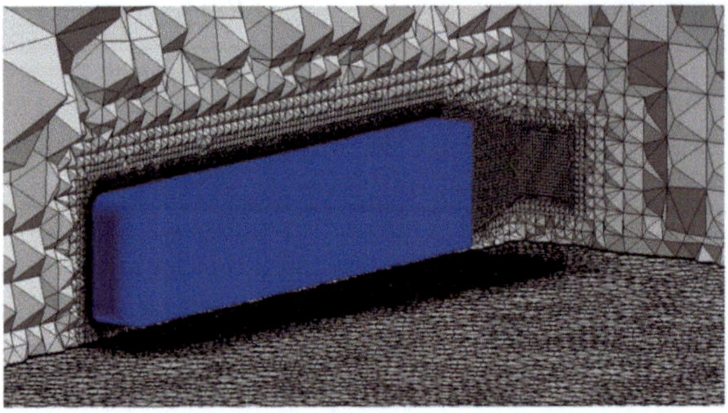

Fig. 4 CFD surface model together with volume mesh

The Reynolds Averaged Navier-Stokes flow equations together with the Realizable $\kappa - \varepsilon$ turbulence model are solved with the commercial package Fluent. There is opted for this type of turbulence model due to the its common usage in the automobile sector, its performance with bluff bodies and its relatively low required computer power, Lanfrit [10]. As near-wall treatment for the boundary layer, the non-equilibrium wall function, is chosen ($30 < y^+ < 300$). The operating pressure of 101325 Pa is defined in one of the outer corners of the computational domain.

The upstream inlet has the velocity inlet boundary condition: an absolute magnitude of 60 m/s in the direction of the flow for 1:15 scaled set-up; 25 m/s for the full-scale situation. The outlet is set as pressure outlet. The actual symmetry plane as well as the other two outer planes (side and upper plane) have the symmetry boundary condition. The floor is defined as a moving wall with a translational velocity as well as a fixed wall or symmetrical plane boundary conditions are simulated.

2.3 Experimental Modeling

The wind tunnel experiments are executed in the atmospheric Low Turbulence Tunnel. This closed circuit wind tunnel has an octagonal test section with a cross sectional area of 2.07 m^2 (width of 1.8 m; height of 1.25 m, Fig. 5) and with parallel wind tunnel walls. The maximum operating velocity of the wind tunnel is 120 m/s; the turbulence intensity can be changed in the range 0.02–0.1 %. The empty test section is calibrated for the wind velocity with the aid of a pitot tube by measuring the dynamic pressure in the centre of the test section and compare it with a static pressure difference between two locations in front of the test section.

The wind tunnel used is not equipped with a moving belt. According to Cooper [4, 5] one can conclude that a fixed-floor with a thinned boundary layer is sufficient for current automotive and commercial vehicle applications. The vehicle model is suspended on a parallel ground board which has an offset of 300 mm with respect to the upper wind tunnel wall and has the same width as the test section, Fig. 5. On the rounded front edge of this ground board develops a new thinner boundary layer.

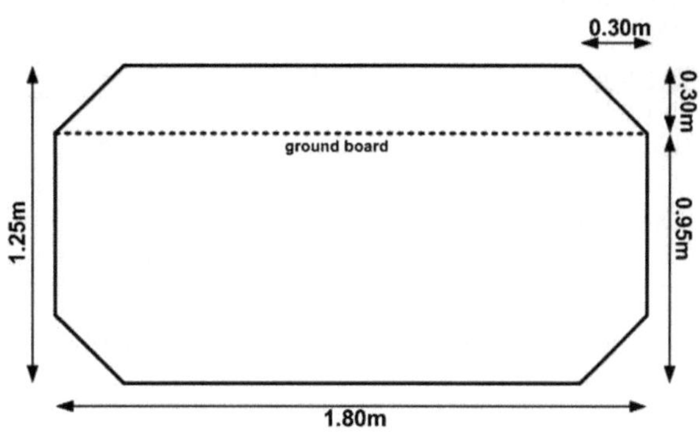

Fig. 5 Cross section low turbulence tunnel

3 Flow Properties of the GETS Model

In this section several flow properties will be analysed and discussed. A comparison will be made between the results of the numerical and the wind tunnel analysis, both for 1: 15 scaled model with a velocity of 60 m/s.

3.1 Force and Pressure Coefficient

An overview of the drag and lift coefficient numerically simulated with different mesh and cell types is given in Table 1. The listed simulations are performed by different students involved in this research project. The difference in the outcome for the coefficients illustrate one has to be careful comparing the absolute drag values.

The bottom half of Table 1 illustrates the corrected (for blockage effects in closed test section) and uncorrected force and pressure coefficient obtained with wind tunnel experiments. A corrected drag and lift coefficient of respectively 0.297 and −0.036 is observed, while a mean rear pressure coefficient of −0.163 is measured.

Comparing the numerical simulations and wind tunnel experiments one can state that the drag and the mean rear pressure coefficients are rather in line which each other, except for the hexahedron mesh. Also a large difference between the simulated and measured lift coefficient can be observed. This is probably due to the fact the wind tunnel is not equipped with a rolling belt to simulate a moving road.

Table 1 Numerical and wind tunnel results: force and pressure coefficients

Numercial results					
Grid type	Scale	Velocity (m/s)	C_D	C_L	$C_{P,base\ mean}$
Tetrahedron [13]	1:15	60	0.3017	−0.1348	−0.1048
Tetrahedron [18]	1:15	60	0.277	−0.136	−0.145
Tetrahedron [18]	1:1	25	0.318	−0.100	−0.133
Polyhedron [13]	1:15	60	0.3023	−0.1412	−0.1194
Hexahedron [17]	1:15	60	0.2399	−0.1635	−0.1483
Wind tunnel results					
	Scale	Velocity (m/s)	C_D	C_L	$C_{P,base\ mean}$
Uncorrected	1:15	60	0.327	−0.080	−0.145
Corrected	1:15	60	0.297	−0.036	−0.163

The numerical simulations resulted in a typical drag breakdown of the GETS model as for a bluff body: 18 % friction drag and 82 % pressure drag, van Leeuwen [18]. The frontal surface, which includes the rounded edges, contributes 41 % which is almost as much as the rear surface: 42 %. The center body only experience friction drag and accounts for 17 % of the total drag of the vehicle.

3.2 Boundary Conditions

With the aid of CFD different boundary conditions for the full-scale GETS model are simulated. Only small relative differences are obtained when a moving ($C_D = 0.318$) or symmetric floor ($C_D = 0.320$) are compared with a stationary wall boundary condition ($C_D = 0.320$). The difference in lift coefficient are larger: −0.101, −0.093 and −0.099 for a moving, stationary and symmetric floor respectively.

Figure 6 illustrate how the velocity profile changes near the floor with different boundary conditions for different downstream locations in the near wake. In case of a symmetric floor no boundary layer is formed, while the boundary layers for the moving and stationary floors differ significantly. The velocity in the wake area is below freestream velocity therefor the flow is accelerated by the floor in the moving wall situation opposed to decelerated in the stationary wall simulation.

Although the wake velocity profiles show good agreement between the moving and symmetric wall conditions, except near the ground, the drag coefficient of the symmetric wall is closer to the fix wall simulation. This is remarkable as the velocity profiles of the fixed and the symmetric wall differ more than expected from the difference in C_D value. For the moving wall simulation the suction peak on the lower curved front face is higher compared to the symmetrical case. This suction peak can be explained by the effect that a boundary layer develops on the wall under the model creating a displacement thickness which accelerates the flow passing through the floor and the model, leading to a higher suction peak.

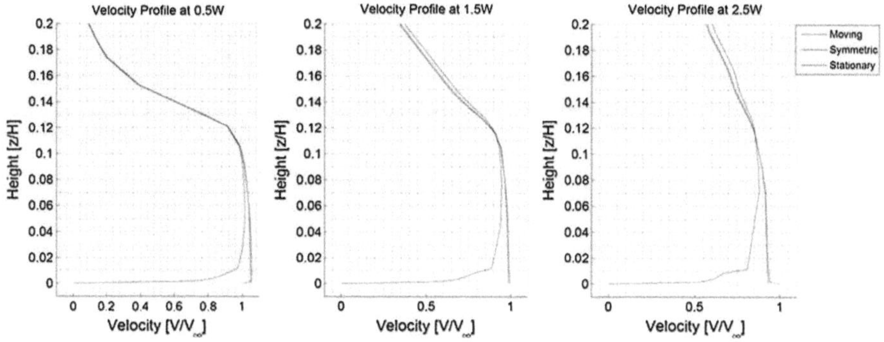

Fig. 6 Rear velocity profiles for different boundary conditions

3.3 Cross Wind Influence

In real case situations a truck is always experiencing cross winds. Cross winds can have a large influence on the total drag C_T and side force C_S due to flow separation on the front-end of the truck.

The force and the mean base pressure coefficients are shown in Figs. 7 and 8 for a certain yaw angle range. The drag coefficient first increases, but remains approximately constant after 6° while the side force coefficient increases. The increased suction peak on the leeward front edge curvature creates a larger forward suction as well as a side ward suction force. The CFD and wind tunnel results follow the same trend, however, the uncorrected data is offset compared to the numerical drag coefficient.

In Fig. 8 the mean base pressure coefficient is illustrated, which decreases with increasing yaw angle, increasing the suction force on the back of the model. With increasing yaw angle the difference between the numerical simulations and the wind tunnel experiments seems to become larger.

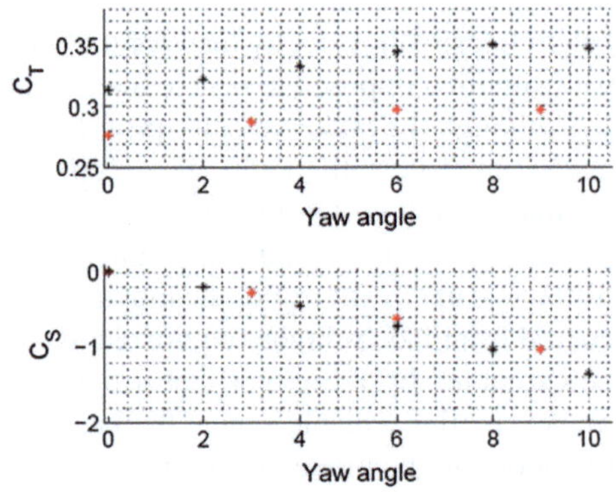

Fig. 7 Drag and side forces in relation with increasing yaw angle

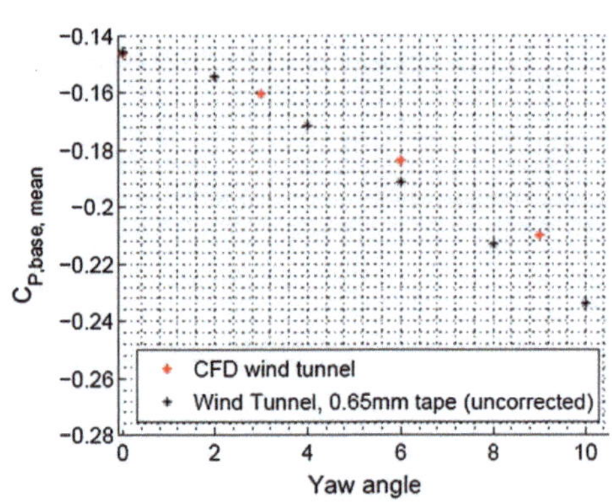

Fig. 8 Rear pressure coefficient in relation with increasing yaw angle

3.4 Boundary Layer Properties

The boundary layer properties on the side surface 15 mm ahead of the rear edge is measured (van Raemdonck, [20]) and compared to simulated values (van Leeuwen, [18]) in Figs. 9 and 10. The latter illustrates different velocity profiles indicating the shape of the simulated, measured and theoretical boundary layers.

The wind tunnel measured values of $\delta,\ \delta^*,\ \theta\ and\ H$ are in close agreement with the theoretical turbulent flat plate equations. The simulated boundary layer on the scaled model is thicker ($\delta/W = 147$) compared to the flat plate estimation ($\delta/W = 110$) and the wind tunnel measured value ($\delta/W = 105$). Also the displacement thickness and momentum thickness are higher compared to the measured and estimated values. The shape factor H is higher in case of the simulated results, however, it must be noted that, as can be seen in the Fig. 10, there is an irregularity in the simulated results for the wind tunnel boundary layer. This irregularity is due to the interface of the hybrid mesh and the interpolation during the data export.

Compared to the 1/15th scale simulation the full-scale simulation has much fuller profile, also indicated by the lower shape factor H. The difference can be explained by

Fig. 9 Wind tunnel model dimensions

	δ/W $(1e^{-3})$	C_f $(1e^{-3})$	δ^*/W $(1e^{-3})$	θ/W $(1e^{-3})$	H
CFD (1:1)	94	1.62	6.6	5.9	1.13
CFD (1:15)	147	2.51	15.6	11.6	1.34
Wind tunnel	105	-	14	10.4	1.3
Flat plate	110	2.70	13.6	10.7	1.278

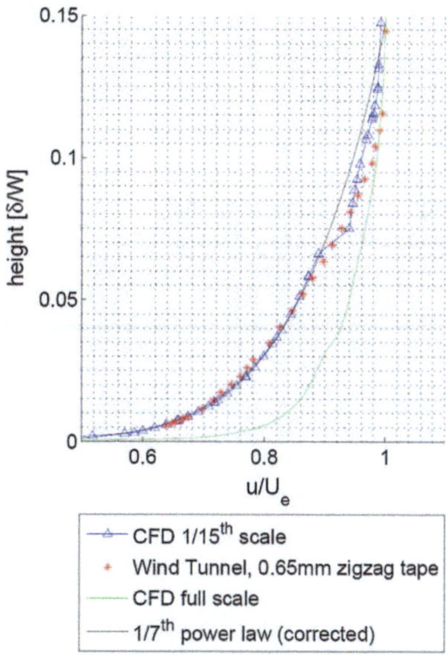

Fig. 10 Boundary layer profiles

the low freestream turbulence intensity of the wind tunnel compared to the full scale simulations. Numerical diffusion leads to thicker boundary layers at high Reynolds numbers.

3.5 Wake Structure and Streamlines

Figures 11 and 12 illustrate the wake structure obtained with the aid of stereoscopic PIV (Particle Image Velocimetry) and CFD simulation respectively, van Dijk [16]. Both figures show the horizontal velocity magnitude and a projection of the path lines. In general one can observer the two counter rotating vortices inducing a negative static pressure in the near wake of the GETS model.

When one compares both wake structures more closely one can observe that the numerical simulation underestimates the thickness of the boundary layers of the model, especially at lower side and the ground plate. This may result in an overestimation of the size of the lower vortex and underestimation of the upper

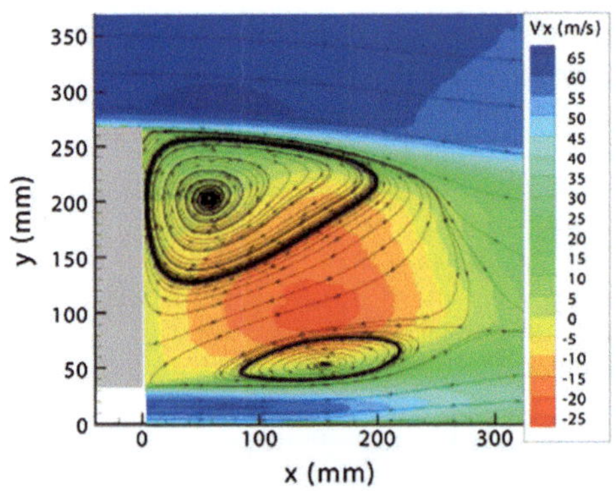

Fig. 11 PIV result of the symmetry plane of the wake

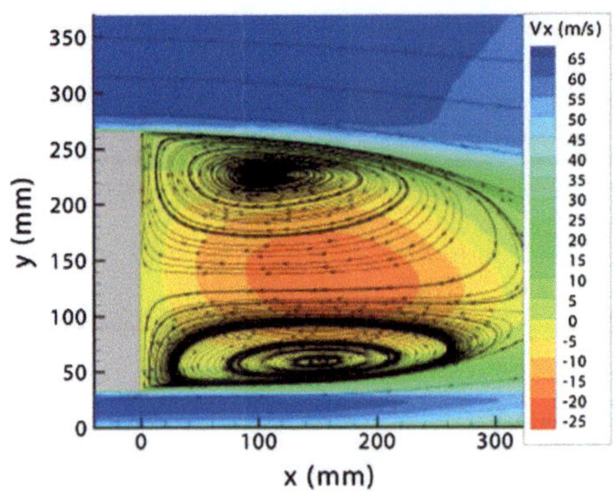

Fig. 12 CFD result of the symmetry plane of the wake

vortex. Clearly visible is that the lower vortex measured in the wind tunnel, see Fig. 11, is smaller than the lower vortex obtained with CFD.

The vertical asymmetry of the two counter rotating vortices is resulting in upstream flow paths that are angle downwards, in stead of horizontally oriented in the numerical simulations. Also the length of the near wake obtained with PIV measurement is much smaller compared with the near wake length resulting from the CFD analysis.

3.6 Rear Pressure Distribution

The rear pressure distribution of the GETS model, obtained with the aid of wind tunnel experiments and numerical simulations are shown in Fig. 13. As can be seen in the contour plots the pressure distribution measured in the wind tunnel differs significantly compared to the simulated distribution obtained with the aid of CFD. There is no pressure maximum close to the center of the base for the wind tunnel results: a more lateral pressure distribution is obtained. Also the minima and maxima of the measured values are completely different compared to the CFD values.

In the case of 6° yaw, see Fig. 14, the flow remains attached on the front of the GETS model, whereas it separates in the wind tunnel data, causing a more asymmetric pressure distribution in the wind tunnel data compared to the CFD results.

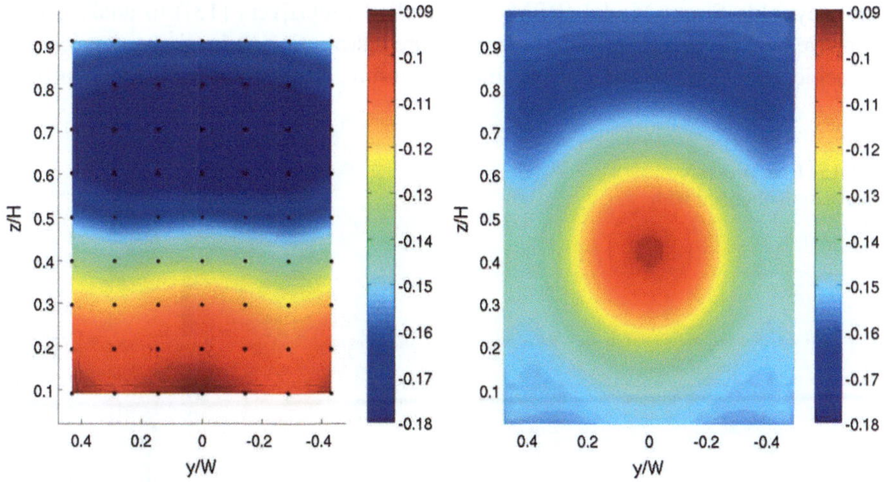

Fig. 13 Rear pressure distribution: (l) wind tunnel experiments and (r) CFD simulation

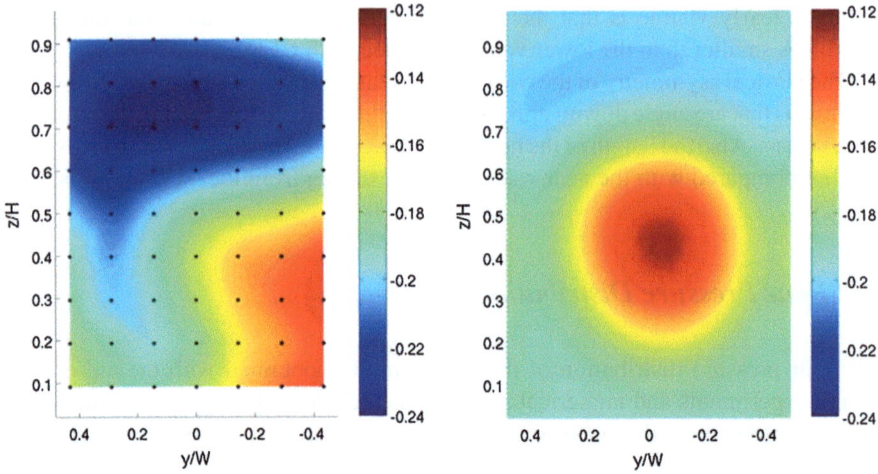

Fig. 14 Rear pressure distribution at 6 yaw angle: (l) wind tunnel test and (r) CFD simulation

3.7 Turbulence Modeling

In the above comparison analysis steady RANS equations are solved with the aid of the realizable $\kappa - \varepsilon$ turbulence model. Two other turbulence models, SST $\kappa - \omega$ and Reynolds Stress Model (RSM), are applied (Schrijvers [13]) to analyse if an improvement in the prediction of rear pressure distribution and wake structure could be achieved. As expected the obtainded pressure distribution with the two other

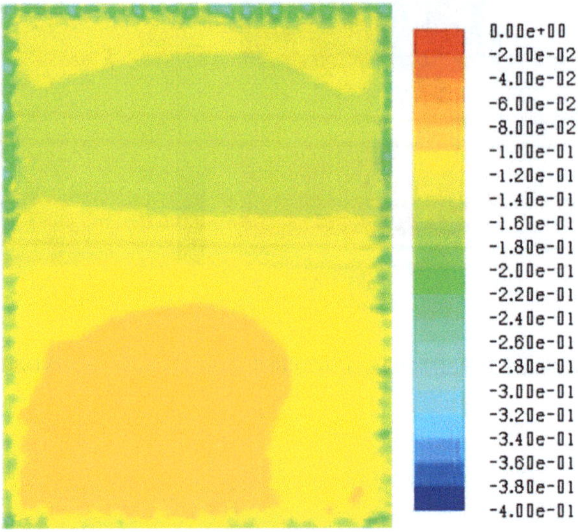

Fig. 15 Rear pressure distribution obtained with LES

turbulence models did not match the rear pressure distribution obtained with wind tunnel experiments.

With the RANS mesh a first attempt is undertaken to conduct a Large Eddy Simulation (LES). The velocity is set on 6 m/s and a time step of 1×10^{-4} s is defined. The time-averaged pressure distribution can be observed in Fig. 15. Although the applied mesh was too coarse for this simulation and a velocity of 6 m/s is set, one can see that the LES rear pressure distribution is more similar to the wind tunnel measurements. Identatical observations are made by Krajnović [9].

4 Design Examples

Different flow properties of the GETS model obtained with numerical simulations and wind tunnel experiments are discussed and compared. In the next part different design examples, to obtain a low drag heavy duty vehicle, will be analysed. Typical flow characteristics like pressure distribution and drag coefficient will be analysed. Numerical simulations (based on RANS) will be compared with wind tunnel tests. Both set-ups are executed with 1:15 scaled models and an inlet velocity of 60 m/s.

4.1 Standard and Stepped Boat Tails

Boat tails are a well known concept to reduce the drag of heavy duty vehicles. In this research project different boat tail configurations (i.e. slant angle, offset distance and tail lenght variations) are tested in the wind tunnel and simulated by Vonk [21].

The standard boat tail with flush side panels showed a drag reduction of 39 % during the wind tunnel experiments and in the numerical simulations for the GETS model. Figure 16a, b display the difference in flow between the wind tunnel results and that of the simulations for the standard boat tail. It can be noted that the location of the upper vortex is comparable as is the main direction of the flow. Although the CFD prospects a flow which is more horizontally directed, where the flow of the PIV analysis shows a flow directed towards the upper inner part of the boat tail. Due to the shadow of the upper boat tail element the PIV results do not show the second vortex which should be, conform the numerical simulations, located inside the cavity. Again at the lower side one can observe lower velocities in the wind tunnel experiments induces a different flow field.

The best performing stepped boat tail only indicated a drag reduction of 10 %, Vonk [21]. Figure 16c and d illustrate the velocity field together with the streamlines for a stepped boat tail configuration. Due to the shadow of the stepped tail, the lower small vortex is not visible. The upper vortex structure is clearly visible in the PIV experiments while the CFD simulaitons has difficulties to indicate the structure. The two counter rotating structures in the near wake of the stepped boat tail are clearly present, both in the PIV results and the numerical simulations. The closure of the

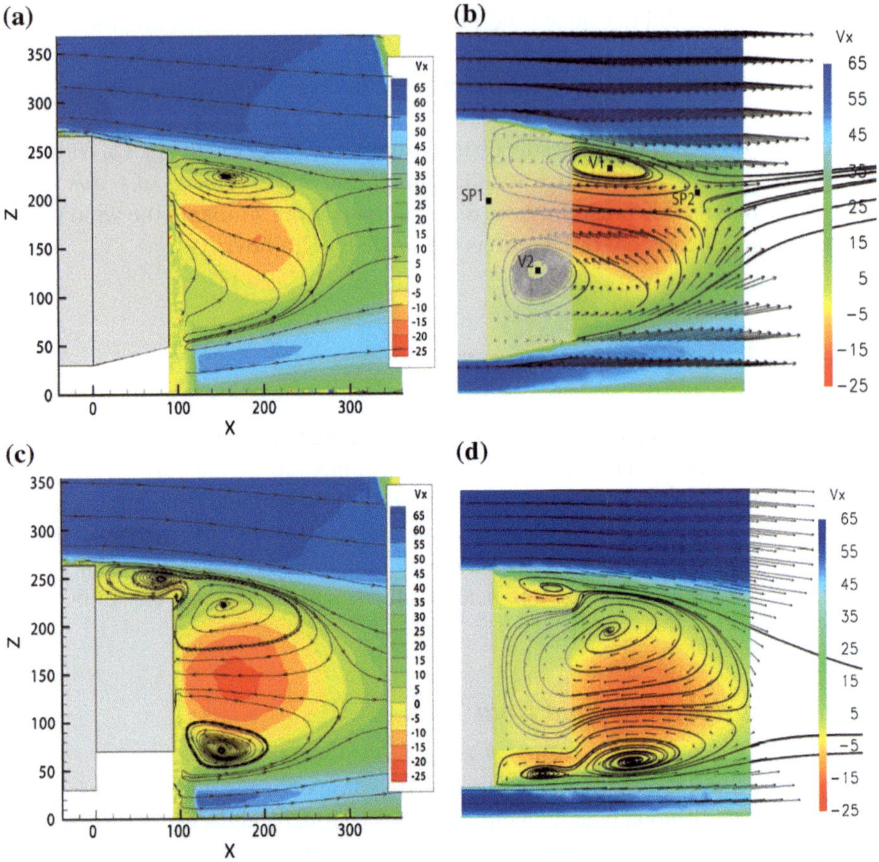

Fig. 16 Streamlines of standard and stepped boat tail on the GETS model **a** PIV result of a standard boat tail. **b** CFD result of a standard boat tail. **c** PIV resuls of a stepped boat tail. **d** CFD result of a stepped boat tal

near wake of the PIV measurement seems to be shorter compared with the numerical simulations. Figure 16c indicates the existance of extra vortex at the rear edge of the stepped boat tail which is not captured with the numerical simulations. This extra vortex is necessary to counteract the two vortices in respectively the back step and the cavity of the tail.

4.2 Tanker Trailer

Beside rectangular shaped models like the GETS model, also cylindrical shaped vehicles are considered in the research project. Tanker trailer transporting for instance liquids or fluids, use cylindrical shaped trailers. The flow behavior around this type of vehicles is analysed by Saat, [12].

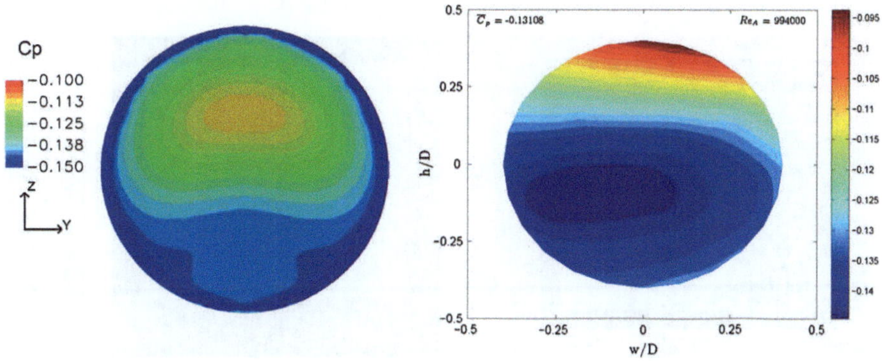

Fig. 17 Rear pressure distribution of a tanker trailer: (l) CFD and (r) wind tunnel analysis

The numerical and experimental pressure distributions at the base of the tank are compared in Fig. 17. As can be seen the general pressure distribution is similar for both analysis tools. The general pressure distribution can be described by relatively low suction at the top of the tank and high suction on the lower. Nevertheless, the difference is that in the numerical C_P-plot the rear stagnation point is located lower on the base, when compared to the measured pressure coefficients. This indicates that in the wind tunnel the lower vortex of the wake is larger in size.

4.3 Combi-Vehicle

Another type of vehicle that has been analysed is the combi-vehilce constisting of a rigid truck together with a drawbar trailer, Buijs [2]. The pressure distribution of the rear surface of the rigid truck and the front surface of the trailer will be compared.

The pressure distributions, see Fig. 18a, b, on the gap surfaces show good resemblances between the wind tunnel measurements and the numerical simulations. The high and lower pressure regions, for both the rigid truck rear (vertical orientation) and trailer front (V-shape orientation) are comparible between the wind tunnel experiments and the numerical simulations. This suggests that the gap flow, for a combi-vehicle is well predicted by numerical analysis, which is not the case for the rear surface of the trailer.

4.4 Rear Shape Modification

For the GETS model a rear shape modification study is executed by Lauwers [11] to analyse different aft concepts and its influence on the drag coefficient. The drag coefficients of the corresponding aft modifications for both CFD and wind tunnel analysis are shown in Fig. 19. The similarity in the trends is clearly notable, when a drag reduction is found with CFD analysis, this is also measured in the wind tunnel.

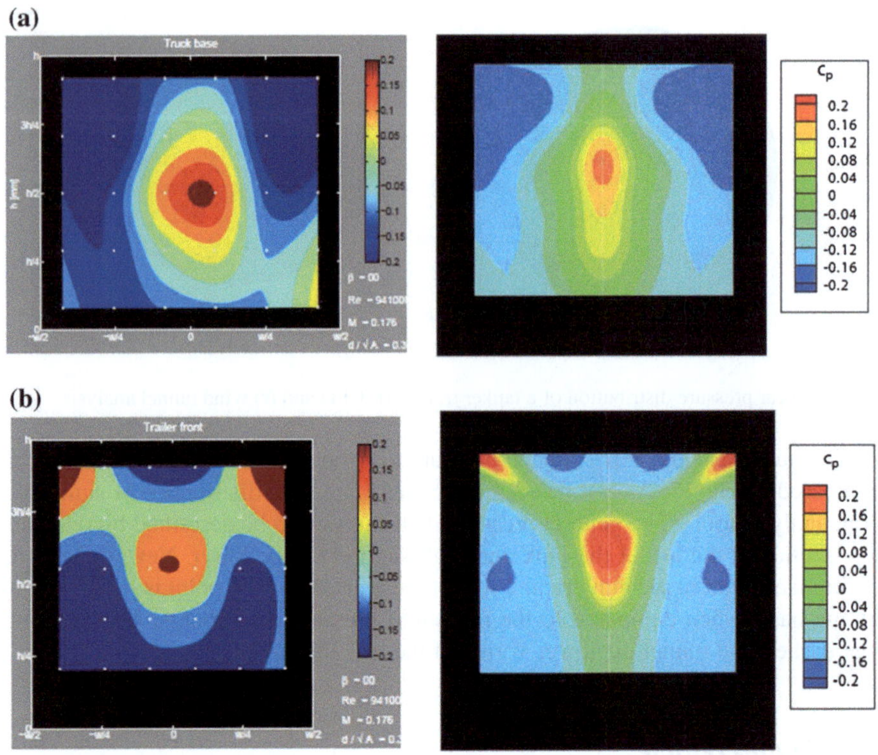

Fig. 18 Pressure distribution truck base (*upper*) and trailer front (*lower*) **a** Pressure distribution rigid truck base (*left*) wind tunnel experiment and (*right*) numerical simulation. **b** Pressure distribution trailer front (left) wind tunnel experiment and (*right*) numerical simulation

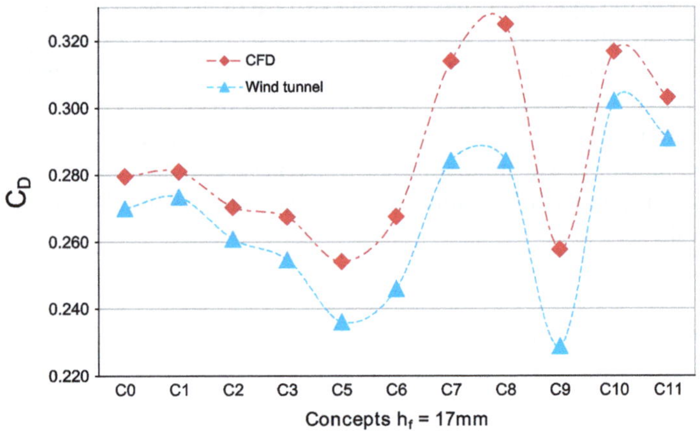

Fig. 19 Comparison of CFD and wind tunnel results for the rear body modifications

The relative improvements of a new concept found by CFD and experimental analysis are very similar. Especially changes in drag coefficients for concept C1, C2, C3, C10 and C11 (all upper aft edge modifications) are almost the same during both methods of analysis. For concepts C5, C6 and C9 (all with the tapered and angled aft surfaces) the CFD analysis predicts a more modest improvement compared to the wind tunnel experiment. The CFD analysis is more pessimistic about concepts C7 and C8 which have both rounded rear side edges and high adverse pressure gradients.

5 Conluding Remarks

A comparison of different flow characteristics of the GETS model is made with the aid of numerical simulations, based on Reynolds Averaged Navier-Stoke equations and wind tunnel experiments.

The absolute numerical obtained values of the drag, lift and mean base pressure coefficient are not exactly in line with the measured coefficients in the wind tunnel. This is mainly due to the numerical modeling and the fact that the underbody flow is not good predicted with the aid of CFD. The drag and side force coefficients of the simulations and measurements agree well for the zero yaw angle case, but the trend agreement decreases for increasing yaw angle. The design example of the aft modifications illustrated that the trend of the different concepts are in good agreement. Only when rear surface modifications imply rear high pressure gradients the numerical simulation overpredict the forces and pressures.

The boundary layer properties, like displacement and momentum thickness and the shape factor, are predicted well with CFD when compared with the scaled analytical equations and wind tunnel experiments. Full-scale and thus high Reynolds number simulations have difficulties to match the theorethical model. Numerical diffusion results in relative thicker boundary layers.

The wake structure obtained from the PIV measurements and the CFD analysis is different in many ways, stating that path lines and flow structures of highly turbulent and separated flow simulated with RANS is insufficient. RANS equations forces the flow in the wake to be steady, while they are highly unstable. This differs from averaging the velocities from an unsteady flow obtained during the PIV experiments. The design example with the standard boat tail illustrated that the resemblance between CFD and wind tunnel experiments is in general positive. The extra vortex at the rear edge of the stepped boat tail is not captured with the numerical simulations. Indicating that detailed flow structures are insufficient simulated with the aid of RANS CFD.

The rear pressure distribution obtained with numerical simulatinos is completely off compared with experimental results. The wind tunnel measurements showed a more lateral pressure distribution while the numerical analysis indicated a central orientated distribution. Applying different turbulence modeling gave no better results. Only LES improved the resemblance with the wind tunnel experiments. The design example with the cylindrical body showed a nicer similarity. Also the pressure distribution in a gap indicated remarquely good similarity between the numerical simulations the wind tunnel experiments.

One has to act carefully when flow patterns and pressure distributions, obtained with RANS simulations of bluff bodies, are considered to design low drag vehicles. The drag and mean pressure coefficients are in good agreement with wind tunnel experiments. If steady numerical simulations are used in the preliminary design phase, only descions based on the relative force and mean pressure coefficients, and not on pressure distributions and flow patterns, should be made before the actual design in the wind tunnel.

Acknowledgments The first author would like acknowledge the contribution of his M.Sc. students to this work: Lennert Buijs, Kiki Lauwers, Patrick Saat, Patrick Schrijvers, Tobias van Dijk, Karl van Ginderdeuren and Arnout Vonk.

References

1. Browand, F., McCallen, R., Ross, J.: Aerodynamics of Heavy Duty Vehicles II: Trucks Busses and Trains. Springer, Berlin (2008)
2. Buijs, L.J.: Numerical and Experimental Analysis on Aerodynamic Solutions for Drag Reduction on Truck-Trailer Combinations. Master's Thesis, Delft University of Technology, Faculty of Aerospace Engineering, The Netherlands (2010)
3. Cooper, K.: The Effect of Front-Edge Rounding and Rear-Edge Shaping on the Aerodynamic Drag of Bluff Vehicles in Ground Approximity. SAE Technical Paper 850288 (1985)
4. Cooper, K.: The Wind Tunnel Testing of Heavy Duty Trucks to Reduce Fuel Consumption. SAE Paper 821285 (1982)
5. Cooper, K.: Bluff-body aerodynamics as applied to vehicle. J. Wind Eng. Ind. Aerodyn. **49**, 1–21 (1993)
6. Freight Best Practice: Aerodynamics of Efficient Road Transport. Department of Transport, UK (2003)
7. Gutierrez, W.T., Hassan, B., Croll, R.H., Rutledge, W.H.: Aerodynamics Overview of the Ground Transportation Systems (GTS) Project for Heavy Duty Drag Reduction. SAE Technical Paper 960906 (1996)
8. Henneman, B.: Modeling of Front Edge Flow Separation on Rounded Bluff Body Bodies Commercial CFD Software, Master's Thesis, Delft University of Technology, Faculty of Aerospace Engineering (2005)
9. Krajnović, S., Davidson, L.: Large-Eddy Simulation of the Flow around a Ground Vehicle Body. SAE Paper, 2001–01-0702 (2001)
10. Lanfrit, M.: Best Practice Guidelines for Handling Automotive External Aerodynamics with FLUENT. Version 1.2. http://www.fluentusers.com (2005)
11. Lauwers, K.: Computational and Experimental Analysis of Trailer Shape Modifications for Drag Reductions. Master's Thesis, Delft University of Technology, Faculty of Aerospace Engineering, The Netherlands (2009)
12. Saat, P.J.W.: Aerodynamic Drag Reduction of Semi-Trailer Tanker Trucks by Means of Tank Base Modification. Master's Thesis, Delft University of Technology, Faculty of Aerospace Engineering, The Netherlands (2010)
13. Schrijvers, P.: Optimizing Towards Minimal Drag of the Aerodynamic Package of a Truck. Master's Thesis, Delft University of Technology, Faculty of Mechanical, Maritime and Materials Engineering, The Netherlands (2010)

14. Society of Automotive Engineers. SAE Wind Tunnel Test Procedure for Truck and Buses. SAE J1252 (1981)
15. Storms, B.L.E.A.: An Experimental Study of the GTS Model in the NASA Ames 7-by 10-FT Wind Tunnel. NASA/TM-2001-209621 (2001)
16. van Dijk, T.N.: Analysis of the Relationship between Front Surface Roughness and the Unsteady Wake behaviour of a Bluff Body. Master's Thesis, Delft University of Technology, Faculty of Aerospace Engineering, The Netherlands (2010)
17. van Ginderdeuren, K.: Design of Guiding Vanes for Bluff Bodies with the aid of Computational Fluid Dynamics. Master's Thesis, Delft University of Technology, Faculty of Aerospace Engineering, The Netherlands (2010)
18. van Leeuwen, P.: Computational Analysis of Base Drag Reduction Using Active Flow Control. Master's Thesis, Delft University of Technology, Faculty of Aerospace Engineering, The Netherlands (2009)
19. van Raemdonck, G.M.R., van Tooren, M.J.L., Boermans, L.: Aerodynamics and Trucks. Platform for Aerodynamic Road Transport, Faculty of Aerospace Engineering—TU Delft, The Netherlands (2009)
20. van Raemdonck, G.M.R., van Tooren, M.J.L.: Time-averaged phenomenological investigation of a wake behind a bluff bofy. In: Bluff Body Aerodynamics and Application VI Conference, Milan, Italy (2008)
21. Vonk, A.F.M.: Stepped Boat Tails on Trucks. Master's Thesis, Delft University of Technology, Faculty of Aerospace Engineering, The Netherlands (2010)

Quantitative Flow Visualization Applied to a Passive Wake Control Problem

A. Landolt, D. Borer, A. Meier and T. Roesgen

Abstract The results of different approaches to quantitative flow visualization with large-scale capability are presented. The techniques were applied on the occasion of a measurement campaign in a medium-sized subsonic wind tunnel that addressed a passive wake control problem on an excursion boat. The obtained data was compared qualitatively with numerical results. Fluorescent tufts attached to the surface in the area of interest were filmed by a digital camera. Subsequent post-processing retrieved the local intensity variance and thus the tuft's movement. This allowed for both, the identification of local flow direction and regions of detached flow on the surface of the model. Helium-filled soap bubbles were tracked using an asynchronous temporal contrast sensor or Dynamic Vision Sensor (DVS). The sensor is recording temporal changes in intensity only and has a high dynamic range which made it possible to track the bubbles in ambient light conditions. The recordings yielded local flow velocity and the streamlines around the model. The pressure field or local flow direction were recorded by tracking hand-held probes in 3D space with a stereo vision system. The system allowed free placement and fast measurements close to the model's complex surface or in its wake. By moving the probe in the volume of interest and with suitable post-processing applied, a quick quantitative assessment of the pressure field and flow topology was possible. The application of the different techniques confirmed the potential of quantitative flow visualization in large-scale testing. The methods complemented each other in the sense that they served to extract surface flow, streamlines and pressure fields.

1 Introduction

Standard wind tunnel tests usually provide measurements of aerodynamic forces and pressures at discrete locations. The information about what is actually happening in the flow field—especially at high Reynolds numbers and in large facilities—can be

A. Landolt (✉) · D. Borer · A. Meier · T. Roesgen
Institute of Fluid Dynamics, ETH Zurich, 8092 Zurich, Switzerland
e-mail: landolt@ifd.mavt.ethz.ch

A. Dillmann and A. Orellano (eds.), *The Aerodynamics of Heavy Vehicles III*,
Lecture Notes in Applied and Computational Mechanics 79,
DOI 10.1007/978-3-319-20122-1_26

difficult to obtain. The standard tufts or smoke streak visualization techniques are easy to use and of great value for the operator's own understanding, but are by nature difficult to record and cannot provide quantitative data.

At the other end of the scale reside highly accurate point measurements (hot wire or laser Doppler anemometry) or field measurement techniques like Particle Image Velocimetry (PIV). Traversing the test volume is at best time consuming, the more so in large facilities or for complex model geometries. PIV hardware costs and setup time are considerable and usually exceed both the actual testing fees and -time by orders of magnitude. Furthermore, the percent-accuracy that would justify time and effort may not always be required. Flexibility, speed and immediate insight into the flow can be more important.

Bömmels et al. [1] discussed possible approaches on the occasion of the first issue of this conference. Along the same line follows the work presented by the MIRA group in a NewScientist article where soap bubbles were tracked with a motion capture system. Although little technical details are given, the system appears to be working at its sensitivity limits due to the bubbles' specular reflection and hence low visibility.

The approach to quantitative flow visualization (also computer aided flow visualization [2]) discussed here can be positioned somewhere in between the traditional visualization techniques like smoke streaks or tufts and the high-precision quantitative methods. The aim is to combine the easy setup and intuitive approach of the traditional visualization techniques (the probe is pointed at where things happen) with the access to quantitative volumetric data for documentation and rating of configurations.

The different methods discussed here were implemented and tested on the occasion of a measurement campaign dealing with a passive wake control problem on an excursion boat. The task was to improve passenger comfort on the upper deck where in certain conditions serving food was impaired by wind gusts. Thorough understanding of the flow topology from wind tunnel tests was mandatory. A CFD study complemented the information about the flow and allowed comparison with the experimental data.

2 Wind Tunnel Facility and Model Geometry

The subsonic wind tunnel of the Institute of Fluid Dynamics at ETH shown in Fig. 1 is of the closed return type with a closed test section of 3 m width and 2.1 m height. The flow can be driven up to a dynamic pressure of 2300 Pa ($\approx$ 64 ms^{-1}) in the test section. The flow velocity is monitored by measuring the static pressure drop over the contraction cone.

The geometry and main dimensions of the ship model investigated are shown in Fig. 2. The model reproduces the structure above the water line and was CNC-machined at 1:25 scale out of foam and Medium Density Fibreboard (MDF). It was

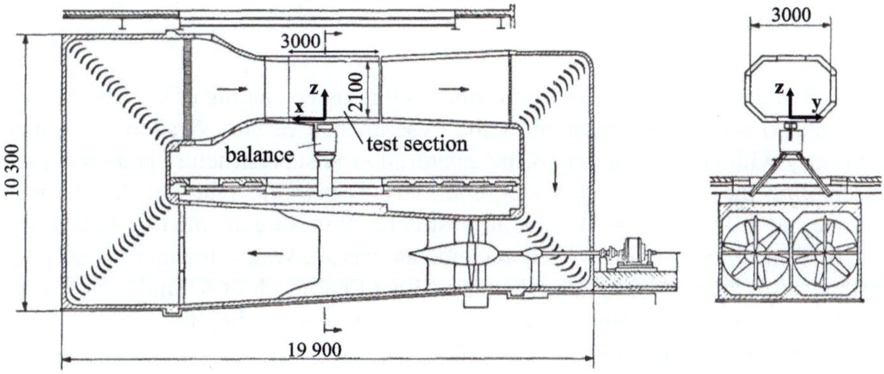

Fig. 1 Closed-loop subsonic wind tunnel of the Institute of Fluid Dynamics, ETH Zurich including the dimensions of the test section and the reference coordinate system

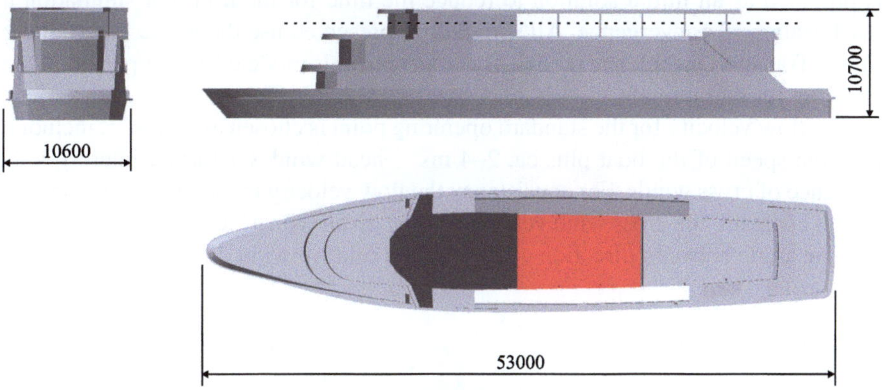

Fig. 2 Geometry of the original including main dimensions in [mm]. The *top view* is a sectional drawing along the *dashed line* in the *side view* to show the area of investigation in *red*

then covered with an epoxy-glass layer and painted black. The model is mounted on the floor and can be rotated to yaw angles between −20° and 20° using a programmable rotary stage. The free-stream velocity V_∞ in the tunnel was set to 15–55 ms^{-1} resulting in a measurement Reynolds-number of 4×10^5–1.5×10^6 based on the width of the model. With a maximum speed of 7.5 ms^{-1} the Reynolds number of the original ship is roughly 3 times higher. Due to the comparatively rough surface of the model and the observed regions of detached flow near the sharp edges of the individual decks, the test were done without tripping the flow.

The area of interest on the sun deck between the bridge and a dividing wall is marked in red in Fig. 2. Initially, the impression was that the flow in the area of interest is governed by the complex vortical structures originating from the bridge's lateral protrusions and the investigations concentrated on this area. However, the flow is in reality driven by the wake of the bluff structure in front. To grant optical access to this region part of the model's MDF roof was replaced with an acrylic glass window.

3 CFD Study

CFD simulations were performed to gain a better understanding of the flow in question and to acquire a validation of the experimental results. A simplified model geometry without the fine details like guardrails and superstructure parts was used, resembling the wind tunnel model as closely as possible. The simulations were carried out with ANSYS CFX on an unstructured mesh generated by ICEM. The standard mesh used has 2.5 million elements overall. Wall refinement is achieved with a prism layer. A simple mesh study with a finer mesh of 6.9 million elements showed reasonable agreement with the standard mesh regarding the flow structures, i.e. the behavior of the flow in the region of interest.

The final simulations are transient with a time step of 0.1 s and simulate up to 50 s in real time (amounting to 500 time steps with 10 internal iterations per time step and a resulting overall RMS residual on the order of 10^{-4}). A steady state simulation is provided as an initial solution to reduce the time for the transient simulation to reach statistical convergence. All simulations presented use the $k - \omega$ based Shear-Stress-Transport turbulence model. The atmosphere is modeled as an open boundary allowing for entrainment.

The flow velocity for the standard operating point is chosen as 10 ms^{-1}, including both the speed of the boat plus ca. 2–4 ms^{-1} head wind. Of further interest is the influence of cross winds. For consistency the flow velocity in the direction of motion is kept constant, the cross wind velocity is then given with 10 $ms^{-1} \times \tan(\alpha)$, where α is the angle between direction of motion and relative wind direction.

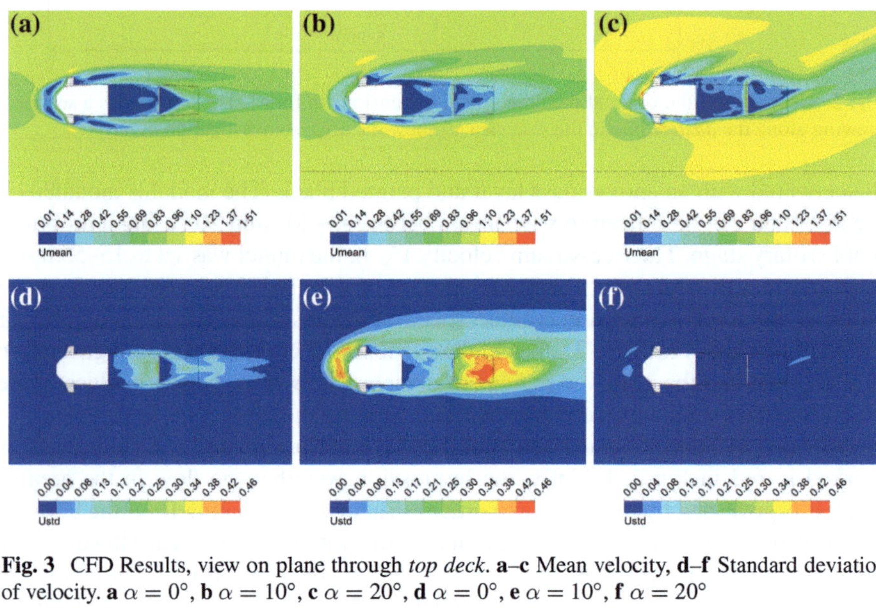

Fig. 3 CFD Results, view on plane through *top deck*. **a–c** Mean velocity, **d–f** Standard deviation of velocity. **a** $\alpha = 0°$, **b** $\alpha = 10°$, **c** $\alpha = 20°$, **d** $\alpha = 0°$, **e** $\alpha = 10°$, **f** $\alpha = 20°$

Figure 3 shows the mean and standard deviation of velocity in the plane through the top deck. The statistics are directly calculated at run time, no intermediate steps are written.

The highest fluctuations in the region of interest occur at $\alpha = 0°$ and $\alpha = 10°$. At higher cross wind angles the flow is much calmer.

4 Fluorescent Tufts

4.1 Experimental Setup

Tufts represent an easy method to visualize flows in the near wall region. Features that can be detected include separation of the flow, turbulent transition in the boundary layer, unsteady flow or recirculation zones. As a qualitative method, it is easy to implement and document.

Still pictures of tufts have the disadvantage of only showing one instant in time. By varying the exposure time it is possible to include transient effects, where the moving tufts smear out over the traversed area at longer exposure, but finding the appropriate time scales of the flow features (turbulence, eddies) can take some time and the results provide limited insight. Recording time series as a video will give a better overview, but it is difficult to document. Here we use a simple method proposed by Crowder [3] to present the results in a single picture.

The tufts are made of UNI-Floss Neon (fishing equipment), which contains fluorescent dye. Illumination is achieved with an ultraviolet neon tube (and an additional ultraviolet spot lamp) for higher contrast against the background of the models surface. A CCD camera with a resolution of 1024×1280 and a frame rate of 15 Hz is used to record the model from the top, with a view through the plexiglass roof window into the region of interest.

4.2 Processing of the Data and Results

Time series of 20 pictures (exposure time 8 ms) were acquired. To compress the results into a single overview the stack of pictures is processed at the pixel level. From grayscale pictures a color coded picture is produced, where the RGB color channels are assigned to different statistical properties. The red channel corresponds to the normalized mean pixel value, and the green channel corresponds to the normalized standard deviation of the pixel values. The blue channel is set to zero.

Figure 4 depicts mean flow and background (red) and transient effects (green) at different yaw angles. For presentation purposes the contrast of the background image was reversed. Clearly visible are the areas of low and high fluctuations, which correspond well with the computed numerical results in Fig. 5. While there is little

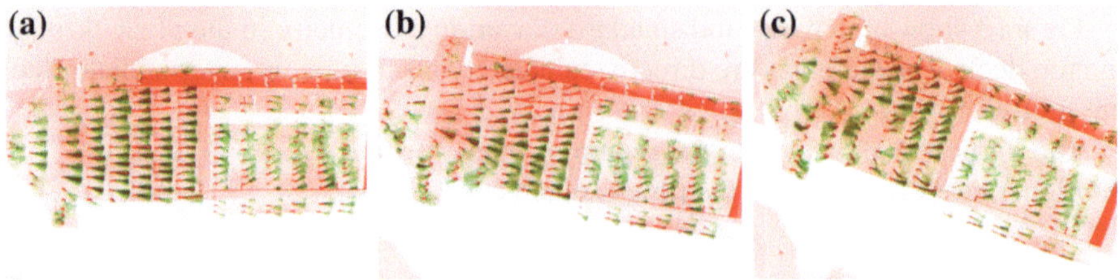

Fig. 4 Tufts on model (insight through plexiglas window): *color coded picture* from time series recording. **a** $\alpha = 0°$, **b** $\alpha = 10°$, **c** $\alpha = 20°$

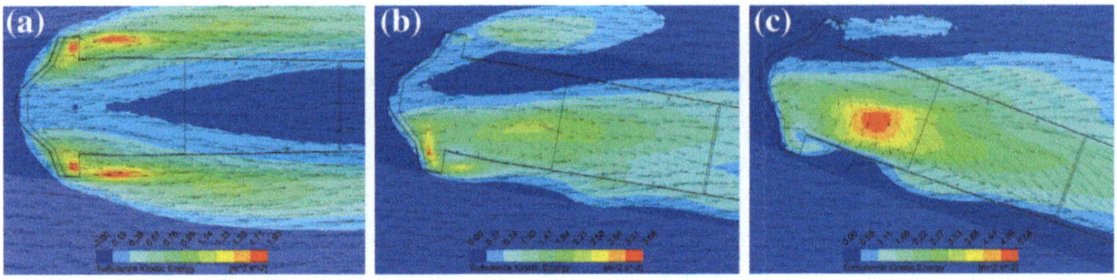

Fig. 5 Turbulence kinetic energy, simulation with CFX. **a** $\alpha = 0°$, **b** $\alpha = 10°$, **c** $\alpha = 20°$

quantitative information in this method, the ease of use and the intuitive display give a good overview over the prevailing flow features and can be used as a starting point for further measurements.

5 Bubble Tracking with a Dynamic Vision Sensor (DVS)

5.1 Working Principle

The Dynamic Vision Sensor (DVS) is an asynchronous temporal contrast CMOS sensor developed at the Institute of Neuroinformatics at the University of Zurich [4]. Local changes in intensity produce spike events on a local pixel level that appear at the sensor output as an asynchronous, time stamped stream of digital coordinates. The result is a sensor with microsecond time resolution and and excellent dynamic range while the amount of data is drastically reduced since no entire images are transmitted. To illustrate the distinction from a full-frame sensor approach Fig. 6 shows the same data set using different time windows for display.

In large scale applications neutrally buoyant Helium filled soap bubbles are the only choice for tracers. The reflective properties of the bubbles require a quality of illumination and background suppression similar to a PIV setup [5]. With the 128×128 pixel DVS sensor mounted in the overhead light of the ceiling, tracking of

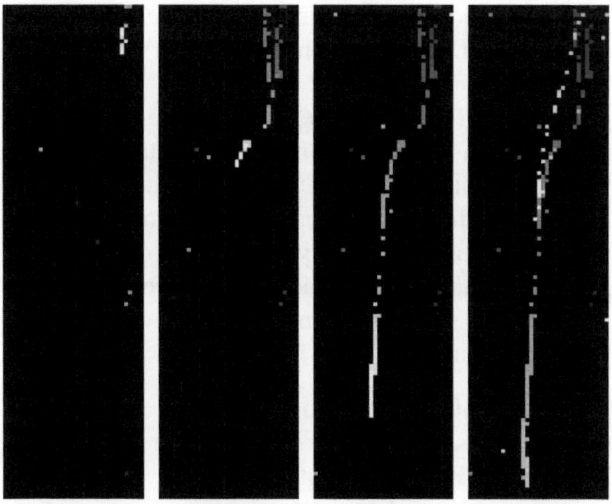

Fig. 6 Display of the identical DVS data using observation time windows of different magnitude to reconstruct full frames

the 3 mm bubbles at 2 m distance was possible using just one powerful spotlight. Due to the volume illumination and the single camera setup, the bubble depth positions could not be resolved.

Background pictures are obtained by slightly shaking the sensor and integrating the results. The bubbles are seeded in a plane at the height of the top deck, the ship is oriented towards the top of picture and the flow is projected downwards.

5.2 Processing of the Data

Data obtained from the DVS is in the form $[x_i, y_i, t_i, p_i]$ where x_i, y_i are the pixel positions, t_i is the timestamp in milliseconds (time of event), and p_i is the polarity of the event (increase or decrease of signal).

Conventional velocimetry methods like PIV or PTV appear to be unsuited for analyzing the obtained DVS data. Due to the low resolution and a viewing area of approx. 1 m^2 the bubble positions are not resolved optically and particle centers cannot be determined. Bubbles appear as streaks of one pixel width and often the streaks are discontinuous. One possible method would be to bin the data in a given time interval and build a picture as done in Fig. 6, which could be used to look for coherent structures (using standard pattern recognition techniques) and to calculate the direction of the streak. This approach is complicated by gaps in the streaks and difficulties in tracing the features back to the corresponding time stamps. Another method would be to find subsets of events which fit a given set of criteria (in this case approximation of a smooth curve). The disadvantage of such an approach is the computational cost for large data sets (here in the order of 10^6 events).

The method used in this paper can be described as velocimetry by POD (Proper Orthogonal Decomposition) of neighborhood events and was chosen for its simplicity. It makes direct use of the obtained data (no binning or picture reconstruction necessary). The goal at the current stage is to provide a simple algorithm to extract the direction of the streak in $x - y - t$ and to implement a validity condition.

One crucial factor in preparing the data is to scale the time axis. This is necessary because Euclidian distances in $x - y - t$ are employed in the method. Each dimension should be processed equally, which is only the case when the time differences between events are in the order of the spatial differences.

Within a subset of N neighboring events a weighted covariance matrix of the relative distances $\mathbf{r}_{i,k} = \mathbf{x}_k - \mathbf{x}_i$ to the current event is computed. The weights are given by a normal distribution $\tilde{\beta}_k = e^{-\frac{|\mathbf{r}_k|^2}{\sigma^2}}$, ensuring a higher contribution of the closer events and a minor influence for outliers. The main direction of the streak (direction with the highest variance) is given by the eigenvector with the highest associated eigenvalue.

A validity criterion is used to filter the results, based on the divergence of the direction of an event to the displacement and direction of its direct neighbors.

5.3 Results

The results in Figs. 7 and 8 give an integral view of the flow using the volume illumination, clearly visible where the bubbles flow over the deck. The low velocity region at the top of the picture is due to the wake of the nozzle of the bubble generator and accelaration of the bubbles when they are seeded into the flow.

An extension of this technique currently under development is tracking bubbles in 3D with a stereoscopic setup.

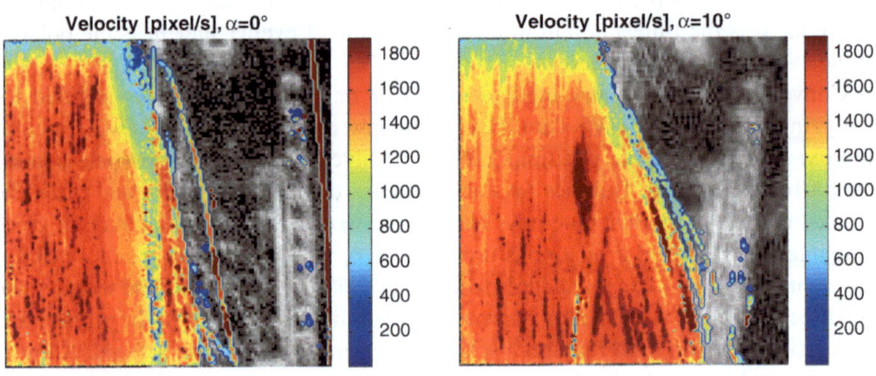

Fig. 7 Average velocity, scale given in pixels/s. **a** $\alpha = 0°$, **b** $\alpha = 10°$

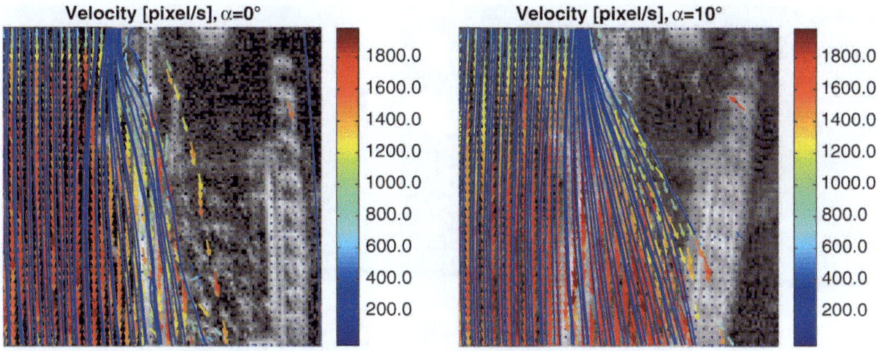

Fig. 8 Streamlines of calculated flow field. **a** $\alpha = 0°$, **b** $\alpha = 10°$

6 Tracking of a Hand-Held Probe

The obvious approach for bringing the traditional hand-guided flow visualization techniques to a quantitative level is to track the probe's position while the operator is investigating the volume of interest and simultaneously logging the measured quantity. This operating mode was presented in the late 1970s and subsequently refined by Crowder et al. [6]. For the present work, two 3D tracking systems were installed in the test section: a stereo setup consisting of two USB cameras and a commercial motion capturing system with four dedicated cameras.

6.1 Experimental Setup

The stereo setup consisted of two monochrome USB CCD cameras with a resolution of 1024×768 pixel equipped with 8.5 mm/F1.5 lenses and mounted on a rail 0.65 m apart (Fig. 9). The cameras were calibrated according to Heikkilä [7] with an reprojection error of 2 mm perpendicular and 8 mm along the optical axis at the working distance of 1.8 m. In the two images, an infrared LED on the backside of the probe was identified and localized by a maximum search, converted to position in space and stored at a rate of 15 Hz. Image and pressure acquisition and data processing were achieved by a dedicated C-code running under Matlab.

The motion capture (mocap) system from Qualisys with four Oqus 3 series cameras was calibrated using the commercial software and calibration package. The system relies on tracking passive markers illuminated by a ring of infrared LED's on each camera. The accuracy of the system as determined during the calibration process was better than 1 mm in all directions. The position of the markers and the analog data were recorded at 200 Hz.

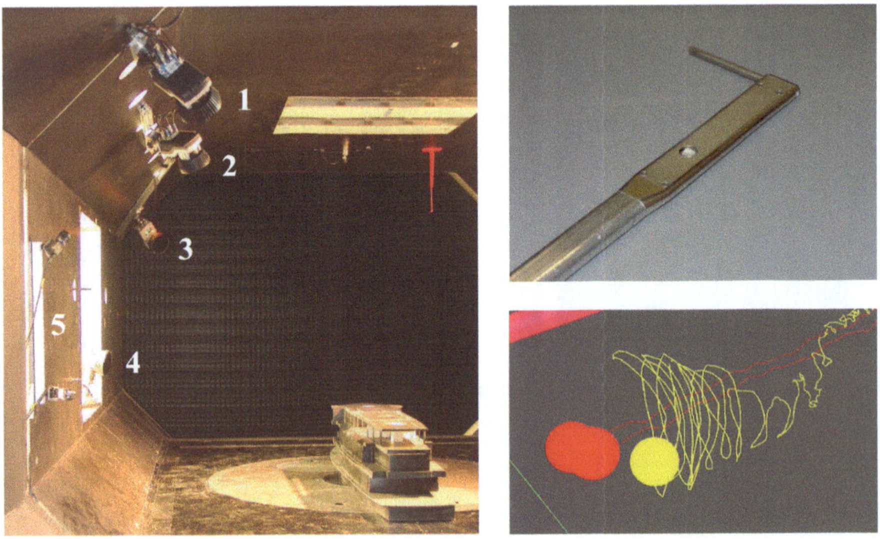

Fig. 9 Arrangement in the test section (*left*), the view is upstream into the flow. The mocap cameras are labeled by *1–4*, the USB camera stereo setup by *5*. *Right*, the Pitot-static probe used (*top*) and an illustration of the flow dynamics captured by the tuft

The measurement volume of approximately 1 m in each spatial direction was similar for both systems. Using the same probe and post processing for both systems the results were identical. Where the two systems differ significantly is in cost, setup and measurement time (i.e. framerate). The latter was 10–15 min with the stereo setup and 2–3 min with the mocap system.

With both tracking systems a Pitot-static probe with a tube diameter of 5 mm and a length of 60 mm was used to record static and dynamic pressure distributions. Two fast pressure transducers are housed in the handle close to the probe head to keep distances short (Fig. 9). At the rear end of the probe head, an infrared LED serves as marker for the tracking. The upper side the probe is equipped with a color LED that can serve the user as feedback by color-coding actual pressure or position. With only one single LED at the back as marker the position of the probe tip can only be estimated from the stereo-setup data by assuming the Pitot to be oriented parallel to the tunnel axis. To work with the mocap system, three passive markers were fitted on the probe head which allowed to calculate the probe's tip position and orientation more accurately.

To record the direction of the flow, a single tuft was fitted with 3 spherical passive markers. Two markers were put on the wand close to the tuft giving position and orientation of the probe and a third, lightweight marker at the end of the tuft from which the instantaneous orientation of the tuft relative to the wand was calculated. Without active markers (LEDs) fitted, this probe was only used with the mocap setup.

Figure 9 illustrates the arrangements and emphasizes the recorded dynamics of the tufts by showing the trace of the tree markers.

The postprocessing of the marker data involved calculating the Pitot probe's tip position and the tufts' position, respectively. The measured pressure and tuft orientation were then interpolated onto a regular grid with 10 mm spacing using standard Matlab routines. The no-slip conditions on the model surface was considered in setting the dynamic pressure to zero at the respective positions during interpolation.

6.2 Results

Figure 10 illustrates the visualization capabilities with access to volumetric experimental data. The distribution of the dynamic pressure for zero yaw angle of the ship is plotted. In the left part, slices through the volume of interest show the wake of the laterally protruding wings of the bridge extending downwards along the structure. On the right hand side, the same slices of the CFD data is plotted. Although the CFD data is based on a full-scale simulation, the agreement is remarkable.

The streamlines calculated from the tuft probe are drawn on the left side of Fig. 11. The data compares favorably to the CFD simulations. Both data sets reproduce the vortical structure of the bridge's wake and allow detailed insights into the flow. It has to be noted that through interpolation the circumferential component of the flow near the vortex core is averaged out in the measured data.

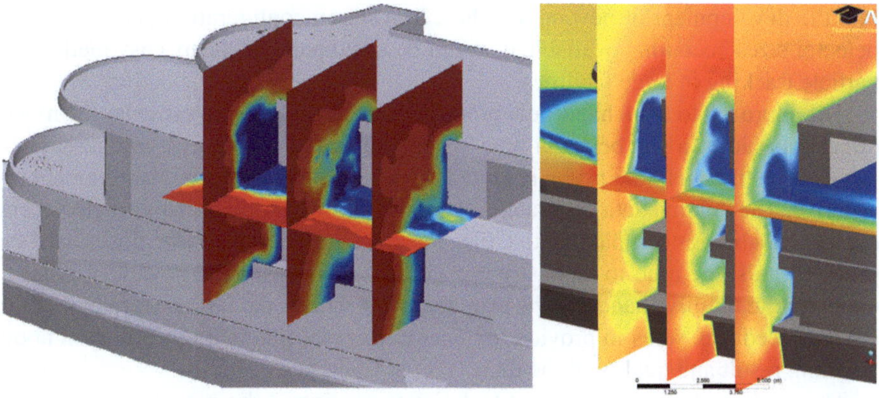

Fig. 10 The distribution of the dynamic pressure showing the wake of the bridge (*left*) recorded with the Qualisys system compared to the CFD data (*right*)

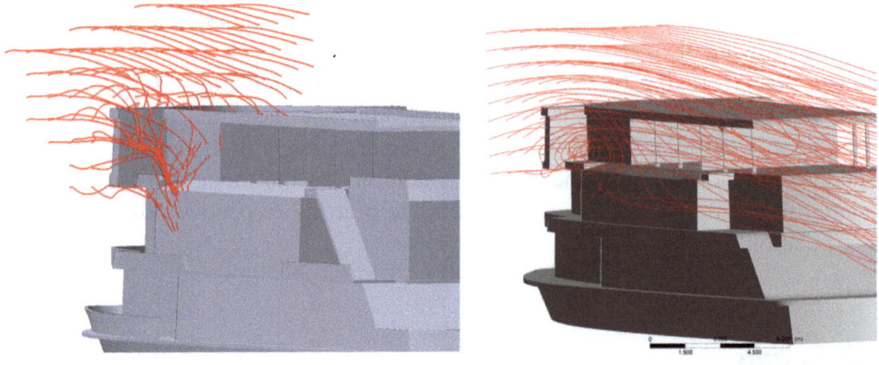

Fig. 11 The streamlines computed using the tuft-probe data from the wind tunnel tests (*left*) and based on the CFD simulations (*right*)

7 Conclusion and Outlook

The aim of the present study was to retrieve quantitative data from traditional visualization methods while maintaining a short setup time and intuitive operation. Recording image time series of tufts attached to the surface of the model allowed to obtain both the mean direction and local variance of the flow. The use of an asynchronous temporal contrast sensor (DVS) allowed to detect Helium filled soap bubbles in ambient lighting and to compute local flow velocity with a tracking algorithm. As one can not think in fixed time slices when handling the asynchronous stream of events the traditional tracking algorithms are not applicable and some work has to be put into the development of new approaches. In a stereo configuration and with larger detector sizes becoming available it would be a comparably cheap, easy method for particle tracking in large facilities.

The approach to track hand-held probes while the operator investigates the model effectively closes a gap between flow visualization and quantitative measurements. Because instinctively the probe is guided to the area of interest its handling is very intuitive. The volume data generated in this way allows for quantitative analysis and display capabilities comparable to CFD. The tracking system and the position of the probe can also be used to reconstruct position and shape of the model. In order to maintain the intuitive handling the primary effort will be the further development of the user feedback so as to provide the operator in real time with information about position, result and quality of the data.

The application of the different techniques confirmed the potential of quantitative flow visualization in large-scale testing. The methods complemented each other in the sense that they investigated surface flow, streamlines and pressure fields. If configuration, installation and operation can be kept fast and intuitive, quantitative visualization methods have to potential to close the gap between traditional flow visualization and measurement techniques.

References

1. Bömmels, R., Machacek, M., Landolt, A., Rösgen, T.: Quantitative flow visualization for large scale wind tunnels. In: McCallen, R., et al. (eds.) The Aerodynamics of Heavy Vehicles: Trucks, Buses, and Trains. Springer, Berlin (2004)
2. Takagi, M.: Applications of computers to automobile aerodynamics. J. Wind Eng. Ind. Aerodyn. **33**, 419–428 (1990)
3. Crowder, J.P.: Tufts. In: Yang, W.-J. (ed.) Handbook of Flow Visualization, Hemisphere Publishing Corporation (1989)
4. Lichtsteiner, P., Posch, C., Delbruck, T.: A 128 $\times$ 128 120 dB 15 μs latency asynchronous temporal contrast vision sensor. IEEE J. Solid-State Circ. **43**(2), 566–576 (2008)
5. Bosbach, J., Kühn, M., Wagner, C.: Large scale particle image velocimetry with helium filled soap bubbles. Exp. Fluids **46**, 539–547 (2009)
6. Crowder, J.P., Watzlavick, R.L., Krutckoff, T.K.: Airplane flow-field measurements. AIAA/SAE World Aviat. Congr. **8**, 1–9 (1997)
7. Heikkilä, J.: Geometric camera calibration using circular control points. IEEE Trans. Pattern Anal. Mach. Intell **22**(10), 1066–1077 (2000)

Author Index

A
Ali, I., 85

B
Boissinot, S., 343
Borer, D., 413
Bruneau, C.-H., 363
Busslinger, A., 61

C
Camosy, M., 287
Catanzaro, C., 99
Cheli, F., 99, 143
Chernoray, V., 253
Creusé, E., 363

D
Davidson, L., 253
Dayan, I., 377
Depeyras, D., 363
Dillmann, A., 161
Duncan, B.D., 343

E
Efraimsson, G., 3
Ehrenfried, K., 161, 175
El-Alti, M., 253
Elofsson, P., 211, 343

F
Frank, T., 196

G
Gilliéron, P., 363
Glas, J., 303
Grossmann, J., 269, 377

H
Henningson, D.S., 3
Herbst, A.H., 3
Hilbert, C.B., 227
Hjelm, L., 211, 253
Horrell, C., 269, 377

J
Jennings, T., 287
Jönsson, M., 21

K
Kehs, J., 269
Kjellgren, P., 253
Kowalczyk, H., 287
Krajnović, S., 311, 327

A. Dillmann and A. Orellano (eds.), *The Aerodynamics of Heavy Vehicles III*,
Lecture Notes in Applied and Computational Mechanics 79,
DOI 10.1007/978-3-319-20122-1

L
Landolt, A., 413
Lauterbach, A., 175
Löfdahl, L., 211
Loose, S., 21, 175

M
Mariano L., 143
Marinko, F., 287
Meier, A., 413
Mercier, G., 343
Mortazavi, I., 363
Muld, T.W., 3

N
Nayeri, C.N., 133, 303
Nyfeler, S., 61

O
Orellano, A., 3
Östh, J., 327

P
Pankajakshan, R., 227
Paschereit, C.O., 133, 303
Paul, J.C., 37

R
Reinke, P., 61
Reynard, A., 287
Rocchi, D., 99, 143
Rodriguez Ahlert, C.-J., 85
Roesgen, T., 413
Rüter, Arnd, 85

S
Schito, P., 99, 143
Schober, M., 133
Seifert, A., 377
Sheridan, J., 113
Smith, A., 269, 377
Söderblom, D., 211
Strangfeld, C., 133

T
Thompson, M., 113
Tian, H.-q, 113
Tietze, A., 133
Tomasini, G., 99, 143
Turney, J., 196

V
van Leeuwen, P.M., 393
van Raemdonck, G.M.R., 237, 393
van Tooren, M.J.L., 237, 393
Visser, K., 269

W
Wagner, S., 161
Whitfield, D.L., 227

Z
Zellmann, C., 133
Zhou, D., 113

MIX
Papier aus verantwortungsvollen Quellen
Paper from responsible sources
FSC® C105338

If you have any concerns about our products,
you can contact us on
ProductSafety@springernature.com

In case Publisher is established outside the EU,
the EU authorized representative is:
Springer Nature Customer Service Center GmbH
Europaplatz 3, 69115 Heidelberg, Germany

Printed by Libri Plureos GmbH
in Hamburg, Germany